To my mentor Frank DiCesare who has given me numerous care, inspiration, and fatherly love since 1987.

MengChu Zhou

To my late father Sankara Narayana who inspired my quest for knowledge.

Kurapati Venkatesh

ABOUT THE AUTHORS

MengChu Zhou received the B.S. degree from Nanjing University of Science and Technology, M.S. degree from Beijing Institute of Technology, and Ph. D. from Rensselaer Polytechnic Institute, NY. He is currently an Associate Professor of Electrical and Computer Engineering at NJIT. His interests include computer-integrated manufacturing, Petri nets, intelligent automation, and multi-lifecycle engineering. He co-authored *Petri Net Synthesis for Discrete Event Control of Manufacturing Systems* in 1993, and edited *Petri Nets in Flexible and Agile Automation* in 1995, both by Kluwer Academic, Norwell, MA. He has published more than 120 journal articles, book chapters, and conference proceeding papers. Dr. Zhou served as Program Chairs of the 9th International Conference on CAD/CAM, Robotics, and Factories of the Future, Newark, NJ, 1993, 1997 IEEE International Conference on Emerging Technologies and Factory Automation, Los Angeles, CA, and 1998 IEEE International Conference on Systems, Man and Cybernetics, San Diego, CA. He is an Associate Editor of IEEE Transactions on Robotics and Automation, and editor of International Journal of Intelligent Control and Systems. He was listed in 1994 for the CIM University-LEAD Award by SME and granted the 1996 H. J. Perlis Research Award by NJIT. He is a Senior Member of IEEE and 1998 President-Elect of the Chinese Association for Science and Technology-USA.

Kurapati Venkatesh is a Senior Technical Staff Member in AT&T. Since 1988, he has been applying Petri nets to investigate a variety of problems in flexible automation, object-oriented software development, communication networks, enterprise modeling, business process reengineering, and neural networks. He has published more than 30 papers in journals and conferences including, International Journal of Production Research, Journal of Manufacturing Systems, IEEE Transactions on Industrial Electronics, Computers and Industrial Engineering, International Journal of Operations and Production Management, and International Journal of Material Processing and Technology. He holds a Doctorate in Mechanical Engineering (1995, New Jersey Institute of Technology) with focus on Petri nets and its applications for modeling, simulation, and control of FMSs using object-oriented concepts. He has an M.S. in Manufacturing Systems Engineering (1992, FAU, Boca Raton); an M.S. (1990, Indian Institute of Technology, Madras) and a B.S. (1988, Sree Venkateswara University, India) in Mechanical Engineering. He is a member of SME, ASME and IEEE.

SERIES IN INTELLIGENT CONTROL AND INTELLIGENT AUTOMATION

Editor-in-Charge: Fei-Yue Wang
(*University of Arizona*)

Vol. 1: Reliable Plan Selection by Intelligent Machines
(*J E McInroy, J C Musto, and G N Saridis*)

Vol. 2: Design of Intelligent Control Systems Based on Hierachical Stochastic Automata (*P Lima and G N Saridis*)

Vol. 3: Intelligent Task Planning Using Fuzzy Petri Nets
(*T Cao and A C Sanderson*)

Vol. 6: Modeling, Simulation, and Control of Flexible Manufacturing Systems: A Petri Net Approach (*M Zhou and K Venkatesh*)

Vol. 7: Intelligent Control: Principles, Techniques, and Applications (*Z-X Cai*)

Vol. 10: Autonomous Rock Excavation: Intelligent Control Techniques and Experimentation (*X Shi, P J A Lever and F Y Wang*)

Forthcoming volumes:

Vol. 4: Advanced Studies in Flexible Robotic Manipulators: Modeling, Design, Control, and Applications (*F Y Wang*)

Vol. 5: Computational Foundations for Intelligent Systems (*S J Yakowitz*)

Vol. 8: Advanced Topics in Computer Vision and Pattern Recognition
(*E Sung, D Mital, E K Teoh, H Wang, and Z Li*)

Vol. 9: Petri Nets for Supervisory Control of Discrete Event Systems:
A Structural Approach (*A Giua and F DiCesare*)

Series in
Intelligent Control and Intelligent Automation
Vol. 6

MODELING, SIMULATION, AND CONTROL OF FLEXIBLE MANUFACTURING SYSTEMS

A Petri Net Approach

MengChu Zhou
Kurapati Venkatesh
New Jersey Institute of Technology, USA

World Scientific
Singapore • New Jersey • London • Hong Kong

Published by

World Scientific Publishing Co. Pte. Ltd.
P O Box 128, Farrer Road, Singapore 912805
USA office: Suite 1B, 1060 Main Street, River Edge, NJ 07661
UK office: 57 Shelton Street, Covent Garden, London WC2H 9HE

Library of Congress Cataloging-in-Publication Data
Zhou, MengChu.
    Modeling, simulation, and control of flexible manufacturing
systems : a Petri net approach / Mengchu Zhou, Kurapati Venkatesh.
      p.   cm. -- (Series in intelligent control and intelligent
automation ; vol. 6)
    Includes bibliographical references and index.
    ISBN 981023029X
    1. Flexible manufacturing systems, ls. etc.   2. Petri nets.
I. Venkatesh, Kurapati.   II. Title.   III. Series.
TS155.65.Z46   1998
670.42'7--dc21   98-45715
                                                                   CIP

British Library Cataloguing-in-Publication Data
A catalogue record for this book is available from the British Library.

Copyright © 1999 by World Scientific Publishing Co. Pte. Ltd.

All rights reserved. This book, or parts thereof, may not be reproduced in any form or by any means, electronic or mechanical, including photocopying, recording or any information storage and retrieval system now known or to be invented, without written permission from the Publisher.

For photocopying of material in this volume, please pay a copying fee through the Copyright Clearance Center, Inc., 222 Rosewood Drive, Danvers, MA 01923, USA. In this case permission to photocopy is not required from the publisher.

This book is printed on acid-free paper.

Printed in Singapore by Uto-Print

PREFACE

Whether it is a manufacturing based company or a service based company, the key to stay at the apex of global competition is to meet the dynamically changing needs of customers. Flexible Manufacturing Systems (FMS) combine both the sophisticated manufacturing equipment and the advanced computer and information technology to impart flexibility to the manufacturing operations, thereby effectively meeting the changing needs of customers.

FMS consist of several types of machines, computers, robots, and automated guided vehicles and are designed to produce a great variety of products. While FMS give a cutting-edge advantage to a manufacturing company through their flexibility, they pose complex problems for their planning, designing, scheduling, controlling, and monitoring. This is primarily due to the inherent nature of FMS that are asynchronous concurrent systems. Modeling tools often help to address the above stated problems in FMS. There is a great demand for integrated tools that address the multifaceted problems in FMS. Unlike the several available tools such as mathematical programming, queuing networks, ladder logic diagrams, commercial simulation packages, Petri nets offer such an integrated solution to address effectively many issues in design and operation of FMS.

In order to address these problems, this book focuses on both the theory and several applications of Petri nets in modeling, analysis, simulation, scheduling, and control of FMS. It not only contributes to the theory of Petri nets by introducing new types of Petri nets such as Augmented Timed Petri Nets and Real-Time Petri Nets but also applies them to address issues related to breakdown handling and real-time discrete event control. It illustrates the use of Petri nets to compare the performance of push and pull systems and to develop object-oriented control software. Basic comparative studies that compare Real-Time Petri Nets and ladder logic diagrams for discrete event control are also presented. Roughly speaking, this book draws a half of material from the second author's doctoral dissertation, and another half from the lecture materials of a graduate level course entitled "Discrete Event Dynamic Systems" at New Jersey Institute of Technology (NJIT) by the first author and some recent research results obtained by his research group at NJIT.

Organization

This book is organized as follows. Chapter 1 discusses the background, motivation, and purpose of this book. Chapter 2 presents an overview of FMS and concludes by emphasizing the importance of integrated modeling tools. It is based on the paper entitled "Flexible Manufacturing Systems: An Overview" in *International Journal of Operations and Production Management*, Vol. 13, No. 4, pp. 26-49, 1994 by Kaighobadi and Venkatesh. Chapter 3 presents Petri nets as an integrated tool and methodology in FMS design. A significant portion of this chapter is based on the article entitled "Petri nets: tool and technology" by Zhou and Zurawski in the March 1994 issue of *IEEE IES Newsletter*, the Newsletter of IEEE Industrial Electronic Society. Some representative papers and reports are surveyed with the focus on the applications of Petri nets in modeling, analysis, performance evaluation, simulation, control, planning, and scheduling of FMS.

The fundamentals of Petri nets are introduced in Chapter 4. Several simple yet interesting manufacturing and assembly examples are used to illustrate the concepts. Chapter 5 presents several important classes of Petri nets in modeling automated manufacturing systems and their application scope. The modeling and synthesis procedures are discussed and many FMS examples are given. Chapter 6 focuses on the analytical performance modeling and evaluation of FMS. The materials presented in Chapters 4-6 are based on several papers including

- ♦ "Modeling, Analysis, Simulation, Scheduling, and Control of Semiconductor Manufacturing Systems: A Petri Net Approach" by Zhou and Jeng, to appear in *IEEE Transactions on Semiconductor Manufacturing*, August 1998.

- ♦ "Petri Nets and Industrial Application: A Tutorial" by Zurawski and Zhou in *IEEE Transactions on Industrial Electronics*, 41(6), pp. 567-583, December 1994; and

- ♦ "Petri Net Synthesis and Analysis of A Flexible Manufacturing Cell" by Zhou, McDermott and Patel in *IEEE Transactions on Systems, Man and Cybernetics*, 23(2), pp. 523-531, March/April 1993.

As the model complexity increases, the use of analytical techniques for performance analysis becomes harder and impossible for many cases. Discrete event simulation becomes an indispensable alternative for the performance analysis. Chapter 7 presents the basic principles of Petri net simulation and explores the use of Petri net simulation tools. It is based on the paper by Venkatesh, Chetty, and Raju, "Simulating Flexible Automated Forming and Assembly Systems," *Journal of Material Processing and Technology*, Vol. 24, pp. 453-462, December 1990.

Chapter 8 demonstrates Petri nets as a powerful tool to investigate the problem often encountered in manufacturing systems management, namely comparing the performance of push and pull systems. It is based on the paper by Venkatesh, Zhou, Kaighobadi, and Caudill, "A Petri Net Approach to Investigating the Performance of Push and Pull Paradigms in Flexible Factory Automated Systems using Petri Nets," *International Journal of Production Research*, Vol. 34, No. 3, pp. 595-620, 1996. Chapter 9 introduces a new type of Petri nets called Augmented Timed Petri Nets for breakdown handling and presents a case study on their use in modeling and analysis of a multi-robot assembly system. It is based on the paper "Augmented Timed Petri Nets for Modeling of Robotic Systems with Breakdowns," *Journal of Manufacturing Systems*, Vol. 13, No. 4, pp. 289-301, 1994 by Venkatesh, Kaighobadi, Zhou, and Caudill.

Another class of Petri nets called Real-Time Petri Nets is introduced in Chapter 10 along with their unique features and how they are different from the various existing classes of Petri nets for real-time control. Chapter 11 presents a case study comparing Real-Time Petri Nets and ladder logic diagrams. It also develops some comparison criteria and analytical formulas to quantify the complexity of models designed by Real-Time Petri Nets and ladder logic diagrams. These two chapters are based on the papers by Venkatesh, Zhou, and Caudill, "Comparing Ladder Logic Diagrams and Petri Nets for the Design of Sequence Controllers Through a Discrete Manufacturing System," *IEEE Transactions on Industrial Electronics*, Vol. 41, No. 6, pp. 611-619, 1994, and "Evaluating the complexity of Petri nets and ladder logic diagrams for sequence controllers design in flexible automation," in *Proc. of Seiken/IEEE Workshop on Emerging Technologies and Factory Automation*, pp. 428-435, Tokyo, Japan, November 1994.

Chapter 12 introduces Petri Nets in the design of object-oriented control software. By combining the existing modeling methods such as Object Modeling Technique Diagrams and Petri Nets, it presents a case study in which Petri Nets are formally applied for the dynamic modeling in the object-oriented software development. The material is based on the paper entitled "Object-oriented design of FMS Control Software based on Object Modeling Technique Diagrams and Petri nets" by Venkatesh and Zhou in *Journal of Manufacturing Systems*, Vol. 17, No. 2, pp. 118-136, 1998. Chapter 13 explores the application of Petri Nets in scheduling of FMS by combining them with heuristic-based search approaches. This chapter is based on several sources including

- The doctoral dissertation research performed by Huanxin H. Xiong under the first author's supervision,
- The paper entitled "Hybrid Heuristic Search for Petri Net Scheduling of an FMS" by Xiong, Zhou and Caudill, in *Proceedings of 1996 IEEE International Conference on Robotics and Automation*, Minneapolis, MN, pp. 2793-2797, April 1996, and

- The paper entitled "Scheduling of Semiconductor Test Facility via Petri Nets and Hybrid Heuristic Search" by Xiong and Zhou, to appear in *IEEE Transactions on Semiconductor Manufacturing*, August 1998.

Finally, Chapter 14 presents the future research and development of Petri Nets in flexible automation.

Acknowledgments

Many people have helped us in completing this book. Especially, we would like to thank Dr. Fei-Yue Wang, the editor of this series for his help, encouragement and support. We would like to thank Professors Reggie J. Caudill, Xiuli Chao, Ernest Geskin, and Dr. Anthony D. Robbi who served as the doctoral committee of the dissertation of the second author of this book. They are also very helpful colleagues of the first author. In particular, Professor Caudill has been our collaborator for many years and served as a co-advisor of the second author's dissertation. We would also like to acknowledge the assistance provided by Dr. H. H. Xiong at Lucent Technologies, Professors N. Ansari, R. Haddad, R. Kane, and G. Thomas at NJIT. The first author would like to thank his wife, Fang Chen, his parents and parents-in-law for their support and love. The second author would like to express his gratitude to all of his family members including his sisters, brothers-in-law, father-in-law, mother-in-law and especially thanks his mother, Sree Lakshmi and his wife, Amita to support the long weekends of time spent in the lab. He would also acknowledge the support of NJIT, Florida Atlantic University, Indian Institute of Technology at Madras where the entire work reported in this book was performed. We are grateful for the support of World Scientific through their professional editorial staff members.

The preparation of this book has been financially supported by the New Jersey State Commission on Science and Technology through the Center for Manufacturing Systems at NJIT and the National Science Foundation under Grant No. DMI-9410386.

MengChu Zhou
New Jersey Institute of Technology, Newark, New Jersey
E-mail address: *zhou@njit.edu*
Website: *http://megahertz.njit.edu/~zhou*

Kurapati Venkatesh
AT&T, Middletown, NJ
E-mail address: *amita@mtmail.att.com*

CONTENTS

Preface vii

Chapter 1. Introduction 1
  1.1. Background, Motivation and Objectives 1
  1.2. Historical Perspective on Manufacturing Systems 2
  1.3. Historical Perspective on Petri Nets 5
  1.4. A Robotic Cell and Petri Net Model 7
  1.5. Summary 14

Chapter 2. Flexible Manufacturing Systems: An Overview 15
  2.1. Introduction 15
  2.2. Definitions of FMS 16
  2.3. Impetus for Change 18
  2.4. Installation, Implementation, and Integration of FMSs 20
  2.5. Applications of FMSs 24
  2.6. Problems in Installation and Implementation of FMSs 25
      2.6.1. Managerial Problems 26
      2.6.2. Technical Problems 27
  2.7. Summary 36

Chapter 3. Petri Nets as Integrated Tool and Methodology in FMS Design 39
  3.1. Introduction 39
      3.1.1. Brief and Informal Introduction to Petri Nets 40
      3.1.2. Advantages of Petri Nets 41
      3.1.3. General Procedure for Applying Petri Nets in FMS Design. 42
      3.1.4. Successful Industrial Applications 43
  3.2. Modeling and Analysis 44

3.3. Performance Analysis and Evaluation 46
3.4. Discrete-Event Simulation 49
3.5. Discrete-Event Control 50
3.6. Planning and Scheduling 55
3.7. Summary 58

Chapter 4. Fundamentals of Petri Nets 59
4.1. Conditions, Events, and State Machines 59
4.2. Formal Definition of Petri Nets 64
    4.2.1. Definition of Petri Nets 64
    4.2.2. Execution Rules of Petri Nets 66
4.3. Properties of Petri Nets and Their Implications 69
    4.3.1. Reachability 70
    4.3.2. Boundedness and Safeness 70
    4.3.3. Conservativeness 71
    4.3.4. Liveness 71
    4.3.5. Reversibility and Home State 72
    4.3.6. Other Structural Properties 73
    4.3.7. Examples 73
    4.3.8. Implications in Flexible Manufacturing 74
4.4. Reachability Analysis Method 75
4.5. Invariant Analysis Method 80
4.6. Reduction Methods 83
4.7. Other Analysis Methods 89
4.8. Summary 90

Chapter 5. Modeling FMS with Petri Nets 91
5.1. State Machine Petri Nets 92
5.2. Marked Graphs 95
5.3. Free-Choice and Asymmetric Choice Petri Nets 99
5.4. Acyclic Petri Nets and Assembly Petri nets 105
5.5. Production-Process Nets and Augmented Marked Graphs 109

Contents   xiii

    5.6. General Modeling Method 118
    5.7. Systematic Modeling Methods 125
        5.7.1. Bottom-up Methods 125
        5.7.2. Top-Down Methods 131
        5.7.3. Hybrid Methods 132
    5.8. Top-down Modeling of an NJIT's FMS Cell: A Case Study 133
        5.8.1. Description of A Flexible Manufacturing Cell 133
        5.8.2. Top-Down Design Process: First Level Net and Modules.. 136
        5.8.3. Final Petri Net Model 143
    5.9. Summary 145

Chapter 6. FMS Performance Analysis 147
    6.1. Deterministic Timed Petri Nets 147
    6.2. Analysis of an FMS Cell: A Case Study 153
    6.3. Stochastic Petri Nets 156
        6.3.1. Exponential Distribution 156
        6.3.2. Definition and Solution of Stochastic Petri Nets 157
        6.3.3. Generalized Stochastic Petri Nets 165
        6.3.4. Other Extensions to Stochastic Petri Nets 166
    6.4. Performance Analysis of a Flexible Workstation 166
        6.4.1. Overview of SPNP 167
        6.4.2. Performance Evaluation of a Flexible Workstation 169
    6.5. Summary 174

Chapter 7. Petri Net Simulation and Tools 177
    7.1. Introduction. 177
    7.2. Discrete Event Simulation 178
        7.2.1. Discrete Event Simulation Procedure 178
        7.2.2. Simulation Models, Schemes and Tools 182
        7.2.3. Simulation Schemes and Tools 185
    7.3. Timed Petri Nets and Token Game Simulation 187
    7.4. Description of the Software Package to Simulate Petri Nets 189

    7.5. Other CASE Tools to Simulate and Analyze PNs 192
    7.6. Summary 194

Chapter 8. Performance Evaluation of Push and Pull Paradigms in Flexible Automation 195
    8.1. Introduction 195
    8.2. Push and Pull and their PN Modeling 197
    8.3. Application Illustration 200
        8.3.1. System Configuration and Assumptions 200
        8.3.2. Statement of the Problem 202
        8.3.3. PNM Formulation and Analysis 202
    8.4. Procedure for PN Modeling and Analysis and Simulation
         results 212
        8.4.1. FMS with the Pull Paradigm 213
        8.4.2. FAS with the Pull Paradigm 215
        8.4.3. FMS with the Push Paradigm 216
        8.4.4. FAS with the Push Paradigm 217
        8.4.5. Simulation Results 217
    8.5. Summary 220

Chapter 9. Augmented-Timed Petri Nets for Modeling Breakdown Handling 223
    9.1. Introduction 223
    9.2. Augmented Timed Petri Nets 225
    9.3. Application Illustration: A Flexible Assembly System 230
        9.3.1. System Description 230
        9.3.2. ATPN Modeling of the System 232
        9.3.3. ATPN Model for Designing the Optimum Number of
               Assembly Fixtures 234
        9.3.4. Simulation and Analysis of the ATPN Model 236
    9.4. Summary 238

Chapter 10. Real-Time Petri Nets for Discrete Event Control 241
10.1. Introduction 241
10.2. Real-time Petri Nets 241
10.3. Real-time PNs and Other PN Extensions for Control 244
10.4. Example: An Automatic Assembly System 248
10.5. A Case Study: An Electro-Pneumatic System 251
10.6. Software Description to Execute Real-time Petri Nets 255
10.7. Summary 257

Chapter 11. Comparison of Real-Time Petri Nets and Ladder Logic Diagrams 259
11.1. Introduction 259
11.2. Comparison Criteria for Control Logic Design by PNs and LLDs 261
11.3. Comparison Through an Electro-Pneumatic System 264
    11.3.1. Sequence Controller Design 264
    11.3.2. Control for Other Sequences 269
    11.3.3. Discussions 272
11.4. Analytical Formulas to Evaluate the Complexity of PNs and LLDs 278
    11.4.1. Logical AND, Logical OR, and Sequential Modeling 279
    11.4.2. Timed Logical AND, Timed Logical OR, and Timed Sequential Models 282
    11.4.3. Other Formulas for Estimating Basic Elements in PN and LLD 284
11.5. Methodology to Use the Analytical Formulas 289
11.6. Illustration of the Methodology Through Examples 292
    11.6.1. An Automatic Assembly System 292
    11.6.2. An Electro-Pneumatic System Without Sustained Signals 293
    11.6.3. An Electro-Pneumatic System With Sustained Signals 296
11.7. Summary 300

Chapter 12. An Object-Oriented Design Methodology for Development of FMS Control Software 303
12.1. Introduction 303
    12.1.1. Background 304
    12.1.2. Literature Review and Motivation 305
12.2. Methodology for FMS Control Software Development 307
    12.2.1. Methodology 307
    12.2.2. Fundamentals of OOD 309
    12.2.3. Object Modeling Technique Diagram as a Static Modeling Tool 311
    12.2.4. Petri Nets as a Dynamic Modeling Tool 314
12.3. Illustration of the Methodology with an FMS 315
    12.3.1. OMT Diagram and PNM of the FMS 317
    12.3.2. Complete Structure of Objects with Their Static and Dynamic Relations 324
    12.3.3. Reusability, Extendibility, and Modifiability of the Design 326
12.4. Summary 328

Chapter 13. Scheduling Using Petri Nets 333
13.1. Introduction 333
13.2. A Brief Review 334
13.3. Petri Net Modeling for Scheduling 337
    13.3.1. Petri Net Model with Traditional Assumptions 339
    13.3.2. Petri Net Model with Finite Buffer Size 342
    13.3.3. Petri Net Model with Multiple Lot Size 342
    13.3.4. Petri Net Model with Material Handling Considered 344
    13.3.5. Petri Net Model for Flexible Routes 345
13.4. Best-First, Backtracking, and Hybrid Search Algorithms 346
    13.4.1. Best First Search and Backtracking Search 346
    13.4.2. Hybrid Heuristic Search Algorithms 349
13.5. Scheduling an Automated Manufacturing System 351

13.6. Modeling and Scheduling of a Semiconductor Test Facility 355
13.7. Summary 363

Chapter 14. Petri Nets and Future Research 365
14.1. CASE Tool Environment 365
14.2. Scheduling Large Production Systems 369
14.3. Petri Nets and Supervisory Control Theory 371
14.4 Application to Multi-lifecycle Engineering Research 372
14.5 Benchmark Studies and Comparisons 375

Bibliography 377

Index 403

CHAPTER 1

INTRODUCTION

1.1 Background, Motivation and Objectives

Flexible and agile manufacturing is of increasing importance in advancing factory automation that keeps a manufacturer in a competitive edge. Flexibility signifies a manufacturing system's ability to adjust to customers' preferences and agility means the system's speed in reconfiguring itself to meet changing demands. Both together make it possible for manufacturers to respond instantly to the market. Design of such flexible and agile manufacturing systems requires tremendous team effort with information, knowledge and expertise from customers, system analysts, designers, and engineers in many disciplines such as manufacturing, industrial, mechanical, electrical, computer, software, and systems engineering. Approaches that allow all the involved persons to communicate and apply effectively at all the design stages have been sought by researchers and practitioners for the past several decades. This book aims to present a unified and consistent approach based on Petri nets (PN) to tackle numerous challenging problems that researchers and engineers encounter in the design of advanced manufacturing systems. The book focuses on the applications of Petri nets to modeling, analysis, simulation, control, and scheduling of flexible manufacturing systems (FMS).

The increasingly expensive hardware and software investment involved in building up a flexible manufacturing system forces engineers and designers to model it and experiment with the model using computers prior to its actual implementation. A model is a mathematical representation of the important features of the system. Conventional modeling tools such as continuous differential and difference equations or inequalities are not sufficient to deal with flexible manufacturing systems that can be characterized as discrete-event driven systems. These systems can be asynchronous, comprising many concurrent components or subsystem interacting in a complex way over time. Petri nets have been developed to serve this modeling purpose. By the manipulation of Petri net models, we hope that new knowledge about the modeled manufacturing systems can be obtained without danger, cost, or inconvenience of manipulating the real system itself. Such Petri net models can be conveniently analyzed with the computer tools available. They can be simulated under various operational policies such as push and pull strategies. They can also be evaluated for different schedules. Furthermore, these models can be used to derive the best schedule by applying such methods as heuristic search. They can serve as a basis for control software development for complex systems.

The main objectives of this book are as follows:
1. To overview flexible manufacturing, agile manufacturing, computer-integrated manufacturing as well as intelligent automation, and review applications of Petri nets as an integrated tool and methodology in FMS design;
2. To present the fundamentals of Petri nets and their properties and introduce special classes of Petri nets, and the modeling methods via several manufacturing systems;
3. To present timed Petri nets and the analytical approaches to their solution for performance evaluation of manufacturing systems;
4. To present Petri net simulation tools and their applications to the performance comparison of a flexible factory automated system with push and pull paradigms;
5. To present and apply augmented timed Petri nets for the simulation of robotic assembly systems with breakdowns;
6. To propose real-time Petri nets for discrete-event control of automated systems and compare with the conventional approaches such as relay ladder logic diagrams;
7. To present a methodology to use Petri nets in the object-oriented design of reusable control software; and
8. To present scheduling approaches using Petri nets and heuristic search algorithms and illustrate their advantages over the previous methods such as mathematical programming and rule-based approaches.

1.2 Historical Perspective on Manufacturing Systems

Production has evolved from labor-intensive production systems, mass-production lines, automated mass-production lines, jobshop, group/cellular manufacturing cells, to flexible manufacturing systems, agile manufacturing systems, and computer-integrated manufacturing systems. They can be shown roughly into four stages in Fig. 1.1.

Labor-intensive production systems remained as a primary way before the 19th century. Highly-skilled craftsmen made a complex product, e.g., watch, by using crude tools and materials. Human energy is a main energy source with limited usage of wind and water power. The invention and development of steam engines in the eighteenth century empowered human being in producing more complex products such as locomotives.

1.2. Historical Perspective on Manufacturing Systems

Mass production arose from the need to supply a large volume of uniform products, such as automobiles, to satisfy the market at the end of the last century. The early production and assembly systems were rigid and expensive, thus impossible to change to handle many variations in product types. Development of electrical devices such as electrical switches and motors led to better control of machines and resulted in machines with certain flexibility. Computer technologies allowed researchers and engineers to develop computer numerically controlled machines and programmable logic controllers to automate manufacturing operations. Recent developments in information technology including computer hardware/software and network techniques make it feasible to achieve the purposes of rapid product prototyping, concurrent engineering, flexible and agile automation and computer-integrated manufacturing.

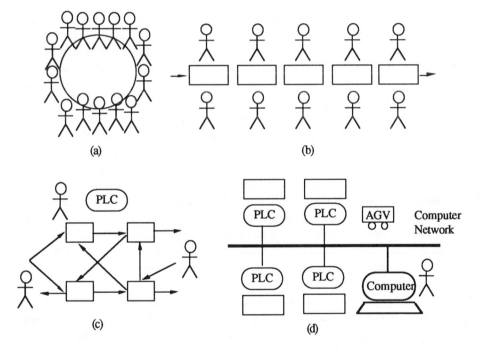

Figure 1.1 Production evolvement (a) Labor-intensive systems, (b) Mass production lines, (c) Job-shops, and (c) Flexible manufacturing systems/Computer integrated manufacturing systems.

The invention and applications of steam engines and other water-driven power generators in the eighteenth century triggered industrial revolution. It freed people from heavy labor work. Meanwhile, the concept of the division of labor was developed and used to increase the productivity. Continuous flow manufacturing systems were introduced and shifted toward mass-production-oriented factories. Each individual was responsible for a small part of a product. Entering the nineteenth century, the communication and transportation infrastructure started to be significantly and increasingly improved for further industrial growth. Electricity was introduced to develop electrically powered machine tools. The internal combustion engines developed around 1900 helped mass-production of products such as automobiles. Large-scale assembly and mass production became very popular over the world. Another revolution lied ahead.

In the late 1940s, a first transistor was invented. Then integrated circuits as well as digital computers were developed in the 1950s. In the manufacturing field, a first numerically controlled milling machines was introduced. In 1967, computer numerically controlled (CNC) machines were developed and put into use. Integration of computer and machines greatly improved the flexibility associated with the machines and manufacturing systems. Fixed automation started to evolve into flexible automation. The fast development of computer and information technologies led to information revolution in many fields including manufacturing. In the early 1970s, computer-integrated manufacturing (CIM) was coined to provide a new concept and direction to grow manufacturing enterprises [Harrington, 1973]. The CIM concept has changed over time from computerized workcell, flexible manufacturing systems, large-scale automation, computer aided design and manufacturing (CAD/CAM), interfacing and communications concepts to the current state: an information system that controls data flow among all the function units in a manufacturing enterprise [Groover, 1980; Rembold et al., 1993; Ranky, 1997]. CIM as a system provides computer assistance to all business functions within an enterprise - from customer needs, product design, order entry to product manufacturing and shipment. The goal of CIM is to reduce the product cycle time, improve the product quality and reliability, increase the productivity, and lower product cost, thus maintain and improve the manufacturing competitiveness in the world.

Implementation of successful CIM systems has taken place in many enterprises. It requires tremendous effort made by both management and engineering personnel in a company. A central technical challenge is the modeling, planning, scheduling, control and monitoring of automated factory systems, in particular, flexible manufacturing systems that are often most precious resources. This book is dedicated to the presentation of Petri net-based methods and tools to resolve this challenge facing to managers and engineers who are responsible for development of FMS and CIM systems. It also offers researchers with fundamentals of Petri nets and related approaches to study design issues of flexible manufacturing systems as well as computer integrated manufacturing systems.

1.3 Historical Perspective on Petri Nets

Petri nets have been originated from Carl Adam Petri's doctoral dissertation work on communication with automata in 1962. He described using a net the casual relationships between events in a computer system. His work came to the attention of A. W. Holt and others of the information System Theory Project of Applied Research, Inc. in the United State. Their work illustrated how Petri nets could be used to model and analyze systems of many concurrent components. Petri's work also came to the attention of The Computation Structure Group at Massachussetts Institute of Technology, led by Professor J. B. Dennis. Several doctoral theses and technical reports were published during the early 1970s. Most of the publications on Petri nets before 1980 were listed in the annotated bibliography of the first book on Petri nets [Peterson, 1981]. The work done in Europe on Petri nets and the published papers are annotated in the second Petri net book [Resig, 1985].

Starting in the late 70's, Petri nets became a very active area, especially in Europe. Annual conferences on Applications and Theory of Petri Nets have been held since 1979 and the proceedings published in the series of Lecture Notes of Computer Science by Springer Verlag. Most of the studies focused on information processing systems in the computer science community. An excellent tutorial paper was given by Professor T. Murata in 1989, which comprehensively presented properties, analysis, and applications of Petri nets and a list of references of significance [Murata, 1989]. Most of the earlier applications and theory of Petri nets aimed to information processing systems. The books [Peterson, 1981; Resig, 1985] and most of the papers were primarily targeted at computer scientists and graduate students.

Researchers with engineering background started their probe into the application of Petri nets in engineering systems particularly automated manufacturing systems in the early 1980's. They found that Petri nets were a powerful tool in describing event-driven systems. These systems may be asynchronous, contain sequential and concurrent operations, and involve conflicts, mutual exclusion and non-determinism. Such systems are termed as discrete event systems or discrete event dynamic systems (DEDS). The book [David and Alla, 1992], which is the first in its kind, presents Petri nets as a modeling tool of discrete event systems. The book [Zhou and DiCesare, 1993] focused on modeling and synthesis methods and discrete-event control of manufacturing systems. Another book [Desrochers and Al-Jaar, 1995] presented a comprehensive review of Petri nets in manufacturing automation and their applications to analysis, performance evaluation, and control of manufacturing systems. Other significant books and tutorials addressing Petri nets and their applications to automation include [Silva, 1985; DiCesare *et al.*, 1993; Proth and Xie, 1996; Silva and Vallette, 1990; DiCesare and Desrochers, 1991; Zhou and Robbi,

1994]. A tutorial on Petri nets and their industrial applications is presented in [Zurawski and Zhou, 1994]. A volume addressing the issues of flexible and agile automation using Petri nets is edited [Zhou, 1995]. Special issues on Petri nets and their applications were published, e.g., 1991 February issue of IEEE Transactions on Software Engineering on "Petri Net Performance Models," 1994 December issue of IEEE Transactions on Industrial Electronics on "Petri Nets in Manufacturing" [Zurawski and Zhou, 1994], and several to be published for IEEE Transactions on Semiconductor Manufacturing and Journal of Advanced Manufacturing Technology. These publications indicate the depth and breadth of the development and applications of Petri nets in the area of manufacturing automation.

While many activities are currently engaged in Petri net applications to the manufacturing automation area, the following foci could be observed since the late 70's [Zhou and Robbi, 1994]:

1. Early interest in Petri nets aroused from the need to specify and model discrete manufacturing systems. The activities in this area started with the Petri net representation of simple production lines with buffers, machine shops, and automotive production systems, and proceeded with modeling of flexible manufacturing systems, automated assembly lines, resource-sharing systems, and recently just-in-time and Kanban manufacturing systems. The work has continued to specify and model plant-wide production systems under different operational policies such as push and pull paradigms.

2. Early research focused on qualitative analysis of PN models of manufacturing systems. Reachability analysis shows whether a system can reach a certain state. Desired sequences of events are validated according to the system requirements. Other PN properties are used to derive the DEDS stability, cyclic behavior, and absence of deadlocks. Deadlock avoidance policies and their evaluation in a PN framework are still a hot research topic.

3. As temporal or quantitative properties become an important consideration, timed PNs are used to derive the cycle time of repetitive and concurrent manufacturing systems. To deal with the stochastic nature of many production operations, stochastic PNs are used to derive the system production rates or throughputs, critical resource utilization, and reliability measures. Their underlying models are Markov or semi-Markov processes. The direct construction of Markov chains is avoided thanks to the conversion algorithms for stochastic PNs.

4. When state explosion problems arise or the underlying stochastic models are not amenable for tractable mathematical analysis, simulation must be conducted for analysis of both qualitative and quantitative properties. Fortunately, PN models can be easily utilized to drive complex discrete event simulation. Several packages based on PNs exist for the simulation purpose. The use of simulation techniques to

evaluate flexible manufacturing systems and a fault-prone robotic assembly system will be explored later in this book.
5. Programmable logic controllers (PLCs) are commonly used in industrial sequence control of automated systems. They are designed through ladder logic diagrams that are known to be very difficult to debug and modify. It is observed that a PLC can be converted into a PN and vise versa for a subclass of PNs. Early work includes the conversion of a PN into a ladder logic diagram for PLC implementation. Direct PN controllers without help of PLCs can also be implemented through either a Petri net interpreter or their corresponding control code. For most cases, additional information to represent the real-time signals and status needs to be incorporated into such PN models. The advantages of PNs include their relative ease to represent and modify the control logic, and their potential for mathematical analysis and graphical simulation to validate a design. It can be proved that the graphical complexity of PNs grows with system complexity less than that of ladder logic diagrams. The comparison results between Petri nets and ladder logic diagrams will be presented later in this book.
6. With a mathematical representation available, designers are able to use PNs for rapid prototyping of a process control system or discrete event control. Virtual factories can be realized through computer graphics using PNs. Stepwise testing and implementation can be achieved by connecting the actual equipment into a PN-based design system, thereby reducing design and development time.
7. Petri nets have been combined with other approaches to achieve various purposes in process planning and scheduling, intelligent control, expert system construction, knowledge representation, and uncertainty reasoning. For example, the correspondence between a PN and an expert system can be established. This can greatly aid in consistency checking of an expert system. Optimal or suboptimal schedulers can be derived with heuristic search algorithms. The research over the past several years demonstrated the efficiency of the PN-based approaches.

1.4 A Robotic Cell and Petri Net Model

We conclude this chapter with an example that illustrates the issues in FMS design and the applications of Petri nets in resolving them. Several Petri net concepts will be introduced and demonstrated. If you do not understand all the concepts and methods discussed here, do not be concerned. They will be more fully developed and explained in more detail later.

Consider a Fanuc machining center located on the factory floor at the Center for Manufacturing Systems at New Jersey Institute of Technology. We start with a

simplified version of the cell and finish with an FMS version. The actual cell is shown in Fig. 1.2. It comprises two workstations (WS), a milling machine WS1 and a drilling machine WS2, and a robot R. Consider a product type that needs only milling. Then the operational specifications for this cell are as follows:
1. To start a cycle, a raw part and the robot must be available.
2. The robot moves a raw part from the incoming conveyor, and loads it at WS1.
3. The milling operation is performed at WS1 while the robot backs off (returns).
4. The robot unloads the part from WS1 and deposits it on the outgoing conveyor, and returns.

The above steps repeat.

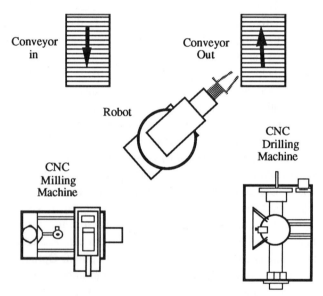

Figure 1.2 Fanuc System Layout.

In Petri net modeling, there are two nodes, places and transitions, represented by circles and bars, respectively. The former are used to represent the status of a resource, e.g., its availability, a process, e.g., its undergoing, or condition, e.g., its satisfaction. The latter is used to model events, e.g., start and end of an operation. A token's presence in a place indicates whether a resource is available, a process is undergoing, or a condition is true. A token is represented by a dot located in the circle representing a place. Multiple tokens often imply availability of multiple resources or the undergoing of operations of several parts. When the conditions for an event become all true, the corresponding transition is enabled and thus can fire. Firing enables the flow of tokens from places to places, implying the change of system status.

1.4. A Robotic Cell and Petri Net Model

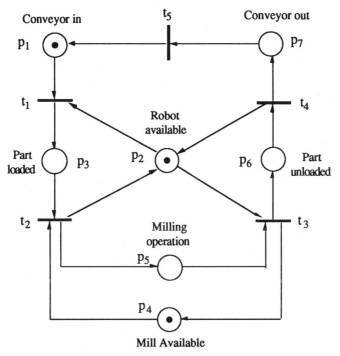

Figure 1.3 Petri Net Modeling.

For the Fanuc system, we determine the initial system conditions, events that can take place, several intermediate conditions and events, and finally the completion of the finished parts and restart the operations. In particular, we model the availability of a raw material in the input conveyor as a place p_1 and robot as p_2. We model the start of loading a part with Robot as transition t_1. Then two directed arcs are introduced linking from p_1 and p_2 to t_1 respectively as shown in Fig. 1.3. They mean that two conditions in p_1 and p_2 have to be met before the event in t_1 can happen. Next, firing t_1, i.e., occurrence of the event "start of loading" allows the robot to enter the status of "loading a part" modeled by place p_3. The arc from t_1 to p_3 is used to reflect this fact. Then the loaded part is available for Mill to initiate a milling operation. Place p_4 models "availability of Mill", and transition t_2 is used to model both "completion of robot's loading" and "start of milling operation". Two arcs leading from p_3 and p_4 to t_2 represent that t_2's being enabled requires two conditions met, i.e., a part being loaded and Mill being available. Two arcs from t_2 to p_2 (Robot available) and p_4 (Milling operation) signify that firing t_2 releases the robot and initiates milling. Applying the similar method, we modeled the remaining portions of the system. Transition t_5 is introduced to model the fact that the system is able to obtain a raw material once there is one product generated. Initially there is one robot and one miller idle. Thus p_2 and p_4 are marked with one token. Place p_1, however, can have one token to represent initially one raw piece available; two or more tokens represent two or more raw pieces available. In practice, the input raw piece may arrive faster than their processing. Then two or more pieces become available.

The dynamic behavior of the system can be observed through the model. When there is only one raw piece in place p_1, the net is executed as follows:

1. Initially only transition t_1 is enabled since only t_1's enabled conditions are met. Firing t_1, i.e., start of loading, removes two tokens in total from p_1 and p_2 and deposits a token to p_3. Places p_1 and p_2 hold no token and transition t_1 is disabled.
2. Only t_2 is enabled and firing t_2, i.e., completion of loading and start of milling, removes two tokens in total from p_3 and p_4 and deposits a token to p_2 and p_5, respectively. Since only t_1 related input and output places, i.e., p_{1-3} are affected, only p_{1-3} related transitions, i.e., t_1 and t_2, are checked.
3. Only t_3 is enabled and firing t_3, i.e., completion of milling and start of unloading, removes two tokens in total from p_2 and p_5 and deposits a token to p_6.
4. Only t_4 is enabled and firing t_4, i.e., completion of unloading, removes a token from p_6 and deposits a token to p_2 and p_7, respectively.
5. Only t_5 is enabled and firing t_5, i.e., releasing another raw material piece, removes a token from p_7 and deposits a token to p_1. The system returns to the initial condition and repeats the above process.

Therefore, the above model accurately describes the system's discrete-event behavior. A strict sequence of the events is followed when only one raw piece is available in place p_1.

Assume that two tokens are assigned to p_1, signifying the availability of two raw pieces initially. Then

1. Initially only transition t_1 is enabled. Firing t_1, i.e., start of loading, removes two tokens in total from p_1 and p_2 and deposits a token to p_3. Place p_2 holds no token and transition t_1 is disabled. However, place p_1 still has a token.
2. Only t_2 is enabled and firing t_2, i.e., completion of loading and start of milling, removes two tokens in total from p_3 and p_4 and deposits a token to p_2 and p_5, respectively.
3. Both t_1 and t_3 are enabled and firing t_3 will continue the system operations smoothly. Firing t_1, however, leads to an undesirable status, a deadlock. Firing t_1 removes two tokens in total from p_1 and p_2 and deposits a token to p_3. At this state, t_1 is not enabled due to the absence of tokens in p_1 and p_2, t_2 is not due to the absence of a token in p_4, and t_3 is not because of no token presence

1.4. A Robotic Cell and Petri Net Model

in p_2, and t_4 and t_5 are neither because of no token in places p_6 and p_7. In other words, Robot is waiting for Mill to start milling operation of the loaded part, while Mill is waiting for Robot to unload its milled part. Such a circular waiting in an automated system is a catastrophe. This is what a designer has to avoid in the design of an automated manufacturing system.

Performance issues arise when operation times are considered. When there is only one token in the model, the cycle time for a product is simply the sum of all the times spent on loading, milling, and unloading plus the conveyor's transportation time in reality. The performance evaluation can be much more complicated when we consider randomness as well as failures of machines and robots. The evaluation of cycle time and productivity is necessary in order to optimize the system performance when several alternatives exist for system configurations, operational policies, etc.

When robot and machine failures are involved, different modeling paradigms have to be used for different faults. Since breakdowns may come any time, their real-time handling has to be effectively modeled. Petri nets can be augmented for this purpose. Furthermore, the optimal operational and/or design parameters of a system may have to be adjusted when breakdowns are taken into account.

Scheduling issues arise whenever there are shared resources or alternative routes in an FMS. Resource sharing happens at two levels in a flexible manufacturing system. At the physical level, robots, automatic guided vehicles, conveyors, and related transportation systems are shared by machines. At the job level, all the flexible machines are shared by different types of jobs. The advantages of Petri net approaches to be fully demonstrated later are that since the model captures all the precedence and concurrent relations among operations, the resulting schedule is automatically deadlock-free, and two level resource-allocation problems are resolved in one step. Many existing approaches such as integer programming cannot handle these without significantly increasing the solution complexity [Xiong, 1996]. Suppose that the FMS in Fig 1.2 produces alternatively Type A parts using only Mill and Type B parts using first Drill and then Mill by using the raw material. Then Fig. 1.4 is the model for the system to produce Type B products only. It models the sequence of loading, drilling, milling, and unloading processes for Type B products. The places such as p_1, p_2, p_4, and p_7, and transition t_5 remain unchanged while others are introduced to model the Type B related events and operations.

In order to obtain a global net model, common places, transitions, and arcs are shared and integrated into the model shown in Fig. 1.5. To guarantee the alternative production of types A and B, a pair of control places, i.e., p_{14} and p_{15} are introduced. Thus initially both t_1 and t_6 are enabled. Firing either one will disable itself without firing the other one. For batch processing, two output places need to be introduced to count the number of Type A and B products generated, respectively as shown in Fig.

1.5. Locating the optimum schedule can be then converted into the search of the optimum path from the initial marking to the desired final marking.

The complexity of a Petri net model becomes a problem from a graphical point of view. Fortunately, its model has its underlying mathematical representations through matrices. In computer implementation, a special structure, e.g., link structure, can be used to reduce the memory requirements for a huge matrix format. The enabling and firing rules can be easily implemented to perform scheduling and discrete event control. The graphical form can be made available in a modular way for a quick visual analysis and help designer debug and maintain the system.

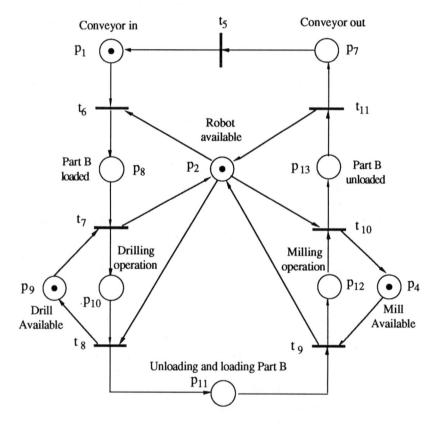

Figure 1.4 Petri net model for Part Type B with the Fanuc cell.

1.4. A Robotic Cell and Petri Net Model

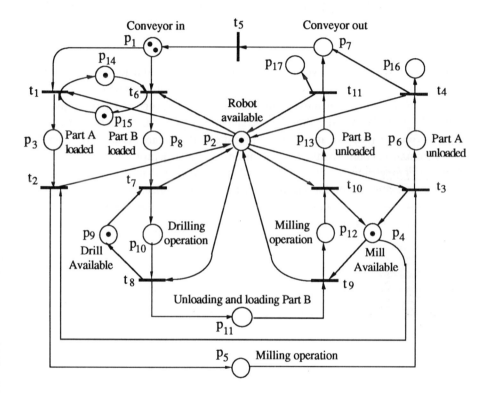

Figure 1.5 A Petri net model of two job types for the Fanuc cell.

1.5 Summary

The flexible manufacturing system (FMS), a job shop of sorts, presents formidable design problems. Formal discrete event system models are useful to predict operational characteristics such as throughput rates and utilization of machines and to perform scheduling and discrete event control of the FMS. The modeling using Petri nets for the Fanuc cell is intuitively conducted and the related issues to system design are discussed. These issues include deadlock avoidance, system behavior modeling, performance evaluation, scheduling, fault modeling and handling, as well as discrete event control.

An FMS features diversity in material movement and job types, requiring flexible automation. Flow shop production can be efficiently implemented by hard automation, with material movement handled by conveyors and fixed-movement robots. In an FMS material flow is mechanized by programmed robots, Automatic Guided Vehicles (AGVs), and people. Since these devices are more costly than their hard automation counterparts, the system design must permit them multiple roles, allowing competition for their services. System designs that allow for concurrent activities and competition for resources are difficult to realize. Discrete event system technology can help the designer predict behavior and tune system performance consistent with economic goals.

CHAPTER 2

FLEXIBLE MANUFACTURING SYSTEMS: AN OVERVIEW

2.1 Introduction

Increasing global competition has made many business leaders and policy makers turn their attention to such critical issues as productivity and quality. Businesses seek new approaches to production processes and manufacturing techniques and explore new boundaries of technology. Buzacott (1995) presents some important features of today's manufacturing environment and speculates on their implications for organizational changes. One of the frequently prescribed remedies for the problem of decreased productivity and declining quality is the automation of factories. More specifically, technologies such as computer integrated manufacturing (CIM), robotics, and flexible manufacturing systems (FMSs) are the focal points of much research and exploration. Flexible manufacturing technologies are a relatively recent addition to the portfolio of automated manufacturing technologies [Kunnathur and Sundararaghavan, 1992]. Such views as shown in the following "official" statements are representative of the new attitude toward advanced technologies:

> *"FMS can help our economic recovery... Flexible manufacturing systems can bring tremendous economic advantages to batch manufacturers. Beyond the attraction of increased efficiency, companies must automate if they are to compete in foreign and domestic markets with companies in Japan, Germany and other foreign countries, which are automating their manufacturing operations vigorously " [Goldhar, 1984]*

Other similar views are held by business leaders on factory automation in general and FMS in particular [Goldhar, 1984]. To achieve strategic competitive benefits of FMSs - improved quality, greater flexibility, and cost reduction, firms must carefully manage the implementation of these technologies [Sanchez, 1996]. Despite rapid world-wide growth of FMS installation [Darrow, 1987], many manufacturers still shy away from these advanced technologies. What has kept the manufacturers from getting serious about flexible manufacturing systems is probably the improper performance evaluation criteria used to evaluate managers [Buffa, 1984]. While there is little disagreement about the necessity of rapid movement toward the automation of manufacturing processes and the establishment of such systems as FMS [Buffa, 1984], there is also a great deal of confusion about the fundamental approaches necessary to do so. The same is true about FMS technological complexities and the barriers to its application, implementation, and particularly, integration of FMS with rest of the operations. The purpose of this chapter is to present an overview [Kaighobadi and Venkatesh 1994] and a survey of research in FMS. The objective is to review the relevant research on:

- The definition of the term "flexible manufacturing system"
- The impetus for change from the traditional manufacturing systems
- The major concepts, analytical models, and application and implementation problems in the FMS

This chapter is organized as follows. In the next section, various definitions of FMSs are discussed. Section 2.3 elaborates the impetus for change from conventional manufacturing systems to FMSs. Section 2.4 deals with the issues related to installation, implementation, and integration of FMSs. Some of the important real-life installations of FMSs and their applications are discussed in Section 2.5. In Section 2.6, several issues related to the installation and implementation of FMSs are presented. The last section of this chapter presents some general conclusions regarding FMSs.

2.2 Definitions of FMS

Despite all of the interest in flexible manufacturing systems (FMSs), there is no uniformly agreed upon definition of the term FMS. The main distinguishing feature of FMS from traditional manufacturing systems is "flexibility" [Gupta and Buzacott, 1989] which does not have the precise definition. One of the most cited definition of FMS is by Ranky [Ranky, 1983], who defines an FMS as a system dealing with high level distributed data processing and automated material flow using computer controlled machines, assembly cells, industrial robots, inspection machines and so on, together with computer integrated material handling and storage systems. In fact, the scope and variety of flexible manufacturing is commonly disputed and are the focus of many research efforts. However, the components and characteristics of an FMS, as described by different authors and

2.2. Definitions of FMS

researchers, are generally as follows [Davis et al., 1989]:
- Potentially independent NC machine tools,
- An automated material handling system, and
- An overall method of control that coordinates the functions of both the machine tools and materials handling system so as to achieve flexibility.

The specific manufacturing situations that would be suitable for the adoption of FMSs were identified as early as 1973. The following are the production situations that are encompassed by FMS [Darrow, 1987]:
- A variety of high precision parts are machined (typically job shop)
- A relatively large number of direct numerical control (DNC) machines are required.
- Some form of automated material handling system (MHS) is used to move the work pieces into, within, and out of the FMS.
- On-line computer control is used to manage the entire FMS under conditions of varying parts production mixes and priorities.

It can be concluded from the above that an FMS involves a number of machine centers and material handling systems integrated by a hierarchy of computer control. Furthermore, FMS is capable of handling flexible routing of parts instead of running parts in straight line through work stations. The term CMS (Computerized Manufacturing System) and Variable Mission Manufacturing (VMS) have also been synonymously used with FMS. In a flexible manufacturing system numerically controlled (NC) machines are controlled by computers; parts are handled by robots; and finished products are carried to specific destination via automatically guided vehicles (AGVs). Tool magazines and automatic tool changing systems are utilized to process the raw-material and as engineering or design changes occur in product, they are incorporated into the computer programs or data base. According to Klahorst (1981), FMS is a group of machines and related equipment brought together to completely process a group of or family of parts and includes the following primary and secondary components:

Primary components:
- Machine tools
- A material handling system
- A supervisory computer control network

Secondary Components:
- Numerical control (NC) process technology
- Spindle tooling

- Work-holding fixtures
- Operations management

Klahorst (1981) indicates that the precise semantics of these components depends on the type of application problems to which the intended FMS will be applied. For instance, the primary components of FMS machining modules may include head changer, machining center, maxi machining center, vertical turning lathe, NC milling, NC turning, and head indexer, while the primary components for typical material handling FMS module would include roller conveyer, in/out shuttle, through shuttle, guided vehicle, shuttle car, and tow line. What seems to be the differentiating point in these definitions and elaborations is where one places the emphasis on flexibility. Some place it on flexibility in changing machine configurations which permits change of product designs without much delay. Others place the emphasis on the flexibility in handling of materials. The former group sees the FMS's potential in adaptability to demand for different product designs and the benefits of low inventory due to made-to-order production, while the latter finds the advantages in maximizing machine utilization [Mullins, 1984]. The nature of *flexibility* in terms of affecting the definition of an FMS is well treated in [Yilmaz and Davis, 1987]. Of course, these two points of emphasis are neither mutually exclusive nor in conflict. Rather they can be supportive of each other. In defining FMS what is important is "understanding of what FMS represents conceptually, and what it means to a company in terms of manufacturing strategy" [Hughes and Hegland, 1983].

2.3 Impetus for Change

The annual growth rate of the U.S. market for FMSs has been significant since 1989. The growth is due to the increased desire for change and increased interest in the automated factory. According to Robert J. Maller of Deere & Corporation, two issues will preoccupy the minds of manufacturing managers in the coming decades: the concern for quality and concern for cost reduction. Manufacturers will be able to pursue two goals which have traditionally been considered as conflicting and irreconcilable; low volume and low cost production in response to rapid market changes [Yilmaz and Davis, 1987]. However, it should be noted that this interest in the automation of factories is not new [Buffa, 1984]. Accordingly, there was a period of interest in automated factories during the 1950s. Then this phase of interest declined and rose again in early 1960s. Another decline followed and by 1973 there was little interest shown in the area. Now, again in the 90s the headlines of the press represent the renewal of interest in automated factory concept. But this time the dream of the factory of the future seems to be closer to reality than ever,

2.3. Impetus for Change

thanks to new technological advancements in computers, robotics, fixtures and other components of advanced manufacturing technologies. The main reason behind this new surge of attention directed to FMS and other forms of the automated factory is increased competition, particularly international. The main incentive is reduced cost in production and adaptability to an ever changing environment. Automated systems such as FMSs, have the potential to improve the position of firms on both counts. Other reasons that account for the renewal of interest in advanced technologies include "truncation of product life cycle, and increasing complexity of products" [Goldhar, 1984]. For example, those companies that have installed FMS have reported the following results [Klahorst, 1981]:

- Benefits related to cost reduction programs (55%)
- Benefits related to market response improvement (30%)
- Benefits related to flexibility in production (15%)

Salomon and Biegel (1984) compare FMS with conventional manufacturing technology under various states of risk and show that FMS provide substantial productivity improvement. Table 2.1 summarizes these findings. The entries in the first column of this table signify the productivity improvement. The subsequent columns compare how machine and work-parts spend their time in the shop of a conventional system and in an FMS using optimistic-pessimistic format. As revealed from the table, FMS outperforms the conventional system in terms of productivity improvement.

The case of unwanted dedicated machinery explains why FMS is a necessity rather than a luxury for some manufacturers in the face of world-wide competition. R&D comes up with the operating principle for a different kind of widget. Sales takes a look and projects requirements of 10,000 every year. And manufacturing, unable to economically produce the new widget on presently installed machinery, tools up to meet the exiting demand by installing the latest in high volume, hard automation machine tools. The widget in our story flops, however, as introductory products often do, and manufacturing faces the question "What do we do with all that dedicated machinery. Apart from these kinds of internal problems, external factors such as changes in demand pattern and consumer tastes, changing regulatory environment and labor force, and changes in competitive policies of other firms, all contribute to the renewed efforts of manufacturers to willingly embrace new options. There seems, however, to be some over-correction of past errors. For example, some companies spend huge amounts of capital on equipment without proper justification or knowledge of how to take advantage of their potentials [Buffa, 1984]. Based on extensive analysis of empirical studies, Yilmaz and Davis (1987) presented some propositions regarding issues of flexibility, productivity, and quality. From their investigation, major findings support that FMS investment leads to reduced labor cost, increased output, decreased manufacturing cost, increased flexibility, and reduced lead time.

Table 2.1 Comparison of how machine and work-parts spend their time in the shop of a conventional system and in an FMS using optimistic-pessimistic format [Salomon and Biegel, 1984].

Parameter	Conventional System Performance	FMS Performance		
		Pessi-mistic	Most Likely	Opti-mistic
Percentage of machine time the machine spends without parts	50	35	20	5
Percentage of machine time when there is a part on the machine	50	65	80	95
Percentage of time that the part is not being worked on while on the machine	70	35	21	7
Percentage of time when the part is being worked on while on the machine	30	65	79	93
Percentage of manufacturing lead time the part spends in moving or waiting	95	92.5	90	85
Percentage of manufacturing lead time when the part spends on the machine	5	7.5	10	15

2.4 Installation, Implementation, and Integration of FMSs

For a recent field study and analysis discussing the issues in FMS installation is presented by [Kunnathur and Sundararaghavan, 1992]. Their report is based on plant visits and extensive interview conducted as a part of field study of five distinct FMS installations. It is estimated that 75% of machined parts produced in the U.S. are in lots of 50 units or less [Gilbert and Winter, 1986]. The need for small lot production has justified installation of the FMS in a number of manufacturing

2.4. Installation, Implementation, and Integration

companies [Harvey, 1984; Attaran, 1992; Jari, 1991; Ram and Yash, 1991; Kakati and Dhar, 1991]. But the installation, implementation and integration of FMS create unique issues and problems. However, as a study reported in 1988 shows [Darrow, 1987] that only 64 FMSs have been installed in the U.S. and 253 worldwide, with the majority being in metal-cutting operations. This number, though small, has helped bring to the surface many issues regarding the installation and operation of FMSs. Subsequently more researchers and practitioners addressed these issues and made the installations of FMS easier and lucrative. Due to the concentrated research efforts in the area of FMS, the number of FMS installations have been sharply increased. A study reported in [Dimitrov, 1990] shows that approximately 750 FMSs are installed in 26 countries. Another case study shows how traditional machine tools can be integrated into FMS [Kwok, 1988].

Suresh (1990) presents a decision-support system to do integrated evaluation of flexible automation investments. His approach considers the factors: (1) a comprehensive investment context for multimachine systems; (2) additional refinements in modeling flexibility in economic evaluation; and (3) a structure to facilitate an integrated physical/financial evaluation. Appropriateness of FMS to a given production environment is extremely important and has to be established before investment commitments are made. To help determine their fitness, the type of FMS has to be considered. Classification of manufacturing systems (with respect to their type, degree of flexibility, and volume-handling capability) helps to determine when and where an FMS is most beneficial. Flexible Manufacturing Systems can be broadly classified into dedicated FMS, sequential FMS, and manufacturing cells. Table 2.2 shows various classes of FMS and the range of production volume for each classification. Klahorst (1981) provides a clear analysis of the installation/integration process of an FMS and provides some insightful guidelines drawn from the experiences of the Kearney and Trecker Company (a major producer of FMS equipment).

Table 2.2 Classification of Manufacturing Systems and FMS.

Type of Manufacturing System	Level of Flexibility	No. of Parts in Product Family	Average Lot Size
Transfer Lines	Low	1-2	7,000 & Up
Dedicated FMS	Medium	3-10	1,000-10,000
Sequential or Random FMS	Medium	4-50	50-2,000
Manufacturing Cell	Medium	30-500	20-500
Stand-alone NC Machine	High	200 & Up	1-50

Some of the questions raised in this analysis with respect to the installation of FMSs are: who should do it, when it should be done, and what the responsibilities of final users are. For instance, Klahorst (1981) argues that since 50% of an FMS project value is related to machines, industrial engineers are the primary people who should be involved in the process of FMS design and installation from the start. The circumstances under which FMS should be installed are obviously significant factors. According to Klahorst (1981), FMS should be installed under the following situations:

- when part size and mass exceed "jib crane" standards.
- when production volume is in excess of two parts per hour.
- when processing needs more than two machine types to complete a work piece.
- when more than five machines are required.
- when phased implementation is planned so that material handling provisions can be considered in the initial phases and bad habits can be avoided from the start.

The conclusion is that the more of the above conditions exist, the more incentive there is for transforming a conventional system into an FMS.

Given the fact that there are numerous uncertainties present in FMS installation, the need for an effective way to uncover the potential problems for the success of FMSs is obvious. A very useful approach to discover the potential problems with a system's operation is to simulate the system and pinpoint such problems. Simulation, in the case of an FMS, will help a) substantially trim installation costs, b) ensure that the design of the system is accurate, and c) spell the FMS design. Some of the major simulation studies and models related to FMS are explored in subsequent sections. The cooperation between users and vendors has been frequently emphasized both by researchers and users/vendors of FMS as a major factor of success in implementation of FMS. For example, Hughes and Hegland (1983) observe that an FMS:

> "requires a level of cooperation and exchange of sensitive business planning information between vendor and user heretofore unheard of in typical capital manufacturing equipment acquisitions." They see the relationship between the vendor and the user as a permanent one and call it "long-term partnership in productivity."

They argue that one of the most important factors contributing to the success of an FMS installation and implementation is the degree of commitment the potential company must make to be successful with FMS. Contrary to common belief, installation of an FMS is not limited to large corporations with vast financial resources. Mullins (1984) reports cases where small firms have installed FMSs that

2.4. Installation, Implementation, and Integration

have been successful. But regardless of size, the magnitude of commitment needed is enormous.

Introducing an FMS into an organization has significant strategic implications, such as replacing "economies of scale" with "economies of scope" [Goldhar, 1984]. The fact that such technologies have strategic implications does not mean that their implementation has to be wholesale installation and that the FMS has to be a full-fledged system with extensive risk involved. FMS can be installed incrementally and such a move can be made through employment of stand-alone machines or the utilization of manufacturing cells. Whether FMS introduction is wholesale or incremental, understanding the characteristics of different manufacturing systems will help recognize the potential problems associated with the installation and implementation. Black (1983) provides a useful and brief explanation of such characteristics and elaborates on different aspects of each manufacturing system that could help decide the implementation approach. Black's work is a ground work for identifying potential issues related to FMS installation. Regardless of size and scope, the question of whether an FMS is useful has significant strategic implications. Jukka and Iouri (1990) provides some guidelines for economics and success of FMSs. From a list of case studies of FMSs, they analyze costs and relative benefits of several hundred FMSs in the world. Primrose and Leonard (1991) provide a helpful overview of flexible manufacturing transfer and present a framework for investment consideration in FMS. They emphasize the significance of financial evaluation in adopting FMS and not relying only on "hopes" for future benefits. They provide guidelines as how to identify benefits related to FMS. In the same venue, Krinsky *et al.*, (1991) provide an analytical model for the evaluation of FMS investment, using the von Neuman-Morgenstern theory of utility together with the mean-variance approach of portfolio analysis. They identify a measure that takes into consideration both the capital cost of the new technology (FMS) and the monetary value of its output.

Analytical model building of FMS is a significant area which has to be cleared before a successful installation, integration, and implementation of the FMS is achieved. Currently, there are numerous studies related to various theoretical aspects of FMS. Stecke (1983) tackles the analytical issues related to production planning problems of FMS. The problems of grouping and loading for FMS production planning are examined in detail and formulated as mixed integer programming problems. Similarly, Buzacott and Yao (1986) review the basic features of FMSs and develop models for determining the production capacity of such systems. These models show the desirability of a balanced work load, the benefit of diversity in job routing if there is adequate control of the release of jobs, and the superiority of common storage for the system over local storage at machines. There are numerous simulation and analytical models dealing with various aspects of FMSs. Among them are: perturbation analysis, queuing networks, artificial intelligence, and more recently Petri nets [Kamath, 1994]. However, the detailed exploration of all these models is beyond the scope of this chapter. Interested readers are referred to the some of the following research studies connected with these models: simulation [Schroer and Tseng, 1985; Rolston, 1985; Glassey and Adiga, 1990; Basnet et. al., 1990; Narayanan et al., 1992]; perturbation analysis [Suri and Dille, 1985]; queuing networks [Suri and Diehl, 1985, Kamath and Sanders, 1991, AT&T, 1992]; artificial intelligence [Dhar, 1991]; and Petri nets [Silva and Valette, 1990; Cecil et al., 1992; Liu and Wu, 1993; Zhou and DiCesare, 1993; Venkatesh, 1995; Stanton, et al., 1996].

2.5 Applications of FMSs

There are numerous reports and case histories about the installation of FMSs. However, accurate statistics about the application of these systems are difficult to obtain. Determining the extent of FMS installations throughout the machine-tool industry is not a clear-cut task and entails many confounding issues both statistically and methodologically [Ranky, 1983].

Despite these impediments, a number of trade and professional journals report case studies of FMS installation and experiences gained through them. Some of the better known firms where either partial or complete FMSs have been installed and are operational include:

- General Electric,
- Ingersall Milling Machine Company,
- GM's Pontiac Division and Saturn plant and locomotive plant in Erie, Pennsylvania,
- Chrysler's Toronto plant,
- Cadillac's Livonia engine plant,
- Ford's Sterling Heights transmission and chassis-axle plant,
- GM's Buick City,
- Hughes aircraft plant in El Segundo, California,
- Pratt and Whitney's plant in Columbus GA, and
- Allen Bradley.

What could be generalized from the reports on FMS is that in almost all cases there have been reported improvements in quality, reductions in labor and inventory costs, and increased responsiveness to the changes in the market place. An interesting report by Kaku (1992) indicates that a number of installed FMSs are "under-utilized" in the sense that the flexibility inherent in the systems installed are

2.6. Problems in Installation and Implementation

being used. The author, in his visit to eight establishments with installed and running FMS, only a few were utilizing the full "flexibility" of their system and the rest used the FMS as a dedicated transfer line.

Cooperative and close relationship with the suppliers has already been discussed as a main factor in successful FMS implementation. Several case studies point to the signs that suppliers of FMS are more welcome now when they approach potential clients than a few years ago. In a round-table discussion reported by Dallas (1984), all participants agreed that FMS would succeed if:

- it is functioning in the right economic context,
- the company's organizational structure has been redesigned to accommodate the special requirements of FMS,
- there is closer cooperation between vendors and users of the technology, and
- the management understands that the rules of the game have changed.

The last requirement is particularly important since it entails the mind-set reorientation of managers with respect to performance evaluation, capital rationing criteria, and human resources management. The more difficult aspects of making FMS work are the issues associated with management and organization of the systems [Klahorst, 1981].

2.6 Problems in Installation and Implementation of FMSs

An FMS poses several problems in its implementation, particularly in its integration. These problems can be classified into two sets of inter-related problems: technical and managerial problems. Technical problems arise due to the complexity of the technology and technical decisions to be made in introducing advanced manufacturing systems. The managerial resistance to change is the primary reason for most of managerial problems and thus add to the complexity of implementing FMSs. Furthermore, managerial problems include "infrastructure" complexity created by such advanced technologies. Even though there are many earlier reports on problems related to FMSs [Harvey, 1984; Attaran, 1992; Anthony, 1991; Ismael, 1991; Kunnathur and Sundararaghavan, 1992], a few explore both the technical and managerial problems. The proposed approach of classifying problems connected with FMS into technical and managerial areas helps to comprehend the prominence of the combined role played by both management and technology in FMSs, and to aid researchers and practitioners in FMSs, in focusing on their specific interest. In this section, first the managerial and then the technical problems are discussed.

2.6.1 Managerial Problems

A number of studies have been devoted to the management of FMSs. National Institute of Standards Technology broadly addresses three main aspects of problems in FMS [Goldhar, 1984]. These are:
- How the control architecture can be simplified,
- Why FMSs are difficult to configure, and
- What can be done to ensure a consistently high level of quality in the products

Harvey (1984) believes that the major barriers to the efficient factory of the year 2000 are financial, human, institutional, and technological. Accordingly, executives must be open to new ways to be able to justify financially investment in the factory of the future. Human and institutional barriers need to be discussed together. Among these barriers are communication and education. The key to technological barriers, according to Harvey (1984), is integrating the individual parts of the factory, which can be accomplished through Computer Integrated Manufacturing (hence the relationship between FMS and CIM). Harvey (1984) provides a four-step procedure as a guideline to building a supportive organizational climate for the factory of the future:

1. Educate senior management,
2. Set goals and develop strategies,
3. Establish a corporate-wide culture, and
4. Develop a unified communication structure

Fear of change is another problem associated with implementation of FMS. Kiesler (1983) deals with this issue under the broad subject of the impact of new technology on the work place. In a broader sense, French (1984) discusses problems and issues concerning the CIM. Several of these issues are applicable to FMS as well. According to French (1984) the following are the major problems faced by manufacturers that may lead to failure in CIM (or FMS) implementation:

- Inadequate measurement system
- Partially obsolete facilities
- Inadequate data base
- User hostility
- Shortage of technical skill
- Incompatibility between systems
- Management generation gap
- Changes in management philosophy

2.6. Problems in Installation and Implementation

- Facilities with mixed processing
- Dynamic volume and mix
- Outdated organization
- Varieties of process options
- Loss of superior/subordinate support

Similarly, Scalpone (1984) considers the lack of education in both management and technical staff to be one of the major factors contributing to the failure of advanced technologies. In summary, the main managerial barriers of successful implementation of an FMS seem to be concentrated in a few major areas. These areas include:

- Lack of top management commitment and support
- Inadequate training and education of the personnel involved
- Improper evaluation of the situation/environment which presumably justified installation of the FMS
- Lack of long-term and committed relationship with the vendors of both raw material and the FMS equipment
- Lack of total commitment to the installation and implementation
- Existence of misconceptions about FMSs (such as FMS being good only for large companies and only applicable to large scale production).

Kunnathur and Sundararaghavan (1992) proposed a strategy for proactive management of automation projects to increase the success of FMS's installation and implementation.

2.6.2 Technical Problems

The smooth and economical functioning of any FMS depends upon the effectiveness of strategies for designing, controlling and monitoring of FMS. This section briefly describes the typical and important technical problems encountered in the successful implementation of FMS. In the following section, selected past and current research concerned with each of these problems are summarized and some suggested solutions are reviewed. Since problems related to FMSs cross the boundaries of manufacturing engineering, industrial engineering, computer engineering, computer science, electrical engineering, and operations management, it is difficult to cover each and every problem in great detail. The approach adopted in this section is based on presenting a synopsis of some selected chapters and major results. However, numerous references are provided for the readers who desire to pursue more detailed coverage of specific problems.

Even though some researchers have broadly divided problems of FMS into planning, designing, and controlling [Suri and Dille, 1985; Stecke 1989], and yet others [Imman, 1991] generally discuss issues relating to FMS implementation, much specific and detailed scrutiny of such problems is still needed. Understanding of the problems associated with the successful implementation of FMS requires a closer look at each and every problem more specifically and in greater depth. In order to address the specific problems in FMS, the following three step approach is adopted:

1. Considering all of the functional subsystems constituting FMS and understanding the prominence of each subsystem.
2. Addressing problems related to each subsystem.
3. Analyzing the impact of each subsystem on the system as a whole.

This section details some of the most common and important functional subsystems present in an FMS:

- CNC/DNC machine tool technology
- Tool management
- Automated material handling
- Communication network to integrate elements present in the FMS (real-time control issues are also included here)

The problems related to CNC/DNC machine tools require deeper understanding of the machine tool technology itself and falls beyond the scope of this book. Therefore, problems specific to CNC/DNC machine tools are not discussed here.

Tool management

The various definitions of tool management available in the literature may be due to diversified aspects by which the tool management is viewed. Some researchers define it in terms of four distinct sets of activities [Gray and Stecke, 1988]:

- stock control and administration,
- functional control,
- handling and transportation, and
- programming.

It is also defined as a strategy which aims at resolving problems related to various tool activities, including: acquisition, storage, database development, selection and allocation, inspection, presetting, delivery, loading, monitoring, replacement, requirement planning, and inventory control of tools [Reddy *et al.*, 1990a]. Despite

2.6. Problems in Installation and Implementation

the various orientations and approaches, there are some common goals in tool management. These goals are discussed in the following paragraphs.

Tooling is one element in FMS that is most prone to change due to external factors and may often cause discrepancy in FMS functioning. In other words, the probability of change in the tooling due to change in one or all of the external requirements (see Section 3), is much greater than the probability of changing any of the other internal elements. The changes in tooling include changes either in the number of each tool type and/or number of different tool types, and/or tool position in FMS (tool crib, magazines, etc.). The extent of individual or combined effect of external factors on tooling changes with respect to its number of tools of each type and the variety of tools can be enormous due to the huge tooling (several number of tools of several tool types) in most of the FMSs. Further, the influence of any external factor on tooling with respect to its position can become more complicated by tool flow from tool crib to inspection, to presetting stations, to tool magazine, and finally to the machine spindle. The constant wear and tear on tools adds to the complexity of the situation mainly when FMS is operating for a long period of time. Hence, tools are more prone to replacements during the long time functioning of FMS. Due to the factors and importance of tool management explained above, tooling is the most dynamic and critical facility in FMS and requires keen attention.

Tool management is a very complicated task and is often stressed by FMS users and researchers [Gray and Stecke, 1988; Reddy et al., 1990a,b]. Despite such complexity, there are successful working FMSs whose performance have been considerably augmented with efficient tool management [Gaymon, 1986; Tomek. 1986]. Tool management also attains paramount significance considering its economic impact, since tooling accounts for 25-30% of the fixed costs of production in automated machining environment [Tomek, 1986; Cumings, 1986]. Several firms have recently developed integrated tool management systems with tremendous encouraging results [Tomek, 1986]. In fact, the centralized tool management has introduced the fifth generation of FMS environment, indicating that tool management is one of the important subsystem in the FMS which influences the whole structure and operation of the FMS [Heywood, 1988].

Owing to the complexities in tool management, Maccarini et al., (1987), have suggested different tool room layouts comparing their performances and identified the parameters which describe better tool room behavior while the production process is being developed. However, the recent literature reveals that there are still many tool management problems to be solved [Reddy et al., 1990b; Zavanella et al., 1990; Venkatesh et al., 1997]. This is mainly due to the lack of comprehensive understanding of tool management which is necessary before attempting any tool related activities such as development of algorithms for tool optimization, development of control software, design of a tool delivery system, and framing of a new storage or tool flow strategy. Modeling of the tool management can be of critical help in this respect and aids in understanding of complex asynchronous concurrent interactions/tasks in tool management [Venkatesh and Chetty, 1992]. A more recent survey on tool management in FMSs is made by Veeramani *et al.*, (1992). They have examined tool management related research efforts in academia and outlined the characteristics of comprehensive tool management systems that are being used in industry.

Automated material handling

Recently, automated guided vehicle systems (AGVSs) have received increased attention by the designers and engineers of automated manufacturing systems [Gould, 1990; Lee and Lin 1995]. An AGVS may consist of multiple automated guided vehicles (AGVs) and computers to control them. AGVs are unmanned vehicles that carry workpieces among the workstations following guide paths. They are usually controlled either by on-board computers or by a central computer. The communication between an AGV and its controller is generally established through dedicated wiring embedded in the floor, although some recent AGVSs utilize wireless communication. AGVSs are widely used in FMSs as they provide flexibility in routing parts among elements present in the system. These systems are highly complex and expensive due to the dynamic environment in which FMS functions. Furthermore, if the AGVS is not efficient, the whole system performance may be impaired by the possible starvation of machines in the system.

A variety of analytical methods have been proposed by researchers for the design and control of AGVSs. Egbelu and Tachoco (1986), Mahadevan and Narendran (1990), Maxwell and Muckstadat (1982), and Wysk *et al.*, (1987) have proposed alternate procedures for estimating the number of AGVs required. Hodgson *et al.*, (1987) devised an analytical control strategy for scheduling AGVs which was extracted from Markov decision process optimal control policies. King *et al.*, (1989) evaluated heuristic control strategies for a system under varying arrival patterns. Bozer and Srinivasan (1989) suggested a tandem configuration for reducing software and control complexity of an AGVS. Malmborg (1990) developed an analytical modeling strategies that can be applied for both design of AGVS and performance evaluation of FMSs. The application of these analytical methods is limited due to their restrictive assumptions and hence they cannot effectively be used for controlling the FMS under investigation. This is because the above mentioned analytical approaches can not predict the dynamic behavior of the AGVS with respect to time. Often simulation studies are preferred over analytical approaches as they provide more insight into the problem and they are not hindered by too many assumptions. Bischak and Stevents (1995) have used simulation to compare the performance of tandem AGV system with that of conventional AGV track systems. Also, a number of simulation studies have been conducted to

2.6. Problems in Installation and Implementation

evaluate the effect of different parameters such as the number of AGVs, the number of pallets, buffer sizes, dispatching rules, bi-directional flows etc. [Egbelu and Tachoco, 1984,1986; Newton, 1985; Ozden, 1988; Vosniakos and Mamalis, 1990]. Vosniakos and Mamalis (1990) have addressed design issues of AGVSs for FMSs such as zone blocking, loading/unloading, and traffic control. Schrocr and Tseng (1985) have demonstrated the use of GPSS in modeling AGV based FMSs. The main limitation of the available simulation methods are that they are time consuming and need large computer memory. Hence, they are limited in their practical usefulness for on-line monitoring of AGVS.

Owing to the limitation of the available analytical and simulation methods, there is a need for a single and versatile tool for addressing both design and operational control issues of AGVSs. Petri Nets (PNs), a powerful modeling tool in the context of FMSs have been successfully used for such analyses. Alanche *et al.*, (1984) modeled AGV movements with Petri nets (PNs) and qualitatively investigated the deadlocks and possible vehicle collisions. Davis *et al.*, (1989) used PNs to formalize rules for allocating and dislocating the zones in an AGVS. Lee and Lin (1995) has used a special class of PNs called, attributed PNs to dynamically predict and avoid deadlocks in real-time operation control of zone-control AGV systems. They used a modular approach to facilitate the construction of PN models of zone-control FMSs. Archetti and Sciomachen (1989) have analyzed an AGVS quantitatively using PNs. Occena and Yokota (1991) modeled an AGV system in a Just-In-Time environment and introduced a new dispatching rule to have better inventory and transport control, compared to the traditional dispatching rules available. It is concluded from their study that traditional dispatching rules such as shortest processing time, first-in-first-served, etc. do not perform well in the FMS environment. In other words, there is no specific rule that performs better in an FMS that is operating under Just-In-Time environment. This is mainly because in such systems, processing on machines starts only when there is a demand and processed parts are dispatched only when the successive machine is ready to take the part from its preceding machines.

Although there is extensive literature on hardware issues, research concerning the design and control issues of AGVSs for FMSs is limited. In order to fully exploit the increased flexibility and adaptability by AGVSs, an in-depth study of the design and control issues is essential. Also, for the realistic analysis of AGV-based FMSs, both the machine scheduling and AGV scheduling have to be simultaneously investigated [Sabuncuoglu and Hommuertzheim, 1986]. Raju and Chetty (1993) used extended PNs, to simultaneously investigate machine scheduling, part scheduling, and AGV scheduling to design and evaluate AGV systems.

Control and communication system development among elements present in the FMS

One of the fundamental building blocks of FMS is data communications and control of devices. Communication development in FMSs is probably the single largest and most troublesome problem area encountered by users and suppliers of factory automation systems. A study by General Motors revealed that upto 50% of the cost for new automation projects was directly attributed to communication and control devices [Balph and Vittera, 1985]. This is because, the complete flexibility and efficiency of any FMS can be realized only when a systematic and reliable control system is developed to coordinate and monitor different activities taking place at the different levels in the system hierarchy [Jones and McLean, 1986; Pimentel, 1989]. Even with adequate automation equipment, an FMS may not live up to its performance potential due to the lack of appropriate integration and control of the FMS operations [Maimon, 1987]. Furthermore, this area of factory automation has become a test bed for establishing various network architectures, protocols, access methods, communication media, and band widths for many computer vendors. Hence, communication and control system development among various elements in FMS is very important. However, development of such systems in FMS is a complicated task because:

- An FMS is a complex distributed processing system where each element in an FMS has a data base and there are several such elements in FMS that have to communicate for the manufacturing of a finished product.
- Communication and compatibility among the equipment of different vendors is critical. These equipment include computers, local area networks (LANs), cell controllers, programmable logical controllers (PLCs), robots, machine tools, and similar digital control devices.

To alleviate the complexity of communication development, many efforts are underway in Europe and the USA to develop a common set of vendor-independent communication protocols which would be usable by all types and brands of factory automation equipment. Towards this aim, Manufacturing Automation Protocol (MAP) is developed. The objective of MAP is to make communication possible between mainframe systems, cell controllers, workstation terminals, programmable logic controllers, material handling systems, robots and other types of factory equipment. MAP has received some criticism by researchers and users because of its complexity. For example, MAP specification is suitable at plant level communications, but many times becomes too heavy at cell level. Other protocols include fieldbus. Typical static data sets in an FMS include configurations of machines, AGVs, and robots and their characteristics and process plans. Common dynamic data sets include status of machines, AGVs, and robots

2.6. Problems in Installation and Implementation

that change with respect to real time. The control algorithms act as an interface between the static and dynamic data sets at different levels of hierarchy in the FMS. Maimon (1987) extensively discussed the tasks of the operational control in such an interface. He also presented a generic hierarchical control system and illustrated the development procedure with an example of flexible manufacturing cell to carry out the FMS integration from the production requirements to the actual operation of the machines.

However, the last task of an operational control system, namely, continuous monitoring and handling of breakdowns is not detailed. For example, there is no mention of about how tool wear is monitored in real time. Tool wear monitoring is essential to achieve uninterrupted machining and to schedule tool replacements when the limit on tool life is reached. The complexity of tool wear monitoring and its significance in the context of FMS functioning is highlighted by Masory (1991) and Venkatesh et al., (1997). Handling of machine and/or robot breakdowns in manufacturing systems is also essential to minimize the system down time and thereby not impairing the system performance. In case of breakdowns, there are many concurrent actions to be taken to bring the system back to its normal functioning.

Furthermore, some design parameters such as the number of assembly fixtures required to achieve maximum production rate may change with and without consideration of breakdowns in the system as addressed later in this book. In addition, some important control and monitoring issues have to be addressed in detail during breakdown handling. However, there are a few reported studies that consider breakdown handling in the analysis of a manufacturing system. For example, the importance of breakdown handling is highlighted and breakdown of work stations is considered in modeling of flexible manufacturing system by Barad and Sipper (1988) and Sheng and Black (1990). However, neither of these papers presents the detailed modeling and analysis of the system with breakdowns. Venkatesh et al., (1994a) presented a detailed breakdown-handling approach using Augmented Timed Petri nets to investigate several design and performance issues of flexible automated assembly systems. By distinguishing the failures at the process-level and production-system level, Adamides et al., (1996) used colored timed Petri nets to present a systemic framework for the recovery of flexible production systems. The important highlight of their research is the separation of the error recovery strategy from the system's dynamics which enables to apply various operational policies without changing the system model.

Jafari (1992) presented an architecture for a shop-floor controller using Petri nets. However, the proposed control system architecture is not applicable to assembly systems. Many authors [Jones and McLean, 1986; Maimon, 1987; Jafari, 1992; Ghosh, 1989] have developed hierarchical control structures for FMS control as it supports information modeling and data abstraction at different levels in the factory. Jones and McLean (1986) proposed a hierarchical control model for automated manufacturing systems, which serves as a research test bed to aid in the identification, design and testing of standards for the automated factories of the future. They also emphasized that a standard factory model must address all of the necessary functional, control, data flow, and interface issues. Recently, McPherson and White (1995) presented dynamic issues in the planning and control of integrated manufacturing hierarchies.

There are a few studies that propose heterarchical control structure for automated manufacturing systems. Duffie *et al.,* (1988) presented a heterarchical control structure after describing the limitations of a hierarchical control structure. They illustrated the heterarchical control scheme with an experimental manufacturing system. Even though the proposed scheme in Duffie *et al.,* (1988) is simple and easy to understand, it can not be easily expanded when the FMS grows in complexity (several machines, robots, and AGVs).

McKay *et al.,* (1995) critically reviewed hierarchical production planning (HPP) and its applicability for modern manufacturing. By presenting the fundamental principles and concepts inherent in the HPP approach, they have suggested that there are a number of modern manufacturing situations where hierarchical production planning approach may not be appropriate. Furthermore, by identifying such situations where HPP is not applicable, they have discussed a number of possible adaptations of control schemes suitable for modern manufacturing. Cell control in FMS is increasingly getting attention from both researchers and practitioners. Xiang and Brien (1995) have presented an excellent overview of cell control research - current status and development trends. Based on the requirements of a cell control system suggested by practitioners, users, and researchers, they have proposed a new set of general requirements for the development of a cell control system. They have also examined the tools such as state tables, state machines, and Petri nets used for modeling the cell control logic.

The control of several devices in FMSs is possible only through communication networks. The major objectives of the communication network in FMS are resource and information sharing. The design of computer networks for FMS is treated by Ghosh (1989). In the same paper, Ghosh developed a methodology for performance evaluation of a network. Ghosh illustrated the design and performance evaluation methodology for three different configurations of networks, namely, direct numerical control (DNC), computer numerical control (CNC), and multi-level control hierarchy with local area network. Zhou *et al.,* (1994), presented an overview of various models for performance evaluation of communication networks in manufacturing. They also presented a Petri net method for modeling and performance of token bus local area networks.

2.6. Problems in Installation and Implementation

Emerging trends and new demands

Previous sections have shown that the research on FMS is vast and growing. In order to help the readers focus on their area of interest, the research reviewed in the present chapter has been classified and the results are shown in Table 2.3.

Table 2.3 Cross-references to research related to problems in FMS.

Problem Area	Related Research References
Managerial Problems	
Installation, implementation, and integration issues	Attaran (1992); Black (1983); Bluementhal (1985); Buzacott (1986); Darrow (1987); Dimitrov (1990); French (1984); Gilbert (1986); Goldhar (1984); Harvey (1984); Hughes (1983); Jari (1991); Jukka (1990); Kaku (1992); Kiesler (1983); Klahorst (1981); Krinsky (1991); Kunnathur and Sundraraghavan (1992); Kwok (1988); Mullins (1984); Primrose (1991); Ram (1991); Scalpone (1984); Stecke (1983); and Suresh (1990)
Technical Problems	
Tool management	Cumings (1986); Gaymon (1984,1987); Gray (1988); Heywood (1988); Maccarini (1987); Reddy (1990a,b,c); Tomek (1986); Veeramani (1992); Venkatesh (1992); and Zavanella (1990)
Automated material handling	Alanche (1984); Archetti (1989); Bischak and Stevens (1995); Bozer (1989); Davis (1989); Egbelu (1984,1986); Gould (1990); Hodgson (1987); King (1989); Lee and Lin (1995); Mahadevan (1990); Malmborg (1990); Maxwell (1982); Newton (1985); Occena (1991); Ozden (1988); Raju and Chetty (1993); Sabuncuoglu (1986); Schroer (1985); Vosniakos (1990); and Wysk (1987)
Control system development among elements present in the FMS	Adamides *et al.;* (1996); Balph (1985); Barad (1988); Duffie (1988); Ghosh (1989); Jafari (1992); Maimon (1987); Masory (1991); McKay *et al.;*(1995); McPherson and White (1995); Jones (1986); Sheng (1990); Venkatesh (1993, 1994c, 1998); Xiang and O'Brien (1995); and Zhou (1994)

Although the references cited in Table 2.3 overlap in various areas of FMS (such as managerial, tool management, etc.), the *primary* focus of the papers reviewed has been used as the classification criterion. Due to the interdisciplinary nature of FMSs, new manufacturing paradigms such as cooperative manufacturing [Buzzacott, 1995] and coordination-intensive manufacturing [Wu, 1995] are emerging. According to these paradigms, various fields such as computer science, business management, physical science, and even social science needs to work cooperatively with the manufacturing science in order to implement successfully the advanced manufacturing technologies. To meet the demands of these emerging paradigms of manufacturing, there is a need for integrated modeling tools that are easy to understand among various people with diverse backgrounds. Also, existing research in the area of FMS indicates the need for developing integrated tools to address not only all the technical problems discussed above but also help solve all the managerial problems discussed earlier. Such integrated tools would be of immense help to integrate the efforts of both technical and managerial personnel which is very essential for the full realization of FMS benefits.

Tools that are useful to specify, model, design, evaluate, control, and monitor FMSs are urgently required. Such tools serve as a common medium among several personnel involved in the above activities. Hence, using such tools the integration between the people and between the tasks of the system can be easily achieved. The integration between people in FMSs is of paramount importance for the success of such systems.

2.7 Summary

In this chapter the literature on definitions of FMS, reasons for change from conventional systems to FMS, installation and implementation issues of FMS, applications issues, and finally problems of FMS were briefly reviewed. The FMS-related problems were categorized into two major areas - managerial and technical. Both managerial and technical problems are discussed along with the earlier research. There is a vast source of materials on the subject, although the interest and the technology of the field is relatively recent. There are a number of predictions and forecasts in regard to the future of FMS. For example, Hughes and Hegland (1983) reported the result of a Delphi study with regard to the future of FMS. This report predicted major advances in various areas of FMSs such as:

- The level of application of automated fixturing and holding devices on numerically controlled machines,
- Reliable and practical sensing strategies for implementing adaptive control in all current metal-cutting operations.

2.7. Summary

- Extension of flexible production systems to machine tool industry,
- In-process adaptive control of surface roughness in machining,
- Non-contact high-speed on-line inspection systems with closed loop feedback to the machine control system, and
- Increased application of diagnostic components in FMSs

Many of these predictions are close to reality or have already been realized. Whenever an FMS is installed the experiences gained through the installation, implementation, and integration of the system are shared by the industry making the growth rate self-accelerating. However, despite these predictions and growth, the myopic views of some managers toward capital expansion in such areas as FMS have caused delay in widespread utilization of these systems [Hays and Wheelwright, 1984]. On the other hand, there are some positive trends.

The use of flexible machining cells will be of primary concern throughout most of the manufacturing industries. Integrated modeling tools that address both technical and managerial problems are essential for the success of manufacturing systems. This chapter has made a solemn attempt to comprehensively highlight the issues and problems related to FMSs starting from their planning to implementation. There is an increasing need for integrated modeling tools for the development of control systems in FMSs. PNs are claimed to be ideal integrated modeling tool in the area of FMSs [Cecil *et al.*, 1992; Proth, 1992; Zhou and DiCesare, 1993; Venkatesh, 1995]. The next chapters presents Petri nets as integrated modeling tools in FMSs, introduces the concepts of PNs, and presents a survey of their applications in FMSs.

CHAPTER 3

PETRI NETS AS INTEGRATED TOOL AND METHODOLOGY IN FMS DESIGN

3.1 Introduction

In view of the capital intensive and complex nature of flexible manufacturing systems (FMS), the design and operation of these systems require modeling and analysis in order to select the optimal design alternative and operational policy. It is well known that flaws in the modeling process can substantially contribute to the development time and cost. The operational efficiency may be affected as well. Therefore, special attention should be paid to the correctness of the models that are used at all planning levels. Petri nets, as a graphical and mathematical tool, provide a uniform environment for modeling, formal analysis, and design of discrete event systems. One of the major advantages of using Petri net models is that the same model is used for the analysis of behavioral properties and performance evaluation, as well as for systematic construction of discrete-event simulators and controllers. Petri nets as a graphical tool provide a powerful communication medium among customers, users, requirement engineers, designers, and analysts. Complex requirement specifications, instead of using ambiguous textual descriptions or mathematical notations difficult to understand by the customer, can be represented graphically using Petri nets. This combined with the existence of computer tools allowing for interactive graphical simulation of Petri nets, puts in hands of the development engineers a powerful tool assisting in the development process of complex systems. As a mathematical tool, Petri nets allow one to perform a formal check of the properties related to the behavior of the underlying system, e.g., precedence relations among events, concurrent operations, appropriate synchronization, freedom from deadlock, repetitive activities, and mutual exclusion of shared resources.

This chapter discusses Petri nets (PNs) as an integrated tool and methodology in FMS design using a layman's language. It includes brief and informal introduction to PNs, and their advantages. It then presents an overview of PN applications to a variety of FMS design fields, i.e., modeling, analysis, performance evaluation, discrete-event simulation, discrete-event control, planning, and scheduling.

3.1.1 Brief and Informal Introduction to Petri Nets

Petri nets were named after Carl A. Petri who created a net-like mathematical tool for the study of communication with automata in 1962. Their further development stemmed from the need in specifying process synchronization, asynchronous events, concurrent operations, and conflicts or resource sharing for a variety of industrial automated systems at the discrete-event level. Mostly computer scientists performed the early studies of Petri nets. Starting in the later 1970's, researchers with engineering background, particularly in manufacturing automation, investigated their possible usage in human-made systems. These systems become so complicated that the continuous/discrete-time systems theory becomes insufficient to handle them.

If we visualize any physical net, we can find two basic elements: nodes and links. Both nodes and links play their own roles. For example, forces could be transferred from one end to another through nodes and links and different nodes/links may bear different forces. Change of either may result in different nets in terms of their properties. A Petri net divides nodes into two kinds: places and transitions. The places are used to represent condition, status of a component, or an operation in a system. They are pictured by circles. The transitions represent the events and/or operations. They are pictured by empty rectangles or solid bars. Two common events are "start" and "end". Instead of bi-directional links in some physical nets, a Petri net utilizes directed arcs to connect from places (called input places with respect to a transition) to transitions or from transitions to places (called output places). In other words, the information transfer from a place to a transition or from a transition to a place is one-way. A two-way transfer between a place and transition is achieved by designing an arc from a transition to a place and another arc from the transition back to the place, called a self-loop.

The places, transitions, and directed arcs make a Petri net a directed graph called Petri net structure. The dynamics is introduced by allowing a place to hold either none or a positive number of tokens depicted by small solid dots. These dots in a place could represent the number of resources, indicate whether a condition is true, or signify whether an operation is ongoing, depending upon what the place models. When all the input places of a transition hold enough number of tokens, an event modeled by the transition can happen, called transition firing. This firing changes the token distribution in the places, signifying the change of system states. Both structural and behavioral properties of Petri nets can be defined and studied. The former is determined by solely the structure or topology of the net; while the latter

3.1. Introduction

also depends upon the initial marking that models the initial system condition. Important properties include boundedness, safeness, liveness, and reversibility. Boundedness implies no overflow of buffer capacity or resource processing capacity. Safeness is a special case of boundedness. Safeness of a place modeling an operation guarantees that there is no attempt to request execution of an ongoing process. Safeness of a place modeling a resource guarantees that the availability of only a single resource. Liveness of a PN model implies freedom from deadlock, i.e., partial or complete system standstill. Deadlock is due to inappropriate allocation of resources or exhaustion of certain types of resources. Reversibility of a PN model guarantees the repetitive and cyclic behavior of the modeled system.

The introduction of tokens to a PN and their flow regulated through transitions allow one to visualize the material, control, and information flow clearly. More importantly, a PN model allows one to perform a formal check of the system properties and behavior, e.g., precedence relations among events, concurrent operations, appropriate synchronization, freedom from deadlock, repetitive activities, and mutual exclusion of shared resources. Such a check prior to a system's implementation is critically important in the development of flexible and agile automated manufacturing systems.

3.1.2 Advantages of Petri Nets

Petri nets as a graphical tool provide a unified method for design of discrete event systems from hierarchical system descriptions to physical realizations. The following advantages make Petri nets a promising tool and technology in implementing a manufacturing automation project:

1. Ease of modeling characteristics of a complex industrial system: concurrency, asynchronous and synchronous features, conflicts, mutual exclusion, precedence relations, non-determinism, and system deadlocks. Petri net models offer excellent visualization of system dependencies. They focus on local information rather global one. Top-down (stepwise refinement) design, bottom-up (modular composition) design, and hybrid methods can be applied to design and construction of Petri net models.
2. Ability to generate supervisory control code directly from the graphical Petri net representation. A Petri net execution algorithm can also be constructed for real-time implementation using either Programmable Logic Controllers (PLC's) or computers.
3. Ability to check the system for undesirable properties such as deadlock and capacity overflow and to validate code by mathematically based

computer analysis - no time-consuming simulations are required for many cases.
4. Performance analysis without simulation for many systems. Production rates, cycle time, resource utilization, reliability, and performability can be evaluated. Bottleneck machines can be identified.
5. Discrete event simulation that can be driven from the model.
6. Status information that allows for real-time control, monitoring and error recovery of FMS.
7. Usefulness for scheduling because the Petri net model contains the system precedence relations among events, concurrent operations, appropriate synchronization, repetitive activities, and mutual exclusion of shared resources, as well as other constraints on system performance.

3.1.3 General Procedure for Applying Petri Nets in FMS Design

How can we utilize Petri nets for a particular situation? This question can be briefly answered as follows:
1. At the requirement and specification stage, Petri nets can be utilized to specify the systems requirements and specification or convert the verbal descriptions into Petri nets in order to validate them. These will also become a basis for later design using Petri nets.
2. After feasibility studies and preliminary system designs are performed, designers need to select several of the most attractive ones based on the general considerations such as the investment, desired production capacity and the service time in the system's life cycle. Then, the selected designs are modeled with Petri nets and their aggregate analysis can be performed. Petri nets with timing information might be applied to simplify the analysis. The aggregate analysis should lead to one or two design candidates.
3. While these designs of the manufacturing systems are further detailed, their PN models should be augmented to represent the change through design and synthesis methodologies. More specific performance measures should be analyzed, e.g., resource utilization, production rates, work-in-process inventory, and product cost. Qualitative analysis should be performed at this stage to guarantee deadlock-free, stable and cyclic system with minimum human intervention. At this step, tractable PN models may be applicable by making reasonable assumptions on the operation times. The results could then be used to make a recommendation

3.1. Introduction   43

for simulation studies or to suggest desired modifications to optimize the designs.
4. After more realistic operation data is collected and included, Petri net simulation can be used to derive more accurate system performance. The results could then be used to make a final recommendation or further modifications if they are not satisfactory.
5. At the implementation stage, discrete-event control code or control software of the flexible manufacturing system is derived and implemented according to the PN model. The PN model also serves as a specification for the system design. It is valuable for the future maintenance, modification or improvement of the system.

3.1.4 Successful Industrial Applications

There have been many applications reported over the past decades. In particular, we noticed the applications of Petri nets and related technologies in the following places:

Europe: In 1975, Grafcet was introduced in France as a European PN like graphic tool standard applicable to all software control systems for industrial automation [David and Alla, 1992; Giua and DiCesare, 1993]. Since then, Grafcet has been adopted by many other countries as an international standard called Sequential Function Charts. It continues to gain many successful applications in a variety of industry worldwide.

Japan: Hitachi developed Petri net-based controllers for sequencing flexible workstation design [Murata, et al., 1986]. They are applied to a number of automation projects. Kobe Steel Ltd. used Petri nets for programming PLC [Sato and Nose, 1995]. All these applications have gained more than two-third reduction in man-months, as compared with that needed in a general-purpose programming environment.

USA: Sequential function charts as the international standard of Grafcet gained popularity among the PLC suppliers and vendors in the States for the design of logic control for many applications. Petri nets are used in system modeling and design in such companies as Applied Materials, Inc. Petri nets have also found their applications in information technology sectors and are used by, e.g., Microsoft Corporation, AT&T, and Digital Equipment Corporation.

The following sections focus on the previous and important work of using Petri nets in modeling and analysis, simulation, performance evaluation, discrete-event control, planning, and scheduling for FMS.

3.2 Modeling and Analysis

Early interest in Petri nets aroused from the need to specify and model discrete-event manufacturing systems. Valette et al. (1982) explored the applicability of Petri nets for flexible production systems by specifying and validating interconnected controllers for a transportation system in a car production system using Petri nets. They showed that Petri nets were applicable to this system and indicated that such a Petri net approach could be based on decomposition and structuring. Furthermore, Valette (1987) indicated that Petri nets were more convenient than other models for concurrency. Alla et al. (1985) employed colored Petri nets to model the same car production system in [Valette et al., 1982]. Every token in a colored Petri net has its own attributes and thus allows different parts and machines to be distinguished among each other. Its colored Petri net model was proved to be deadlock-free and bounded. They revealed the possible benefit of using colored Petri nets over ordinary Petri nets, i.e., their conciseness made it possible to describe a complex FMS.

Martinez et al. (1985, 1986) used colored Petri nets to model a transfer line consisting of two machines and three buffers, as well as a transportation system consisting of an automatic guided vehicle (AGV), three loading and one unloading stations. They briefly discussed the characteristics of a "good model," including boundedness, liveness, and reversibility. Other specific properties included mutually exclusive places and finiteness of synchronous distance. Note that synchronous distance between two transitions measures the number of a transition's firings without firing another one given an initial marking. Narahari and Viswanadham (1985) used Petri nets to model two manufacturing systems: a transfer line with three machines and two buffers, and an FMS with three machines and two part-types. Boundedness and liveness were analyzed for both systems using invariant methods, and the significance of boundedness, liveness, and reversibility in manufacturing systems was explored. They also presented a systematic bottom-up modeling approach. Furthermore, they demonstrated the usefulness of colored Petri nets in automated manufacturing systems [Viswanadham and Narahari, 1987]. Gentina and Corbeel (1987) proposed structured adaptive and structured colored adaptive Petri nets including inhibitors, self-modifying arcs, etc. to facilitate Petri net modeling of production systems and their usefulness was illustrated through design of an FMS hierarchical control system.

Krogh and Sreenivas (1987) proposed the concept of *essentially decision freeness* and presented Petri nets to deal with real-time resource allocation for manufacturing processes. Krogh and Beck (1986) proposed the modified Petri nets to model and simulate a robotic assembly system. They also made a significant contribution to bottom-up synthesis in the sense that (a) their bottom-up approach can

3.1. Modeling and Analysis

guarantee safeness and liveness of the resulting net, and (b) they extended common place or transition sharing concepts to common path sharing. Thus they avoided using invariants for the verification of safeness and liveness of a Petri net in [Narahari and Viswanadham, 1985] and reachability analysis methods. Valavanis (1990) introduced extended Petri nets in which each of tokens, places, transitions, and arcs differs from the other so that more information can be carried in a net model. An FMS containing two workstations, a robot, and input and output stations is modeled and simulated.

Wang and Saridis (1990) used Petri nets to describe a coordination model of intelligent machines. The coordination structure was developed to describe activities and connections among dispatchers and coordinators, each of which is specified as a Petri net transducer capable of translating languages (task plans). Therefore, some system properties can be analyzed using PN models.

Ferrarini (1992) proposed an incremental approach to logic controller design using Petri nets. Three connection methods, i.e., transition sharing, self-arcs, and inhibitor arcs, are investigated when they are used to couple subsystems specified by state machine-like Petri nets. His further work [Ferrarini, 1994 and 1995] included the theoretical development and CAD tools for logic controllers. The application example is a super-elevated conveyor allowing four elevators to carry parts from four production lines to power driven rollers.

Zhou and DiCesare (1991) proposed parallel and sequential mutual exclusions to model resource sharing problems in FMS. The hybrid synthesis approach [Zhou et al., 1992a] enables one to develop a Petri net with good behavioral properties for sophisticated manufacturing systems. Thus the analysis based on enumeration of all the markings can be avoided. The approach has been applied to an FMS that consists of two different machining workstations with robotic loading and unloading, a robotic assembly workstation, a materials movement system, raw material and final product inventory storage, and an automated storage and retrieval system [Zhou and DiCesare, 1993]. Buffers are important resources in flexible manufacturing. Various types of buffers are discussed and modeled with Petri nets in [Zhou and DiCesare, 1996]. The buffers and PN models are classified as random order or order preserved (first-in-first-out or last-in-firs-out), single-input-single-output or multiple-input-multiple-output, part type and/or space distinguishable or indistinguishable, and bounded or safe.

Cao and Sanderson (1992) mapped AND/OR nets into Petri nets and then searched the task sequences from the Petri net representations of a robotic system. Fuzzy Petri nets [Cao and Sanderson, 1995] are developed to derive effectively a correct task sequence. Zurawski and Dillon (1993, 1994) proposed the concept of functional abstraction and used it to model manufacturing systems with Petri nets. The correctness of the model is verified via temporal logic.

Just-in-time (JIT) production is a very important philosophy to achieve agile manufacturing. Its basic idea is to produce the right quantity of products at the right time at the right place. As a means of material transportation and production information exchange between workstations, Kanbans are used to implement the JIT philosophy in JIT production systems. In a JIT system, the demand for completed parts from a station depends only on the needs of the next station, which are communicated through Kanbans posted on a bulletin board. Di Mascolo, et al. (1991) provided a unified PN model for Kanban systems. A basic PN model was presented for a three-stage Kanban system. Jothishankar and Wang (1992) modeled a two-station Kanban system using stochastic Petri net and the optimal numbers of Kanbans were determined corresponding to varying demands. Generalized stochastic Petri nets were also used to model Kanban systems with and without workstation failures [Ajmone Marsan et al., 1995].

Modeling and analysis of the photo area in an Integrated Circuit (IC) manufacturing system was performed by using Petri nets in [Jeng et al., 1996]. Azzopardi et al. (1996) synthesized an object-oriented Petri net model of a semiconductor testing plant. Kim and Desrochers (1997) investigated semiconductor fabrication processes to analyze and predict plant capacity, on-line turn-around time, machine utilization, and work-in-process. Timed Petri nets with inhibitor functions are adopted as a modeling tool. Based on technology process flows and fabrication facilities in a floor, the models can be generated in a structural and automatic way. Simulation method is then used to evaluate the systems. In [Jeng et al., 1997], several fundamental PN modules were proposed and discussed for IC wafer fabrication processes. They are used to model a 0.44 μm 4MB DRAM fabrication process in the etching area. The obtained PN model contains 218 places and 260 transitions.

Modeling a manufacturing system including the error recovery blocks was performed in [Zhou and DiCesare, 1989] using PNs. Hierarchical time-extended PNs were proposed to model complex FMS for their integrated modeling, control and diagnostics [Ramaswamy and Valavanis, 1994, 1996]. A powerful modeling tool based on PNs for FMS with breakdowns will be discussed in Chap. 9.

3.3 Performance Analysis and Evaluation

After time was introduced into Petri nets, one can use timed Petri nets to conduct temporal performance analysis, i.e., to determine production rates of systems, resource utilization, and the like. It is also possible to detect a bottleneck in an FMS or to determine optimal buffer size, optimal pallet distribution, etc. as indicated by Dubois and Stecke (1983). In fact, Dubois and Stecke were the first to apply timed

Petri nets to describe, model, and analyze production processes. In their research, deterministic time variables were assumed, and a Petri net-based simulation method was utilized to find the minimum cycle time and to identify the bottleneck machine for an FMS with three machines and three part-types in a fixed route.

In the *1985 International Workshop on Timed Petri Nets*, many research projects in the field of timed Petri nets for performance analysis were presented, and various applications were reported for analyzing the performance of computer systems, communication protocols, and manufacturing systems [Bruno and Biglia, 1985]. More importantly, some significant software packages were introduced, for example, DEEP [Dugan *et al.*, 1985], and ESP [Cumani, 1985]. Chiola (1987) presented a user-friendly software package for analysis of generalized stochastic Petri Nets (GSPN), whose improved version is called GreatSPN. GreatSPN is powerful compared with the other existing tools in the sense that it can accept various time variables, inhibitors, and random switches. It can also be used as a simulation tool.

To evaluate the performance of job-shop systems under deterministic and repetitive functioning of a production process, Hillion and Proth (1989) applied a special class of timed PNs called timed event-graphs or marked graphs for an FMS with three machines and three job types. In such nets, no choice is allowed. In other words, the routes of jobs in a jobshop system are fixed before the entire system is modeled using event-graphs. The number of jobs in process is nearly minimized using integer programming, while the system still works at its maximal productivity. The upper and lower bounds of the cycle time of stochastic marked graphs can be derived if each delay's mean time is known [Campos *et al.*, 1992]. More accurate bounds can be obtained if each delay's derivation is also known. [Xie, 1994]. Hanisch (1994) introduced and applied PNs with timed arcs to batch process control.

Viswanadham and Narahari (1988) have provided an excellent introduction to the use of generalized stochastic Petri nets in analyzing the system performance of automated manufacturing systems. They used a software package they developed to evaluate two representative systems: a manufacturing cell with multiple material handling robots, and an FMS with three machines and two part-types. Wang, and Jiang (1995) presented commonly used stochastic Petri net models and solution methods and applied them to a number of communication and manufacturing systems.

All the existing software tools mentioned above can be used to analyze the performance of automated manufacturing systems. For example, Desrochers (1989, 1995) and Al-Jaar and Desrochers (1990) studied transfer lines and production networks using SPNP and GreatSPN. They also formulated a modular approach to evaluating the performance of automated manufacturing systems.

To demonstrate applicability of and accuracy achieved by stochastic Petri nets for FMS, Watson and Desrochers (1991) performed three representative case-studies from different sources using SPNP and GreatSPN. The work showed that the performance analysis results obtained using Petri nets agreed with those obtained by simulation, queuing theory, and probability theory.

SPNP was used to analyze the performance of manufacturing systems that have fixed routing and produce limited types of parts. Zhou et al. (1990) first developed a stochastic Petri net modeling approach such that desired behavioral characteristics can be preserved. Then they analyzed deadlock-free and deadlock-prone manufacturing systems. The results showed that supervisory controllers with freedom from deadlock are better than deadlock-prone controllers in designing real-time resource-sharing distributed systems. Zhou and Leu (1991) presented Petri net modeling and performance analysis of a flexible workstation, AT&T FWS-200, for automatic assembly of printed circuit boards. Two control implementation methods are evaluated for the performance when robot arms are subject to failure. Some details will be presented later in Chaps. 5 and 6. To determine the holding and storage cost in a just-in-time production system and find the optimal number of Kanbans, Jothishankar and Wang (1992) modeled and investigated a two-station Kanban system using stochastic Petri nets. SPNP software was used. Zhou *et al.* (1993) modeled an FMS with three machining centers, four robots, and an inspection station using SPNP. To improve the system productivity, several system structures are explored given a certain amount of the capital investment. Choices include upgrading either of three machines, automating the inspection procedure, and purchasing an additional robot. The original Petri net is extended to accommodate the new structures and the resulting stochastic Petri nets are evaluated with the help of SPNP software.

Recent developments in performance analysis of automated manufacturing systems have led researchers to investigate FMS with possible deadlocks [Viswanadham *et al.*, 1990]. One reason for this investigation is that it can be extremely difficult to design a deadlock-free controller for such complex systems with numerous job types. Deadlock-avoidance methods are evaluated through stochastic Petri nets for a five-robot-five-assembly system in [Zhou, 1995].

Although the underlying models of stochastic Petri nets are still Markov processes, the Petri net approaches preclude the direct construction of a state space required by Markov analysis, which is often too large for human being to manipulate.

Ajmone Marsan *et al.* (1995) evaluated a Kanban system with workstation failures using GSPN. They also discussed a push production system containing three machines, a Load/Unload station, and a material handling system that is implemented using either conveyors for continuous transportation or an Automatic Guided Vehicle (AVG). Two product types can be produced in their example system. The applications of GSPN to concurrent programs, computer architectures, and random polling systems are also discussed in [Ajmone Marsan *et al.*, 1995].

3.4 Discrete-Event Simulation

PNs were used for the discrete-event simulation of FMSs for a wide variety of purposes. Venkatesh *et al.*, (1990) developed PN models that can be used for the simulation of flexible automated forming and assembly systems. The models were simulated to check some of the qualitative properties of the system such as absence of deadlocks, liveness, and repeatability. Venkatesh *et al.*, (1992) also used PN models for simulation and scheduling of robots in a flexible factory automated system operating with Just-In-Time principles. The system considered has four machine cells, four robots, and an AGV system. They have determined the utilization of robots and machines and production rate by varying the robot scheduling policies. Righini (1993) presented modular PNs for simulation of flexible production systems. He has introduced the concept of Petri subnet and presented an algorithm that allows automatic composition of subnets into a larger PN model. In order to account for the varying time duration of activities such as processing of parts, setting-up of machines, conveying of parts among machines in FMSs, stochastic PNs were applied for simulation and performance evaluation of FMSs [Chan and Wang, 1993].

Petri nets were applied to model and emulate a highly-concurrent pick-and-place machine that is widely used in mass production of printed circuit boards [Sciomachen, *et al.*, 1990]. Different operating scenarios are modeled and emulated and the system throughput rates are obtained. The use of Petri nets allows one to generate emulation code automatically. Camurri and Coglio (1997) introduced and applied colored Petri nets-based architecture for simulation of industrial plants.

Several researchers and practitioners have used PNs for simulation and performance evaluation of Automated Guided Vehicle Systems (AGVS). Raju and Chetty (1993a) have developed extended timed PNs and applied them for the design and evaluation of AGVS in FMS. The proposed methodology is elucidated through an example FMS that has three machining centers, one pallet station, and an AGVS. Raju and Chetty (1993a) have considered three types of scheduling namely job scheduling, vehicle scheduling, and machine scheduling. For this purpose Raju and Chetty (1993b) also presented a dynamic scheduling algorithm to determine the operating policies under varying FMS operation conditions. The system considered was an FMS with six machining centers with input and output buffers for each center; a cell with three machines, input/output buffer, and a robot; and a load/unload station connected by an AGVS. Archetti and Sciomachen (1987) presented the development, analysis, and simulation of PN models by applying them to AGVS.

Wang and Hafeez (1994) used stochastic PNs for the planning of an AGV system such that there are no conflicts or deadlocks for the vehicles. The obtained PN models were then simulated to quantify the performance of the AGV system for comparing two different AGV traffic management methods. In this study, as an example, two FMS - one using a conventional AGV system and another using a tandem AGV system were considered. In both of these systems there are three machines and one load/unload station. Hsieh and Shih (1994) proposed an easy and quick method for the development of an AGVS model using modularized floor-path nets that are based on PNs.

PNs were applied for simulation of FMSs to model breakdowns that occur in shop-floor and their effect on some of the design parameters (buffer sizes, number of assembly fixtures) and performance measures (production rate, utilization of robots and machines) [Venkatesh *et al.*, 1994a]. The application of augmented timed PNs, a special class of PNs for modeling breakdown handling is discussed in detail in Chap. 9. Similar to the activation arc concept presented in [Venkatesh *et. al.*, 1994a], Ramaswamy *et al.*, (1997) have developed a type of PNs called hierarchical time-extended PNs to model complex manufacturing systems. These models help in studying issues such as static priority scheduling, dynamic failure recognition, and rescheduling in FMSs. The PN modeling concepts presented in this paper were illustrated by using an example of an assembly line that consists of three machines, three buffers, and two robots. PNs are also shown as a powerful simulation tool to investigate the problem often encountered in manufacturing systems management, namely comparing the performance of an FMS operating under *push* (Just-In-Case) and *pull* (Just-In-Time) paradigms by varying several parameters such as processing times, number of AGVs, lot sizes of products, etc. [Venkatesh *et al.*, 1996]. The application of PNs for this purpose is discussed in detail in Chap. 8.

3.5 Discrete-Event Control

Around 1980, various researchers independently studied Petri net-based methods to develop a programmable logic controller (PLC). Silva and Velilla (1982) presented a comparative study of various implementations and indicated that one benefit of using Petri net approaches is that both software and hardware errors can be prevented in the PLC design phase due to the possibility of early validation of controllers.

To put such methods to use, Courvoisier *et al.* (1983) developed a prototype system for designing a colored Petri net-based PLC and deriving control code. This system is used for designing a system with two AGVs sharing a common path. Valette *et al.*, (1983) demonstrated PNs as a tool for formal specification and validation of the control procedures used in the implementation of programmable logic controllers in industrial automated systems. The system considered is a concrete production station that consists of four aggregate stocking hoppers, a conveyor, a

3.5. Discrete-Event Control

weighing machine, a mixer, a counter, and a timer. They developed a prototype of a programmable logic controller that controls the concrete production station using a PN. Valette *et al.*, (1983) and Chocron and Cerny (1980) used PN description languages that were input to the translator or compiler. The translator or compiler generates the control tables that drive the actuators.

To effectively handle the complexity of FMS control, Valette et al. (1985) proposed a hierarchical control structure that comprised local control level, coordination level, and monitoring and real-time shop scheduling level. Communication among different PLCs was attained by a local area network in this type of architecture. Independently, Merabet [1986] implemented a three-level control structure for an FMS. The system considered was a flexible manufacturing cell using a multibus/multiprocessor M-68000 based structure. The cell consists of a milling machine tool, a robot, tools, grippers, parts, and sensors.

One of the successful applications of Petri net approaches to system control was reported by Murata et al. (1986). First, Petri nets were modified into Control Nets. Then Control Nets were used to design station controllers for sequencing control. The station coordinators for monitoring and diagnosis were implemented using IF-THEN rules. Two objectives, i.e., flexibility and maintainability, were achieved when such a Petri net approach was used. In fact, their research produced the Hitachi commercial systems, i.e., Station Controller. Three systems are considered to demonstrate the application of "Station Controller". The first system is a parts assembly station which has an automatic robot gripper changer, a robot, automatic tool changer, a screw-driver machine, a press machine tool, pallet transportation station, a boring machine tool, a main control panel, a Control-net editor/monitor, and a HMCS 68000 micro-controller. The second one that uses Station Controller is a microcomputer-based industrial controller that has an operator console, a Station Controller graphic editor, two robots, and peripheral equipment. The third system is an automatic warehouse system that consists of a loading conveyor, an unloading conveyor, a part carrier, two lane conveyors, a warehouse, and a stacker crane.

To accommodate changes in FMS, Menon *et al.* (1988) developed a system for configuring coordination controllers in FMS using colored Petri nets. Also, they presented a pilot Petri net workstation controller.

Crockett *et al.*, (1987) presented an application independent manufacturing workstation controller that uses PNs to describe the sequencing information. They implemented a PN based controller, without need of a PLC, for a machining workstation consisting of a Cincinnati Milacron 5VC machining center and parts handling station. The PN based workstation controller runs on VAX 11/760 system, the CATIA design NC code runs on IBM 4341 that communicates with the IBM PC-

XT machine controller, and the IBM data acquisition and control adapter communicates with discrete interface board that interfaces with the Cincinnati Milacron 5VC machining center. The machine controller communicates with the machining center via a serial interface board using RS232. To deal with more complicated manufacturing systems, Kasturia *et al.* (1988) extended Crockett's direct implementation method to colored Petri nets. Their proposed controller is implemented in real-time on a multi-level manufacturing system that consists of two workstations. To demonstrate the applicability of such a method to FMS, Zhou *et al.* (1992) developed a real-time Petri net controller for a flexible manufacturing system which consisted of three workstations, three robots, four conveyors in a material transfer system, and an automatic storage and retrieval system. Its systematic design and implementation work is reported in more detailed in [Zhou and DiCesare, 1993].

Boucher *et al.*, (1989) reported the application of PNs for the real time production control of an automated manufacturing cell. They also presented some of the differences between a PN controller and a programmable logic controller that uses ladder logic diagrams for the same application. The manufacturing cell considered consists of a CNC lathe, a robot, an input buffer, an output conveyor, an IBM-PC XT cell host controller. The CNC lathe and robot have their own controllers that are dedicated to programmed operations and communication with the cell host controller. The cell host controller communicates with the rest of the elements in the cell using an input/output interface board. Teng and Black (1989) designed an expert control system for manufacturing cell control using PNs to generate the cell control algorithm. They presented the structure of the control system and the application of PN by considering a cell that consists of a CNC milling machine, two decouplers (buffers), a CNC lathe, an input station, and an output station.

Jafari (1992) developed an architecture for a shop-floor controller using colored PNs that are suitable for modeling complex FMS with similar components, e.g., machines, robots, and controllers performing generic manufacturing activities. The system considered was a hierarchical shop-floor control system with one shop-floor controller controlling several cell controllers and one material handling system controller. Each cell controller in turn controls several workstation controllers and one robot controller. Huang and Chang (1992) used PNs for specification and modeling of an FMS. Then based on the cell controller specification they developed a colored timed PN for control of an FMS. PNs were also applied for monitoring and fault detection in the shop floor [Srinivasan and Jafari, 1991]. Automated generation and verification of control programs were discussed in [Krogh *et al.*, 1988].

D'Souza and Khator (1993) presented PN control models for deadlock and conflict detection and performance evaluation in computer-integrated assembly cells. These control models may be also applied to debugging controller programs prior to

3.5. Discrete-Event Control

beginning a production run. The computer-integrated assembly cell used to demonstrate the application of PNs consists of five assembly workstations with each station having input and output buffers, an automated storage and retrieval system, a transporter to move parts among workstations, a programmable electronic controller, and a host computer. D'Souza and Kelwyn (1994) used PNs to develop a control model that was generated from a list of programmable logic controller events. The PN model is then used to debug the programmable logic controller programs by testing it for deadlocks that may occur due to the concurrent nature of activities in an automated machining cell.

Jackovic et. al., (1990) presented an approach to the modeling of the highest level control of flexible manufacturing cell using PNs. They decomposed the PN model of the cell into individual robot nets and machine nets. The cell considered in their study consists of two robots, three machine tools - a milling machine, a lathe, a press, two auxiliary buffers, two transporters, bin for rejected parts, visual TV system, and a set of sensors. Martinez and Silva (1985) developed a language for the description of concurrent systems modeled by colored PNs. The considered methodology was applied to the design of a system of coordination between elements of an FMS. The resulted interpreted colored PN was described using the programming language developed. The system considered was a part of a bigger FMS producing automobiles and consisted of six workstations, one work place for each workstation, and a transfer bench with two roller tables - one designed to load and other to unload. Lewis et al. (1993, 1994) presented an elegant and promising design method for sequencing controllers of discrete event manufacturing systems. Matrix-based equations are formulated and applied. Different conflict and deadlock resolution methods can be efficiently incorporated in their design framework. Hanisch et al. (1996) proposed a modular modeling, controller synthesis and control code generation framework based on condition/event systems. Hanisch et al. (1997) proposed timed condition/event systems to model PLC behavior in order to verify the control code. Westphal and Renganathan (1997) presented a method for state dependent observer synthesis of supervisory controllers of stochastic discrete event dynamic systems

More recently, a new type of PNs called real-time PNs (RTPNs) are used for sequence control of FMSs [Venkatesh et al., 1993; Venkatesh and Ilyas, 1995; Venkatesh et al., 1994b]. RTPN models can be used not only for control but also for the simulation of FMSs [Venkatesh, 1995]. The system considered in these studies using RTPNs is an electro-pneumatic system with four pneumatic pistons that are operated by spring loaded five port two way solenoid valves. Each piston has two normally open limit switches that sense the start and stop motion of the piston. The system was controlled by an RTPN model running on an IBM compatible PC386.

The system also consists of a digital input and output interface to communicate with the PC, solenoids, and the limit switches; on/off push button switches to start and stop the motion of the pistons; and a green and a red light to monitor the functioning of the system. Chap. 10 presents the fundamentals of RTPNs in detail along with several methods of using PNs for control by various researchers. Chap. 11 presents the comparison between RTPNs and the more traditionally used ladder logic diagrams for discrete event control.

PNs have also been integrated with object-oriented methodologies for the development of control software for FMSs [Venkatesh et. al., 1994c, Venkatesh, 1995, Stulle and Schmidt, 1996]. The FMS considered by Venkatesh et. al., (1994c) consists of four machine cells, four input and output buffers, four robots, and four AGVs. Stulle and Schmidt (1996) presented object-oriented PNs for modeling of FMSs by constructing a reusable OOPN that models a typical production cycle. The OOPNs thus modeled are then interconnected in a predefined way to model FMS similar to the way shown in [Venkatesh et al., 1994c]. Wang (1996) has combined PNs and object-oriented concepts to develop object-oriented PN cell control models. More recently, Venkatesh and Zhou (1998) have introduced PNs in a widely popular and formal object-oriented modeling methodology. The uniqueness of their approach is to use PNs for dynamic modeling of the system as opposed to statecharts that are commonly used in many of the formal object-oriented methodologies [Booch, 1994; Rumbaugh et al., 1991]. The application of PNs for object-oriented software development of FMS control software is discussed in detail in Chap. 12.

Petri nets are also used in designing robotic control software [Caloini et al., 1998]. High-level timed Petri nets are applied in the development process of such control software. Building blocks including flags, toggles, shared memories, clocks, boards, sensors, and channels are proposed and discussed. They reported their experience in using Petri nets for the design and verification of a prototype of a new robot control system based on a commercial system. The important advantages of using such a Petri net method include its capability of describing concurrent and real-time systems, and a formal basis that prevents nonambiguities and supports powerful dynamic analysis techniques.

Comparing Petri nets with the industrial dominant approaches such as relay ladder logic diagrams is very important topic that is of both practical and theoretical interest. A portion of the work is presented in Chap. 9. Comparison through a large scale industrial system was performed in [Zhou and Twiss, 1995, 1998]. It is a water pollution treatment facility near Boston, MA, which consists of nine clarifier channels, nine skimmer pipes, and two scum boxes. Each individual system can be operated in the local and remote modes. Each pipe is equipped with a motorized actuator and position limit switches and can change to different positions.

3.6 Planning and Scheduling

The use of Petri nets to evaluate the schedules of FMS is widely known due to the work [Carlier and Chretienne, 1988; Hillion and Proth, 1989]. Their applications to determination of optimal or sub-optimal schedules for FMS have emerged as an important research area only until recently due to the computational complexity involved.

Shih and Sekiguchi (1991) presented a timed Petri net and beam search method to schedule an FMS. Beam search is an artificial intelligence technique for efficient searching in decision trees. When a transition in a timed Petri net is enabled, if any of its input places is a conflicted input place, the scheduling system calls for a beam search routine. The beam search routine then constructs partial schedules within the beam-depth. Based on the evaluation function, the quality of each partial schedule is evaluated and the best is returned. The cycle is repeated until a complete schedule is obtained. This method based on partial schedules does not guarantee global optimization.

Hatono *et al.* (1991) employed the stochastic Petri nets (SPN's) to describe the uncertain events of stochastic behaviors in FMS, such as failure of machine tools, repair time, and processing time. An FMS was modeled in a hierarchical structure using both continuous-time and discrete-time SPN's in which deterministic times and random times of exponential and normal distributions are used. They developed a rule base to resolve conflicts among the enabled transitions. The performance of the on-line rule-based scheduling system is evaluated by simulation. The example system deals with three product types with fixed routes by using three machines, limited-sized buffers, load/unload stations, and an AGV. The failure rate and mean repair time of each machine are known. The variance of machining and transportation times is known. Fixed machine setup time is included.

Shen *et al.* (1992) presented a Petri net-based branch and bound method for scheduling the activities of a robot manipulator. To cope with the complexity of the problem, they truncate the original Petri net into a number of smaller size subnets. Once the Petri net is truncated, the analysis is conducted on each subnet individually. However, due to the existence of the dependency among the subnets, the combination of local optimal schedules does not necessarily yield a global optimal or even near-optimal schedule for the original system. Zhou, Chiu and Xiong (1995) also employed a Petri net based branch and bound method to schedule flexible manufacturing systems. In their method, instead of randomly selecting one decision candidate from candidate sets (enabled transition sets in Petri net based models), they select the one based on heuristic dispatching rules such as SPT. The generated schedule is transformed into a marked graph for cycle time analysis.

To simulate and schedule FMS with AGVS, Raju and Chetty (1993b) developed a special type of PNs called Priority Nets. They investigated three types of scheduling, i.e., job scheduling, vehicle scheduling, and machine scheduling. A dynamic scheduling algorithm was proposed to determine the operating policies under varying FMS operation conditions. The system considered was an FMS with six machining centers with input and output buffers for each center; a cell with three machines, input/output buffer, and a robot; and a load/unload station connected by an AGVS with different numbers of AGVs.

Lee and DiCesare (1993, 1994a, 1995) presented a scheduling method using Petri nets and heuristic search. Once the Petri net model of the system is constructed, the scheduling algorithm expands the reachability graph from the initial marking until the generated portion of the reachability graph touches the final marking. Theoretically, an optimal schedule can be obtained by generating the reachability graph and finding the optimal path from the initial marking to the final one. But the entire reachability graph may be too large to generate even for a simple Petri net due to exponential growth of the number of states. Thanks to the proposed heuristic functions, only a portion of the reachability graph is generated. Three kinds of heuristic functions are presented. The first one favors markings that are deeper in the reachability graph. The second one favors a marking that has an operation ending soon. The last one is a combination of the first and the second ones. These three heuristic functions do not guarantee the admissible condition [Pearl 1984], thus the proposed heuristic search algorithm does not guarantee to terminate with an optimal solution. In [Lee and DiCesare, 1994b], the PN approach and heuristic search method are used to investigate the integrated scheduling and control of FMS using AGVs. Both centralized and distributed AGV systems are studied. In a centralized system, all AGVs have their home position in the loading area. They are dispatched to serve one or more machines and come back to the home position after the tasks are finished. In the latter, each of machines and loading/unloading station has its own AGV at its complete disposal. The AGV returns to its owner machine after it finishes the delivery commanded by its machine. The example system includes three machines, one robot, and five AGVs and deals with four jobs with different batches.

Sun *et al.* [1994] developed a timed-place Petri net model for an FMS based on its transportation model and process-flow model. A limited-expansion A* algorithm, developed from A* search algorithm, is proposed to schedule the part processing. It trades the schedule optimality for reduced memory requirement and lower algorithm complexity. It is applied to scheduling three jobs, each with three operations with limited routing flexibility, for a target FMS that contains a load/unload station, automated storage/retrieve system, AGVs, a robot, a lathe, a miller, and a machine center.

3.7. Planning and Scheduling

Li et al. (1995) investigated scheduling and re-scheduling of flexible and agile manufacturing systems using timed PNs. In their system, parts are transported by a number of AGVs and processed by the machines. Engineers have to schedule AGVs to transport parts cooperatively to satisfy a machining schedule that is free from jamming and collision, and generate quickly a new schedule in case of AGV troubles. Two deterministic timed Petri net models are constructed for machining and transporting events, respectively, resulting in a hierarchical net model for the entire system. Machining and transporting priority functions are defined in order to resolve the conflicts on line. They are determined to achieve job balancing, the least idle time, etc. A simulation study is performed for the system consisting of three machines, four AGVs, and three input and output buffers to produce four job types in a required quantity. The simulation shows that excellent scheduling and rescheduling results are obtained with the proposed priority functions.

Proth and Minis (1995) discussed planning and scheduling problems and solution techniques for both cyclic and non-cyclic production systems using Petri nets and algorithmic methods. For cyclic systems, marked graphs as a special class of Petri nets are used as a basis to model the physical and decision-making systems. The proposed scheduling algorithm is able to generate a near-optimal schedule that maximizes productivity while minimizing work-in-process in deterministic cases. Non-cyclic production systems arise to satisfy the short-term (e.g., one day) planning requirement. Petri net models are constructed and several new concepts are formulated to facilitate the discussion of their scheduling problem. A scheduling algorithm is proposed to minimize the sum of inventory and backlogging cost for non-cyclic production systems. It can generate a near-optimal schedule. In addition, a modular approach is proposed to develop Petri net models to facilitate the formulation of and solution to management problems in manufacturing systems.

Zuberek (1995; 1996) investigated and evaluated different fixed schedules for a flexible manufacturing cell using timed Petri nets. Uzma et al. (1995) explored the use of timed PN to investigate anticipatory scheduling for FMS. Both a robotic cell and an AGV-based FMS are used in their study. The anticipatory scheduling is compared with other fixed or optimized schedules under certain parameter conditions. The simulation results illustrate the robustness of the anticipatory scheduling policy when the system parameters are subject to change.

To formulate aperiodic scheduling problems, Jeng and Lin (1997) presented two classes of time-place PN models, i.e., symmetrical and asymmetrical nets. The former is used to model job-types with a fixed route and the latter with alternative routes. Such properties of resultant PN model as boundedness and freedom from deadlock are guaranteed. Thus the heuristic search algorithms [Chen and Jeng, 1995] based on approximate solutions of PN state equations to evaluate the performance index can be used without risk leading to an infeasible schedule. Due to the guaranteed properties, backtracking steps can be eliminated compared with other heuristic search algorithms in [Lee and DiCesare, 1994; Xiong, 1996], thereby reducing the memory requirements significantly.

3.7 Summary

This chapter presents Petri nets as a modeling tool and methodology for their various applications in modeling and analysis, simulation, performance evaluation, discrete-event control, planning, and scheduling of FMS. There are frequently many technical sessions of papers devoted to Petri nets and manufacturing automation in the following conferences:
>IEEE International Conference on Systems, Man and Cybernetics,
>IEEE International Conference on Emerging Technologies and Factory Automation,
>IEEE International Conference on Robotics and Automation,
>International Conference on Robotics, CAD/CAM, and Factories of the Future,
>etc.

Other major conferences on theory and applications of Petri nets include:
>International Workshop on Petri Nets and Performance Models, and
>International Conference on Applications and Theory of Petri Nets (formally European Workshop on Applications and Theory of Petri Nets).

Papers in the subject of Petri nets and FMS often appear in the below journals:
>IEEE Transactions on Systems, Man and Cybernetics
>IEEE Transactions on Robotics and Automation
>IEEE Transactions on Automatic Control
>International Journal of Production research
>Journal of Manufacturing Systems
>International Journal of Flexible Manufacturing Systems
>International Journal of Computer Integrated Manufacturing,
>etc.

The references we cited in this chapter are merely a small number of technical contributions in the area of Petri nets and FMS design. Many more significant contributions can be found in the above mentioned conference proceedings and journals, as well as other sources.

CHAPTER 4

FUNDAMENTALS OF PETRI NETS

The concepts of conditions, events, and state machines are discussed and used to motivate the definition of Petri nets. Their formal definitions are introduced in this chapter. A variety of important extensions are also discussed toward the convenience in their use and the increase in their representation power. Petri net properties related to resource capacity overflow, stability, deadlock-freeness, and cyclic behavior are discussed. The fundamental analysis methods for these properties are introduced.

4.1 Conditions, Events, and State Machines

The concept of an event is closely related to the conditions for such an event to take place. For example, in a robotic assembly system, a robotic arm has to be ready to pick up a component and a component is ready for pick-up before the event "pick-up" or "start a pick-up operation" occurs. Once an event happens, a new status may result. After a "pick-up" event, the robot arm may hold a component if the pick-up operation is successful and otherwise it either waits for the further instruction or makes another try. Therefore, the simplest mechanism to describe the relationship between conditions and events is a state-transition diagram, formally a finite state machine or automaton. Here all the conditions of components in a system together at a time instant represent a system status, or state. A system state completely describes its current status of the entire system. Occurrence of an event at a state may change the system's state to another. A state diagram uses a circle to represent a state and an arc from one state to another to represent an event. An arc in a diagram is called a transition, and a circle is also called a node. It is in fact easy to build up the state machine model especially when only one or few active resources are involved. A general procedure is as follows:
1. Determine all the possible events and states;
2. Draw and label a circle for each state; and
3. At each state, decide all possible events; for each permissible event, draw and label an arc with the event from the current state to the next.

This process ends after all the states are considered.

Example 4.1: Build up a finite state machine model for a single robot assembly system. The robot performs pick-up and insertion operations on a Printed Circuit Board (PCB) with the assumption that the components are always available and normal operations are considered only. Two state machines can be defined:

A. Model "pick-up" and "insert" as two events and two states are "being ready to pick-up" and "having a component ready to insert". At state "Robot is ready to pick-up" labeled with 0, there is only one event, *pick-up*. Draw one arc from state 0 to state 1, i.e., "Robot is ready to insert". Similarly, an arc labeled with event *insert* is introduced from State 1 to State 0. The complete state diagram is shown in Fig. 4.1(a).

B. Model "pick-up" and "insert" as operations and their starts and ends as events. This leads to four events in total, i.e., *start pick-up, end pick-up, start insertion*, and *end insertion*. Four states can be used to describe the robot, i.e., ready to pick-up, performing pick-up, ready to insert, and performing insertion. Then by adding the arcs to represent events, we obtain the complete state diagram in Fig. 4.1(b). ◊

For this system, two versions of the models have different numbers of nodes and arcs, or states and events. The second one describes the system in more detail than the first one does.

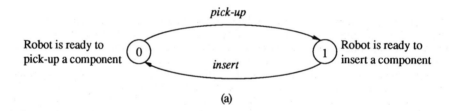

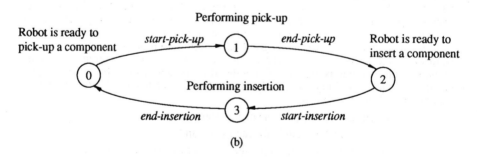

Figure 4.1 (a) Two-state two-event and (b) four-state four-event automata for a single robot assembly system.

4.1. Conditions, Events, and State Machines

Example 4.2: Build up a finite state machine model for a dual robot assembly system. The same assumptions as those in Example 4.1 are applied to the two robots. We can again have two models.
- A. It contains four events and four states as shown in Fig. 4.2(a). The state is defined as a two-tuple (R_1, R_2) where $R_i = 0$ if Robot i is ready to pick-up a component and 1 if ready to insert, i = 1 and 2.
- B. It contains eight events and sixteen states as shown in Fig. 4.2(b). The state is defined as a two-tuple (R_1, R_2) where $R_i = 0$ if Robot i is ready to pick-up a component, 1 if performing pick-up, 2 if ready to insert, and 3 if performing insertion, i = 1 and 2. ◊

Based on the above two examples, we can obtain the following observations.

Observation 1. The number of events of the system is the sum of the events of each individual resource in the model if we assume that only one of them happens at any time instant. The number of events in Fig. 4.2(a) equals 4 and in Fig. 4.2(b) equals 8. In other words, the number of events is linearly increased. The number of states, is however, grows exponentially with the number of states of each individual resource in the model, which is referred to as state explosion. The number of states in Fig. 4.2(a) equals $2^2 = 4$, and it becomes $4^2 = 16$ in Fig. 4.2(b).

Observation 2. The model in Fig. 4.2(a) dose not express the concurrent operations of two robots but that in Fig. 4.2(b) does implicitly. In particular, two states (1 3) and (3 1) in Fig. 4.2(b) implies that one robot is performing pick-up and another insertion. Thus concurrent operations may be implied in certain states in a state diagram model. If one wants to model a system's concurrency, then the model in Fig. 4.2(a) is not appropriate.

Most systems containing multiple robots have to satisfy additional constraints except the precedence relations among events that robots have to follow. For example, two robots are not allowed to enter the pick-up area nor the PCB area simultaneously to avoid the collision. To incorporate these constraints into the state machine model, one has to disable events or disconnect the arcs in state diagrams, which may be present in the unconstrained model. Take the model in Fig. 4.2(a), to accommodate the above constraints, the arcs from state (1 0) to (1 1) and (0 1) to (1 1) are removed. State (1 1) becomes not allowable, which implies that both robots have picked up components and are ready to insert. The new model is shown in Fig. 4.3(a). For the model in Fig. 4.2(b), we notice that at states (0 1), (0 2), (1 0) and (2 0), we cannot start any robot's pick up operations, therefore the resulting arcs and states (1 1) (1 2) (2 1) (2 2) have to be removed. At states (2 3) and (3 2), we cannot start another insertion, thus the arcs from these two states to (3 3) together with state (3 3) are removed. The correct model is shown in Fig. 4.3(b). It is noted that the model in Fig. 4.3(b) can be obtained from the scratch by listing the feasible states and derive the permissible events at each state and then linking states together.

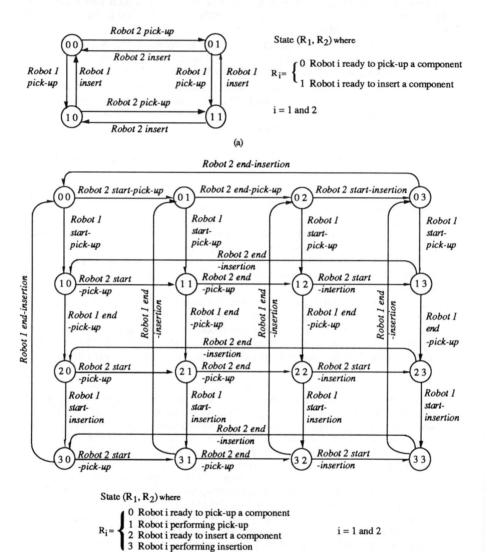

Figure 4.2 Two state machine models for a dual robot assembly system.

4.1. Conditions, Events, and State Machines

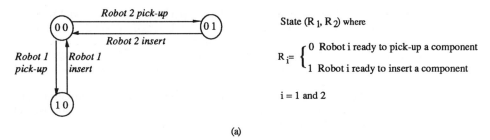

(a)

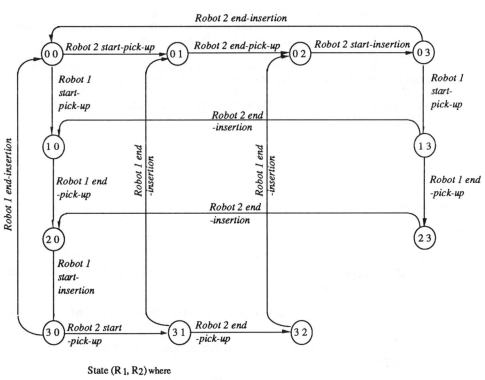

(b)

Figure 4.3 Two state machine models for the dual robot system with additional constraints incorporated.

Observation 3. It would be a tedious job to build up a state machine model because one has to define first the system state that is global, and then enumerate all the states and feasible events at each state. A state explosion problem is immediately encountered for a complex system's design and analysis. The advantage is that the complete information of a system is also obtained once its state machine model is obtained.

Observation 4. The states, events, precedence relations, conflicting situations are explicitly represented but synchronization concepts, concurrent operations, and mutually exclusive relations are not.

A formalism that can overcome some of the limitations in the state machine modeling is extremely desired to handle complex systems. It should use local states rather than global states, thereby avoiding the state enumeration problems in the modeling stage. It can explicitly represent conditions and events together, precedence relations, conflicting situations, synchronization concepts, concurrent operations, and mutually exclusive relations. The invention of Petri nets by Carl A. Petri in his 1962 doctoral dissertation met the above demand. The next section presents a formal definition and several examples. Then the properties and analysis methods of Petri nets are discussed.

4.2 Formal Definition of Petri Nets

4.2.1 Definition of Petri Nets

A marked Petri net (PN) $Z = (P, T, I, O, m)$ is a five tuple where
1. $P = \{p_1, p_2, ..., p_n\}$, $n>0$, is a finite set of places pictured by circles;
2. $T = \{t_1, t_2, ..., t_s\}$, $s>0$, is a finite set of transitions pictured by bars, with $P \cup T \neq \emptyset$ and $P \cap T = \emptyset$;
3. $I: P \times T \rightarrow N$, is an input function that defines the set of directed arcs from P to T where $N = \{0, 1, 2, ...\}$;
4. $O: P \times T \rightarrow N$, is an output function that defines the set of directed arcs from T to P;
5. $m: P \rightarrow N$, is a marking whose i^{th} component represents the number of tokens in the ith place. An initial marking is denoted by m_0. Tokens are pictured by dots.

The four tuple (P, T, I, O) is called a PN structure that defines a directed graph structure. Introduction of tokens into places and their flow through transitions enable one to describe and study the discrete-event dynamic behavior of the PN, thereby the modeled system. The above defined net is called an ordinary Petri net. When $I(p, t) > 1$, the number $I(p, t)$ is also named as the weight of the arc or the number of directed arcs from p to t. The same applies to $O(p, t)$ that is named the

4.2. Formal Definition of Petri Nets

weight of the arc or the number of directed arcs from t to p. I and O represent two n×s non-negative integer matrices. Subtraction of I from O yields an incidence matrix denoted by

$$C = O - I.$$

A PN can be alternatively defined as (P, T, F, W, m) where F is a subset of $\{P \times T\} \cup T \times P\}$, representing a set of all arcs, and W: F → N, defines the multiplicity or weight of arcs.

If I(p, t) = k (O(p, t) = k), then there exist k directed arcs connecting place p to transition t (transition t to place p). If I(p, t) = 0 (O(p, t) = 0), there exist no directed arcs connecting p to t (t to p). A single directed arc is used when k=1. For k>1 cases, either k parallel arcs are used to connect a place (transition) to a transition (place), or more often a single directed arc labeled with its multiplicity or weight k is used. When m(p)=j>1, instead of j dots in p, we often write j in place p to represent j tokens.

Example 4.3: A marked Petri net shown in Fig. 4.4 and its formal description is given as follows:

$P = \{p_1, p_2, p_3\}$
$T = \{t_1, t_2\}$
$I(p_1, t_1) = 1, I(p_1, t_2) = 0; O(p_1, t_1) = 0, O(p_1, t_2) = 0;$
$I(p_2, t_1) = 1, I(p_2, t_2) = 0; O(p_2, t_1) = 0, O(p_2, t_2) = 1;$
$I(p_3, t_1) = 0, I(p_3, t_2) = 1; O(p_3, t_1) = 1, O(p_3, t_2) = 0;$
$m = (1\ 1\ 0)^\tau.$

Input and output functions can be represented as matrices, i.e.,

$$I = \begin{pmatrix} 1 & 0 \\ 1 & 0 \\ 0 & 1 \end{pmatrix} \text{ and } O = \begin{pmatrix} 0 & 0 \\ 0 & 1 \\ 1 & 0 \end{pmatrix}.$$

The incidence matrix $C = O - I = \begin{pmatrix} -1 & 0 \\ -1 & 1 \\ 1 & -1 \end{pmatrix}.$

Its alternative definition of the net can also be derived. This exercise is left to the reader. ◊

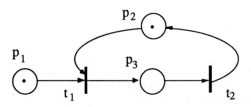

Figure 4.4 A Petri Net Example in Example 4.3.

Given a Petri net diagram, an intuitive way to obtain the input matrix is as follows. Starting from place p_i, $i = 1, 2, ..., n$, if there are directed arcs from p_i to t_j, $j = 1, 2, ...,$ and s, enter in the ith row and jth column an entry with a positive number equal to the number of the directed arcs and otherwise zero. Therefore, if $I(p_i, t_j) \neq 0$, p_i is called an input place of transition t_j. To obtain the output matrix, starting from transitions t_j, $j = 1, 2, ...,s$, if there are directed arcs from t_j to p_i, $i = 1, 2, ..., n$, enter in the ith row and jth column an entry with a positive number equal to the number of the directed arcs and otherwise zero. Similarly, if $O(p_i, t_j) \neq 0$, p_i is called an output place of transition t_j. A reader is encouraged to obtain I and O for Example 4.3 using the described way.

4.2.2 Execution Rules of Petri Nets

The execution rules of a PN include enabling and firing rules:

1. A transition $t \in T$ is *enabled* if and only if $m(p) \geq I(p, t)$, $\forall\ p \in P$; and
2. Enabled in a marking m, t *fires* and results in a new marking m':

$$m'(p) = m(p) - I(p, t) + O(p, t), \forall\ p \in P.$$

Marking m' is said to be (*immediately*) *reachable* from marking m. The enabling rule states that if all the input places of transition t have enough tokens, then t is enabled. This means that the conditions associated with the occurrence of an event are all satisfied, then the event can occur. Note that these conditions form an AND relation from the logic viewpoint. The firing rule says that an enabled transition t fires or an event occurs. Its firing can be viewed as two separate stages. First, remove the required number of tokens from each of its input places and the quantity equals the number of arcs between the input place and fired transition t. In the above equation, the item -I(p, t) states this fact. Second, deposit tokens into each of t's output places and the number of tokens equals the number of arcs from t to the corresponding output place. This is reflected by +O(p, t) item in the equation.

Example 4.4: In Fig. 4.4, transition t_1 is enabled since $m(p_1) = 1 = I(p_1, t_1)$ and $m(p_2) = 1 = I(p_2, t_1)$; t_2 is not since $m(p_3) = 0 < 1 = I(p_3, t_2)$. Firing t_1 removes one token from p_1 and p_2, respectively; and deposits one token to t_1's only output place p_3. The result of firing t_1 is shown in Fig. 4.5(b). The new marking is m' = $(0\ 0\ 1)^\tau$. At m', t_2 is enabled since $m'(p_3) = 1 = I(p_3, t_2)$. Firing t_2 results in m" = $(0\ 1\ 0)^\tau$. At this marking no transition is enabled. Note that t_1 is not at m" since $m"(p_1) = 0<1 = I(p_1, t_1)$ although $m"(p_2) = 1 = I(p_2, t_1)$. ◊

One significant extension to increase the modeling power of Petri nets is achieved by adding the zero testing ability, i.e., the ability to test whether a place has no token. An inhibitor arc is used to implement this ability. The inhibitor arc connects an input place to a transition, and is pictorially represented by an arc terminated with a small circle. Note that the small circle means "not" in digital circuits.

4.2. Formal Definition of Petri Nets

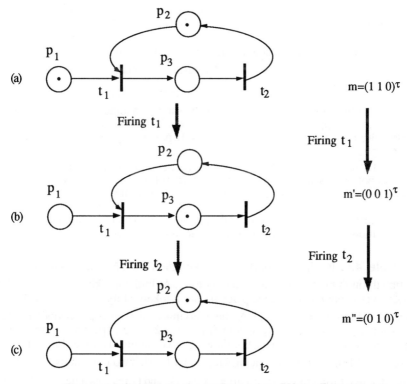

Figure 4.5 Evolution of markings of a PN in Example 4.4: (a) $m = (1\ 1\ 0)^\tau$, (b) $m' = (0\ 0\ 1)^\tau$, and (c) $m'' = (0\ 1\ 0)^\tau$.

The presence of an inhibitor arc connecting an input place to a transition changes the transition enabling conditions. In the presence of the inhibitor arc, a transition is regarded as enabled if each input place, connected to the transition by a normal arc (an arc terminated with an arrow), contains at least the number of tokens equal to the weight of the arc, and no tokens are present on each input place connected to the transition by an inhibitor arc. The transition firing rules are the same for normally connected places. The firing, however, does not change the marking in the inhibitor arc connected places. The inhibitor arc can be generalized to one with multiplicity. Then when the number of tokens in the place is less than the number of inhibitor arc's multiplicity, the transition is allowed to be enabled by its regular input places.

It can formally introduced into the PN definition by adding an inhibitor function $H: P \times T \rightarrow N$. A transition $t \in T$ is regarded as being *enabled* if and only if

$$m(p) \geq I(p, t) \text{ and } m(p) < H(p, t), \forall p \in P.$$

The transition firing rule remains the same. It should be indicated that when $H(p, t)=1$, then $I(p, t)=0$ since, otherwise, t would never be enabled according to the above rule. For cases that $H(p, t)>1$, $I(p, t)<H(p, t)$ if any such normal directed arc (p, t) exists.

Theoretically, an inhibitor arc becomes necessary only when there are places that can potentially obtain an infinite number of tokens and the system needs the knowledge of those places' token numbers. Such arcs are, however, often used to facilitate modeling of many practical systems. A reader is referred to [Ferrarini, 1993] for its extensive use in his controller design methods. The following example shows the execution of a Petri net when such an inhibitor arc exists.

Example 4.5: In Fig. 4.6(a), the upper portion of the PN is affected by the lower portion through the inhibitor arc from p_3 to t_1. At initial marking $m_0 = (3\ 0\ 1\ 0)^\tau$, transition t_1 is not enabled since place p_3 has a token. Only t_3 is enabled. Firing t_3 reaches marking $m_1 = (3\ 0\ 0\ 1)^\tau$. At this marking, t_1 is now enabled since p_3 has no token or $m_1(p_3) = 0 < H(p_3, t_1)$ and $m_1(p_1) = 3 > 2 = I(p_1, t_1)$. Transition t_4 is also enabled since $m_1(p_4) = 1 = I(p_4, t_4)$. Either t_1 or t_4 can fire at a time. Firing t_4 returns to $m_0 = (3\ 0\ 1\ 0)^\tau$. Firing t_1 generates $m_2 = (1\ 1\ 0\ 1)^\tau$. Two transitions, t_2 and t_4, are enabled at m_2. Firing t_2 returns to $m_1 = (3\ 0\ 0\ 1)^\tau$ and t_4 reaches a new marking $m_3 = (1\ 1\ 1\ 0)^\tau$. Both t_2 and t_3 are enabled.at m_3 Firing t_2 returns to the initial marking $m_0 = (3\ 0\ 1\ 0)^\tau$ and t_3 to $m_2 = (1\ 1\ 0\ 1)^\tau$. ◊

A self-loop is a pair of places such that $I(p,t) = O(p, t)$ with p is marked with at least $I(p, t)$ tokens. For the net in the above example, we can find out its equivalent PN by using a self-loop between p_4 and t_1 as shown in Fig. 4.6(b). The reason behind this is that when p_3 is marked, p_4 is not and vice versa. Therefore, p_4 is also called p_3's complement place. A reader can find that the PN in Fig. 4.6(b) exhibits the same behavior as described as above. In other words, both nets in Fig. 4.6 generate the same state machine representation.

4.3. Properties of Petri Nets and Their Implications

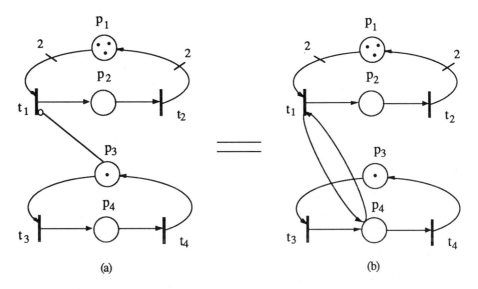

Figure 4.6 (a) A PN model with an inhibitor arc and (b) its equivalent PN with a self-loop.

Other methods to enhance the modeling power of Petri nets include assigning 1) priorities over the transitions, and 2) timing delays to transitions, places, and/or arcs. The latter will be discussed later in Chapter 6. The former can achieve the same effect as inhibitor functions. It is easy to see that removing the inhibitor arc from p_3 to t_1 in Fig. 3(a) but assigning t_3 a higher priority to fire when both t_1 and t_3 are enabled changes no behavior of the net.

4.3 Properties of Petri Nets and Their Implications

Petri nets as a mathematical tool possess a number of properties. These properties, when interpreted in the context of the modeled manufacturing system, allow one to identify the presence or absence of the functional properties of the system. Two types of properties can be distinguished, behavioral and structural ones. The behavioral properties are those that depend on the initial state or marking of a Petri net. The structural properties, on the other hand, do not depend on the initial marking of a Petri net. They depend on the Petri net topology or structure only. In this section, we provide an overview of some of the most important, from the practical point of view, structural and behavioral properties. They are reachability, boundedness, safeness, conservativeness, liveness, reversibility and home state. Additional structural properties include repetitiveness and consistence. Their implications in flexible manufacturing are also discussed. Other properties such as coverability, persistence, synchronic distance, and fairness and their meanings are discussed in [Murata, 1989].

4.3.1 Reachability

Given a Petri net $Z = (P, T, I, O, m_0)$, marking m is *reachable* from marking m_0 if there exists a sequence of transitions firings that transforms m_0 to m. Marking m' is said to be immediately reachable from m if firing an enabled transition in m leads to m'. R is used to represent the set of all reachable markings from m_0. Sometimes, $R(m_0)$ or $R(Z, m_0)$ is also used to represent this set such that different initial markings and nets are indicated.

Reachability is a behavioral property since it depends upon the initial marking. An important issue in designing a manufacturing system is whether the system can reach a specific state, or exhibit particular functional behavior. In general, the question is whether the system modeled with a Petri net exhibits all desirable properties, as specified in the requirement specification, and no undesirable ones.

In order to find out whether the modeled system can reach a specific state as a result of a required functional behavior, it is necessary to find such a sequence of firings of transitions that would transform marking m_0 to m, where m represents the specific state, and the sequence of firings represents the required functional behavior. It should be noted that real systems may reach a given state as a result of exhibiting different permissible patterns of functional behavior. In a Petri net model, this should be reflected in the existence of specific sequences of transition firings, representing the required functional behavior, which would transform m_0 to the required m. The existence in the Petri net model of additional sequences of transition firings that transform m_0 to m indicates that the Petri net model may not be exactly reflecting the structure and dynamics of the underlying system. This may also indicate the presence of unanticipated facets of the functional behavior of the real system, provided that the Petri net model accurately reflects the underlying system requirement specification [Zurawski and Zhou, 1994].

4.3.2 Boundedness and Safeness

Given a Petri net $Z = (P, T, I, O, m_0)$ and its reachability set R, a place $p \in P$ is *B-bounded* if $m(p) \leq B$, $\forall\ m \in R$ where B is a positive integer, and $\in$ means "belong to" and $\forall$ means "for any". Z is B-bounded if each place in P is B-bounded. *Safeness* is 1-boundedness.

In many cases, the specific value of the bound B is not concerned. Thus a place p is *bounded* if it is B-bounded for some B; Z is bounded if all places are bounded.

Z is *structurally bounded* if Z is bounded given any finite initial marking m_0.

Places are frequently used to represent storage areas for parts, tools, pallets, and AGVs in manufacturing systems. They may also represent availability of resources. It is important to be able to determine whether proposed control strategies prevent from the overflows of these storage areas and capacity overflow of these resources. The concept of boundedness of a Petri net is used to identify the existence of overflows in the modeled system. When a place models an operation, its safeness guarantees that the controller have no attempt to initiate an ongoing process. The

4.3. Properties of Petri Nets and Their Implications

concept of boundedness is often interpreted as stability of a discrete manufacturing system particularly when it is modeled as a queuing system.

4.3.3 Conservativeness

A Petri net $Z = (P, T, I, O, m_0)$ is *conservative* with respect to a vector w if there exists a vector $w = (w_1, w_2, ..., w_n)^\tau$, and $w_i > 0$, $i = 1, 2, ..., n$, such that $w^\tau m = w^\tau m_0$, $\forall\ m \in R$ where R is the reachability set. Z is *strictly conservative* if Z is conservative with respect to $w = (1, 1, ..., 1)^\tau$, or $\sum_{i=1}^{n} m(p_i) = \sum_{i=1}^{n} m_0(p_i)$, $\forall\ m \in R$.
Since this property does not depend on an initial marking, it is a structural property.

In real systems, the number of resources in use is typically restricted by the financial as well as other constraints. If tokens are used to represent resources, the number of which in a system is typically fixed, then the number of tokens in a Petri net model of this system should remain unchanged irrespective of the marking the net takes on. This follows from the fact that resources are neither created nor destroyed, unless there is a provision for this to happen. For instance, a broken tool may be removed from a manufacturing cell, thus reducing the number of tools available by one. Strict conservativeness implies that the number of tokens is conserved. From the net structure point of view, the number of input arcs to each transition has to equal the number of output arcs from it. Conservativeness implies that the weighted sum of tokens remains the same for each marking. In real systems resources are frequently combined together so that certain task can be executed. Then they are separated after the task is completed. For instance, a robot picks up a part, loads it to a machine for processing, and subsequently unloads it from the machine. One token may represent several resources combined. This token may be used to create multiple tokens (one per resource) by firing a transition with more output arcs than input ones.

A Petri net $Z = (P, T, I, O, m_0)$ is *partially conservative* if there exists a vector $w = (w_1, w_2, ..., w_n)^\tau$, and $w_i \geq 0$, $i = 1, 2, ..., n$, and $w \neq 0$ such that $w^\tau m = w^\tau m_0$, $\forall\ m \in R$.

4.3.4 Liveness

A transition t is *live* if at any marking $m \in R$, there is a sequence of transitions whose firing reaches a marking that enables t. Equivalently, t is live if at any marking $m \in R$, there is a transition firing sequence that includes t. A Petri net is live if every transition in it is live. A Petri net is *structurally live* if there is a finite initial marking that makes the net live.

A transition t is *dead* if there is $m \in R$, such that there is no sequence of transition firings to enable t starting from m. A Petri net contains a *deadlock* if there is $m \in R$ at which no transition is enabled. Such a marking is called a dead marking.

Deadlock situations are as a result of inappropriate resource allocation policies or exhaustive use of some or all resources. In automated manufacturing, many resources are shared. These resources can be machines, material handling systems including such components as robots and AGVs, and storage areas or buffers. For example, robots are shared by several machines; machines by different or same jobs; and buffers by raw, intermediate and final products. In such a resource sharing environment, the following four conditions may hold simultaneously, thus leading to a deadlock [Coffman et al., 1971]:

1. Mutual exclusion: a resource is either available or allocated to a process that has an exclusive access to this resource.
2. Hold and wait: a process is allowed to hold a resource(s) while requesting more resources.
3. No preemption: a resource(s) allocated to a process cannot be removed from the process, until it is released by the process itself.
4. Circular wait: two or more processes are arranged in a chain in which each process waits for resources held by the process next in the chain.

For instance, in a flexible manufacturing system, a deadlock occurs when the input/output buffer of a machining tool holds a pallet with already machined products, and another pallet with products to be machined has been delivered to the buffer. Assuming that the buffer can hold one pallet only at a time, and an automated guided vehicle (AGV), for instance, has a space for one pallet, a deadlock occurs. The pallet with machined parts cannot be moved from the buffer to the AGV. The pallet with parts to be machined cannot be moved from the AGV to the buffer. In this example, all four conditions hold, with the buffer and AGV space for pallets regarded as resources. Unless there is a provision in the control software for deadlock detection and recovery, a deadlock situation, initially confined to a small subsystem, may propagate to affect a large portion of a system. This frequently results in a complete standstill of a system. Additional examples of deadlock can be referred to [Banaszak and Krogh, 1990; Hsieh and Chang, 1994; Viswanadham et al., 1990; Zhou and DiCesare, 1993; Xing et al., 1995].

Liveness of a Petri net means that for any marking m reachable from the initial marking m_0, it is ultimately possible to fire any transition in the net by progressing through some firing sequence. Therefore, if a Petri net is live, then there is no deadlock. Note that the liveness can be too strict to represent some real systems or scenarios. In particular, a system's initialization may be modeled by some transition(s) that may fire a finite number of times and then become dead. After the initialization period, the system may exhibit a deadlock-free behavior, although the Petri net representing the entire system is not live as specified above. Fortunately, we can always treat the initialization as an exception in FMS design.

4.3.5 Reversibility and Home State

A Petri net $Z = (P, T, I, O, m_0)$ is reversible if $\forall\ m \in R(m_0)$, $m_0 \in R(m)$. Marking $m' \in R(m_0)$ is called a home state if $\forall\ m \in R(m_0)$, m' is reachable from m. A Petri net is structurally reversible if there is a finite initial marking that makes the net reversible.

4.3. Properties of Petri Nets and Their Implications

Reversibility of a Petri net means that for any marking reachable m from m_0, m_0 is also reachable from m. Many systems are required to return from the failure states to the preceding correct states. Thus this property is important in error recovery in a manufacturing system. Also, this property guarantees the cyclic behavior of the system that should be true for all the repetitive manufacturing systems. This requirement is closely related to the reversibility and home state properties of a Petri net. Reversibility is a special case of the home state property, i.e., if the home state m' = m_0, then the net is reversible. Also, if a net contains a deadlock, then the net is not reversible.

Boundedness/safeness, liveness, and reversibility do not imply each other for general cases. The examples can be seen in [Murata, 1989].

4.3.6 Other Structural Properties

A PN is said to be repetitive if there exists a finite marking m_0 and a firing sequence S from m_0 such that every transition occurs infinitely often in S.

A PN is said to be *consistent* if there exists a marking m_0 and a firing sequence S from m_0 back to m_0 such that every transition occurs at least once in S.

4.3.7 Examples

Example 4.4 (Continued): The net shown in Fig. 4.4 is safe since each place holds at most one token. It is not conservative since one token in p_1 is consumed for good. It is neither live nor reversible since the net contains a deadlock, i.e., at marking $(0\ 1\ 0)^\tau$, no transition is enabled. Note that this deadlock results from the exhaustion of the tokens in place p_1. Only markings $(0\ 0\ 1)^\tau$ and $(0\ 1\ 0)^\tau$ are reachable from the initial marking $(1\ 1\ 0)^\tau$. Others are not. The Petri net is structurally bounded. It is, however, structurally neither live nor reversible since given any finite initial markings, a) a finite number of tokens in place p_1 will be exhausted through firing transitions t_1 and t_2 if p_2 or p_3 is initially marked; or b) if none of p_1 and p_3 is initially marked, the net has no transition enabled initially. ◊

Example 4.5 (Continued): The net shown in Fig. 4.6(a) is bounded since place p_1 is 3-bounded and all three other places are safe. The net is conservative with respect to the following weighing vector $(1\ 2\ 1\ 1)^\tau$. It is not strictly conservative though. The net is also live and reversible. These can be verified based on all the markings this net can reach. A marking that p_1 has no token is not reachable from the initial marking m_0 = $(3\ 0\ 1\ 0)^\tau$ since under all the reachable markings, p_1 holds at least one token. A state or marking with exactly three places having one token is reachable from m_0. In fact there are two markings satisfying this, i.e., $m_2 = (1\ 1\ 0\ 1)^\tau$ and $m_3 = (1\ 1\ 1\ 0)^\tau$. The net is also structurally bounded, live, and reversible. ◊

4.3.8 Implications in Flexible Manufacturing

Based on the two interpretations of places and transitions as shown in Table 4.1, and the modeling convention with Petri nets explained in Table 4.2, we can summarize the implications of the Petri net properties in Table 4.3 [Zhou and Jeng, 1998].

Table 4.1. Two interpretations of places and transitions

PN Elements	First Interpretation	Second Interpretation
Places	Resource status and operations	Resource status and conditions
Transitions	Start and/or end of operations, processes, activities, and events	Operations, processes, activities, and events
Directed arcs	Material, resource, information, and/or control flow direction	

Table 4.2. Modeling convention with Petri nets

Concepts in Manufacturing	Petri net modeling
Moving or production lot size	Weight of directed arcs modeling moving or production Kanban
Number of resources, e.g., AGVs, machines, workstation, and robots	The number of tokens in places modeling quantity of the corresponding resources
Capacity of a workstation	The number of tokens in places modeling its availability
Work-in-process	The number of tokens in places modeling the buffers and operations of all machines
Production volume	The number of tokens in places modeling the counter for or the number of firings of transitions modeling end of a product
The time of an operation, e.g., setup, processing, and loading	Time delays associated with the place or transition modeling the operation
Conveyance or transportation time	Time delays associated with the directed arc, place, or transition modeling the conveyance or transportation
System state	Petri net marking (plus the timing information for timed Petri net)
Sequence, concurrency, conflict, resource-sharing, etc.	The Petri net modules to be discussed in Chapter 5

4.4. Reachability Analysis Method

Table 4.3. Petri net properties and their meanings

PN Properties	Meanings in the Modeled Manufacturing System
Reachability	A certain state can be reached from the initial conditions
Boundedness	No capacity (of, e.g., buffer, storage area, and workstation) overflow
Safeness	Availability of a single resource; or no request to start an ongoing process
Conservativeness	Conservation of non-consumable resources, e.g., machines and AGVs
Liveness	Freedom from deadlock and guarantee the possibility of a modeled event, operation, process or activity to be ongoing
Reversibility	Re-initialization and cyclic behavior
Repetitiveness	Existence of repetitive operations/activities/events for some marking
Consistency	Existence of cyclic behavior for some marking

4.4 Reachability Analysis Method

Starting from the initial system condition or state, it is desired to derive all the possible states the system can reach, as well as their relationship. The resulting representation is called reachability tree or graph. Then all the behavioral properties discussed above can be discovered if the number of states is finite. The method toward enumeration of all states or markings in Petri nets for analysis of their properties is called a reachability or coverability analysis method.

There are two strategies to generate and label all the markings in a marked Petri net: depth-first and breadth-first. In the depth-first strategy, starting from the initial marking, identify all the enabled transitions, firing one of them (randomly) resulting in a "new" marking. If it is an old or dead marking, stop exploring it, come back to the marking generating it, and continue with those unfired transitions. Otherwise, at this new marking, again identify all the enabled transitions, firing one of them resulting in a "new" marking. Continue the above process until all the enabled transitions are fired and all the markings are generated if the number of markings is finite.

The second strategy, breadth-first, identifies all the enabled transitions, firing all of them resulting in "new" markings. For each marking, if it is an old or dead marking, go to the next marking. Otherwise, identify all the enabled transitions and fire them to generate all the "new" markings, and then go to next marking until the markings at the same level are exhausted. Then start the next level of markings.

The resulting representation consists of nodes that are markings and the arcs. An arc links from one node to another and is labeled with its corresponding fired transition. The representation is termed as a reachability or coverability tree. Elimination of duplicate markings from the tree leads to a reachability or coverability graph. Both generation strategies lead to an identical a reachability graph. In fact, given a net and its initial marking, its reachability graph is unique.

To illustrate these two strategies, we use the Petri net example in Fig. 4.6(b).

Example 4.6 (Continued): Starting from $m_0 = (3\ 0\ 1\ 0)^\tau$, only transition t_3 is enabled. Firing t_3 generates $m_1 = (3\ 0\ 0\ 1)^\tau$ as shown in Fig. 4.7(a). At m_1, t_1 and t_4 are enabled. Firing t_1 generates $m_2 = (1\ 1\ 0\ 1)^\tau$. At m_2, t_2 and t_4 are enabled. Firing t_2 generates an old marking, i.e., m_1. Then firing t_4 reaches a new marking $m_3 = (1\ 1\ 1\ 0)^\tau$. At m_3 both t_2 and t_3 are enabled. Firing t_2 generates an old marking m_0 and t_3 reaches m_2. The only transition remaining unfired is t_4 at m_1. Its firing generates an old marking m_0. In summary, the use of the depth-first strategy leads to the following sequences of markings and transitions: m_0-t_3-m_1-t_1-m_2-t_2-m_1; m_2-t_4-m_3-t_2-m_0; m_3-t_3-m_2; m_1-t_4-m_0. Its equivalent reachability graph is shown in Fig. 4.7(b).

Now let us generate the same tree by using breadth-first strategy. Starting from m_0, only transition t_3 is enabled. Firing t_3 generates $m_1 = (3\ 0\ 0\ 1)^\tau$. At m_1, t_1 and t_4 are enabled. Firing t_1 generates $m_2 = (1\ 1\ 0\ 1)^\tau$ and t_4 reaches an old marking m_0. At m_2, t_2 and t_4 are enabled. Firing t_2 generates an old marking m_1 and firing t_4 reaches a new marking $m_3 = (1\ 1\ 1\ 0)^\tau$. At m_3 both t_2 and t_3 are enabled. Firing t_2 generates an old marking m_0 and t_3 another old one m_2. Thus, the breadth-first strategy leads to the following sequences of markings and transitions: m_0-t_3-m_1-t_1-m_2 and t_4-m_0; m_2-t_2-m_1 and t_4-m_3; m_3-t_2-m_0 and t_3-m_2. ◊

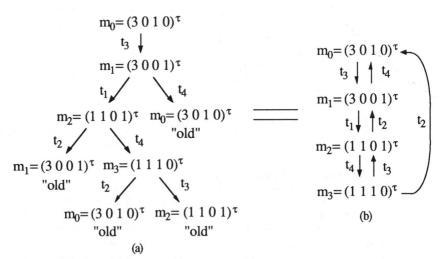

Figure 4.7 (a) Reachability tree and (b) reachability graph.

4.4. Reachability Analysis Method

When a Petri net is unbounded or there are infinite number of states in the modeled system, then the above procedure will continue forever. To keep the tree finite, the symbol ω is introduced. It can be understood as the infinite and satisfies

$\omega > k$, $\omega \leq \omega$, and $\omega \pm k = \omega$ for any finite integer k.

The following algorithm is used to generate a finite tree.

Algorithm 4.1:
 1.0) Let the initial marking m_0 be the root of the tree and tag it "new"
 2.0) While "new" markings exist, do the following:
 3.0) Select a "new" marking m;
 3.1) If m is identical to another marking in the tree, then tag m "old", and go to another "new" marking;
 3.2) If no transitions are enabled in m, tag m "deadend";
 4.0) For every transition t enabled in marking m do the following:
 4.1) Obtain the marking m' that results from firing t in m;
 4.2) If on the path from the root to m, there exists a marking m" such that $m'(p) \geq m"(p)$ for each place p, and $m' \neq m"$, then replace $m'(p)$ by ω for each p wherever $m'(p) > m"(p)$";
 4.3) Introduce m' as a node, draw an arc from m to m' labeled t, and tag m' "new".
 4.4) Remove tag "new" from m.

Note that in Step 3.1, markings in the tree do not include those labeled with "new" node. Step 3.1 can be modified into:

If m is identical to a marking on the path from the root to m, then tag m "old", and go to another "new" marking.

Then the resulting reachability tree may contain more nodes.

Example 4.7: Algorithm 4.1 is applied to the PN model of a two-machine production line with the infinite intermediate part storage. The initial marking is $m_0 = (1\ 0\ 0\ 1\ 0)^\tau$. Following Steps 1.0)-4.4), we find the only enabled transition t_1 whose firing leads to a marking $m_1 = (0\ 1\ 0\ 1\ 0)^\tau$. The resulting marking is neither identical to nor greater than m_0. Then it is labeled "new". It is the only "new" marking at present. Following the above steps generates a marking $m' = (1\ 0\ 1\ 1\ 0)^\tau$ by firing t_2 at m_1 This marking $m' \geq m_0 = (1\ 0\ 0\ 1\ 0)^\tau$ and $m(p_3) > m_0(p_3)$. Thus according to Step 4.2), change $m'(p_3)$ into ω. At $(1\ 0\ \omega\ 1\ 0)^\tau$ that is only "new" and labeled m_2 in Fig. 4.8(b), there are two transitions enabled, i.e., t_1 and t_3. Firing t_1 leads to $(0\ 1\ \omega\ 1\ 0)^\tau$ that is not identical to any markings along the path. Although it is greater than $(0\ 1\ 0\ 1\ 0)^\tau$, there is no need to execute Step 4.2) since its third component is ω already.

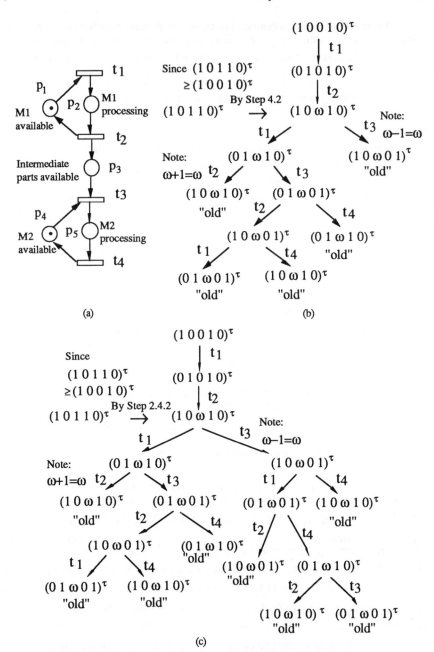

Figure 4.8 (a) Petri net model (b) Finite reachability tree.

4.4. Reachability Analysis Method

Firing t_3 leads to $(1\ 0\ \omega\text{-}1\ 0\ 1)^\tau = (1\ 0\ \omega\ 0\ 1)^\tau$ due to the property of $\omega\pm k = \omega$ for any finite k. This marking is neither identical to nor greater than any marking along the path. At present, there are two "new" markings $(0\ 1\ \omega\ 1\ 0)^\tau$ and $(1\ 0\ \omega\ 0\ 1)^\tau$. At $(0\ 1\ \omega\ 1\ 0)^\tau$, there are two enabled transitions t_2 and t_3. Firing t_2 leads to $(1\ 0\ \omega\text{+}1\ 1\ 0)^\tau = (1\ 0\ \omega\ 1\ 0)^\tau$ that is identical to a previous marking on the path, thereby labeled "old". Firing t_3 leads to $(0\ 1\ \omega\ 0\ 1)^\tau$ that is "new". Again there are two "new" markings, $(0\ 1\ \omega\ 0\ 1)^\tau$ and $(1\ 0\ \omega\ 0\ 1)^\tau$.

At $(0\ 1\ \omega\ 0\ 1)^\tau$, t_2 and t_4 are enabled. Firing t_2 results in $(1\ 0\ \omega\ 0\ 1)^\tau$ that is neither identical to nor greater than any marking along the path, thereby labeled "new". Firing t_4 results in $(0\ 1\ \omega\ 1\ 0)^\tau$ that is identical to a marking in the path, thus labeled by "old".

Now there are two "new" markings, $(1\ 0\ \omega\ 0\ 1)^\tau$ and $(1\ 0\ \omega\ 0\ 1)^\tau$ that are in fact the same although they represent different "positions" in the tree. At marking $(1\ 0\ \omega\ 0\ 1)^\tau$ (most newly added one), t_1 and t_4 are enabled. Firing either one leads to an "old" marking as shown in Fig. 4.8(b). Now select the only remaining "new" marking $(1\ 0\ \omega\ 0\ 1)^\tau$. It is however identical to the just explored marking already in the tree. Thus label it "old". Since there is no "new" marking left, the algorithm halts. The complete reachability tree is shown in Fig. 4.8(b).

Using the modified Step 3.1, we cannot stop at marking $(1\ 0\ \omega\ 0\ 1)^\tau$ during the above process. Expanding it leads to the reachability tree shown in Fig. 4.8(c). ◊

Note that the depth-first, breadth-first, or combined strategies can be implemented by selecting different markings for the next exploration. In the above example, the first strategy was applied.

Once we generate the reachability tree, the following conclusions can be made.

1. The net is bounded if and only if ω does not appear in any node. We can find the maximum number of tokens in a place over all markings, say, B, then the place is B-bounded. Similarly, if B is the maximum component over all markings, then the net is B-bounded. A place p is unbounded if there is a marking m in the tree such that m(p) is ω.

2. The net is safe if and only if each node of the tree contains only zeros and ones.

3. If no deadend contains ω, the number of different deadends in the tree is the number of dead markings of the net. If a deadend marking has ω, then the net has an infinite number of dead markings.

4. A transition is dead if it does not appear as an arc label in the tree.

5. If given any two nodes in the reachability tree without symbol ω, there is a directed path and all transitions are present, then the net is live.

6. If there is a directed path from any node to the initial marking in the reachability tree without symbol ω, then the net is reversible.

Example 4.6 (Continued): Based on the reachability tree as shown in Fig. 4.7(a), the net in Fig. 4.6(b) is bounded, live and reversible. ◊

Example 4.7 (Continued): Based on the reachability tree as shown in Fig. 4.8(b), the net in Fig. 4.8(a) is unbounded. Places except p_3 are safe. ◊

The reachability analysis method is a fundamental approach to Petri net analysis. Algorithm 4.1 can be applied to the Petri net with inhibitor arcs. The limitation of this method is the combinatorial state explosion problem. The method is applicable to analysis of behavioral properties only. Since symbol ω is used in generating a finite reachability tree for an unbounded Petri net, its liveness and reversibility cannot be decided based on the appearance of the tree. Such examples can be seen in [Peterson, 1981; Wong et al., 1993].

4.5 Invariant Analysis Method

Arcs describe the relationships among places and transitions and can be represented by two matrices. By studying linear equations based on the execution rule and matrices, one can find subsets of places over which the sum of the tokens remains unchanged. One may also find that a transition firing sequence brings the marking back to the same one.

Mathematically, denote $C = O - I$ called an incidence matrix. Then execution rule is described as the following state equation:

$$m_k = m_{k-1} + Cu_k, \quad k = 1, 2, \ldots$$

where m_k is a marking immediately reachable from marking m_{k-1}, u_k is an $s \times 1$ column vector with only one entry being one and the others zeros and called the k-th firing vector. If transition t_i fires at the k-th firing, then the i-th position of u_k is 1 and other positions are filled with 0's. The i-th column of C represents a change of a marking as a result of firing transition t_i.

The positive integer solution x of $C^\tau x = 0$ is called a *P-invariant*. By multiplying a transposed P-invariant x^τ to both sides of the state equation, we obtain

$$x^\tau m_k = x^\tau m_{k-1} + x^\tau Cu_k, \quad k = 1, 2, \ldots$$

Since $C^\tau x = 0$, thus $x^\tau C = 0$,

$$x^\tau m_k = x^\tau m_{k-1}, \quad k = 1, 2, \ldots$$

Therefore, $x^\tau m_k = x^\tau m_0 =$ constant.

4.5. Invariant Analysis Method

The P-invariants can be explained intuitively in the following way. The non-zero entries in a P-invariant represent weights associated with the corresponding places so that the weighted sum of tokens on these places is constant for all markings reachable from an initial marking. These places are said to be covered by a P-invariant, denoted by ‖x‖.

The integer solution y of $Cy = 0$ is called a *T-invariant*. Suppose that firing a sequence of transitions leads m_0 back to m_0. Let the i-th element of the aggregate firing vector u be the number of t_i's firing times in the sequence. Vector u is also called firing count vector. Then

$$m_0 = m_0 + Cu$$

Clearly $Cu = 0$ and u is a T-invariant.

The non-zero entries in a T-invariant represent the firing counts of the corresponding transitions that belong to a firing sequence transforming a marking m_0 back to m_0. A T-invariant indicates all the transitions that comprise the firing sequence transforming m_0 back to m_0, and the number of times these transitions appear in this sequence. It does not specify the order of transition firings, however.

Invariant findings aid in net analysis for some PN properties. For example, if each place in a net is covered by a P-invariant, then it is bounded. However, this approach is of limited use since invariant analysis does not include all the information of a general PN. It is applicable to ordinary Petri nets only.

P- and T-invariants can be obtained by solving the linear equations: $C^\tau x = 0$ and $Cy = 0$. This is shown in the following example.

Example 4.8: Find the incidence matrix of the Petri net shown in Fig. 4.6(b) and its P- and T-invariants.

$$I = \begin{pmatrix} 2 & 0 & 0 & 0 \\ 0 & 1 & 0 & 0 \\ 0 & 0 & 1 & 0 \\ 1 & 0 & 0 & 1 \end{pmatrix}$$

$$O = \begin{pmatrix} 0 & 2 & 0 & 0 \\ 1 & 0 & 0 & 0 \\ 0 & 0 & 0 & 1 \\ 1 & 0 & 1 & 0 \end{pmatrix}$$

The incidence matrix

$$C = O - I = \begin{pmatrix} -2 & 2 & 0 & 0 \\ 1 & -1 & 0 & 0 \\ 0 & 0 & -1 & 1 \\ 0 & 0 & 1 & -1 \end{pmatrix}$$

Solving

$$C^\tau x = 0$$

we obtain

$$x_2 = 2x_1 \text{ and } x_4 = x_3$$

Let free variables $x_1 = x_3 = 1$, we have a P-invariant $(1\ 2\ 1\ 1)^\tau$. Letting $x_1 = 1$ and $x_3 = 0$, and $x_1 = 0$ and $x_3 = 1$, we obtain $(1\ 2\ 0\ 0)^\tau$ and $(0\ 0\ 1\ 1)^\tau$, which are also P-invariants.

Solving

$$Cy = 0$$

we obtain

$$y_2 = y_1 \text{ and } y_4 = y_3$$

Let free variables $y_1 = y_3 = 1$, we have a T-invariant $(1\ 1\ 1\ 1)^\tau$. Letting $y_1 = 1$ and $y_3 = 0$, and $y_1 = 0$ and $y_3 = 1$, we obtain two $(1\ 1\ 0\ 0)^\tau$ and $(0\ 0\ 1\ 1)^\tau$, which are also T-invariants.

Since the net is covered by a positive P-invariant, the net is bounded. We can also find a sequence of transitions firings, e.g., t_3, t_1, t_2, t_4. Its firing brings the initial marking back to the initial one. This sequence has the firing count vector $(1\ 1\ 1\ 1)^\tau$ that is identical to the obtained T-invariant. ◊

The incidence matrix C plays a very important role in analyzing the structural properties of a Petri net. Some of the results are summarized in Table 4.4.

Table 4.4 The conditions for structural properties of a PN where $C = O - I$ is the incidence matrix and x is an integer vector.

Properties	Necessary and Sufficient Conditions
Structurally Bounded	$\exists\, x > 0,\ x^\tau C \leq 0$
Conservative	$\exists\, x > 0,\ x^\tau C = 0$
Partially Conservative	$\exists\, x \gneq 0,\ x^\tau C = 0$
Repetitive	$\exists\, x > 0,\ Cx \geq 0$
Partially Repetitive	$\exists\, x \gneq 0,\ Cx \geq 0$
Consistent	$\exists\, x > 0,\ Cx = 0$
Partially Consistent	$\exists\, x \gneq 0,\ Cx = 0$
Neither live, nor consistent	$\exists\, x \gneq 0,\ x^\tau C \lneq 0$ (sufficient condition only)

Example 4.9: Consider the net in Fig. 4.6(b). We have obtained its incidence matrix in Example 4.8. Since $x = (1\ 2\ 1\ 1)^\tau > 0$ satisfies $C^\tau x = 0$, the net is structurally bounded and conservative. It is also repetitive and consistent due to the existence of $y = (1\ 1\ 1\ 1)^\tau$ satisfying $Cy = 0$. ◊

4.6 Reduction Methods

By simplifying a subnet or structure while preserving the concerned properties, one is able to derive the properties of a complex Petri net. The limitation lies in that reducible subnets or structures may not exist or are difficult to find. Some rules are hard to apply. A set of easy-to-use reduction rules is given in Fig. 4.9, including:

Rule a: Fusion/augmentation of series places as shown in Fig. 4.9(a).

Rule b: Fusion/augmentation of series transitions as shown in Fig. 4.9(b).

Rule c: Fusion/augmentation of parallel places as shown in Fig. 4.9(c).

Rule d: Fusion/augmentation of parallel transitions as shown in Fig. 4.9(d).

Rule e: Elimination/addition of self-loop places as shown in Fig. 4.9(e).

Rule f: Elimination/addition of self-loop transitions as shown in Fig. 4.9(f).

It can be proved that these rules preserve the boundedness, liveness, and reversibility when they are applied to reduce or augment a Petri net. Safeness is guaranteed by Rules a, c, d, and f, as well as by Rules b and e when $k = j = 1$.

Example 4.10: An example to use some of these rules is given in Fig. 4.10(a)-(d).

1) After fusing t_1 and t_2 into t_{12} by Rule b, we obtain Fig. 4.10(b).
2) After eliminating self-loop place p_1 by Rule e, we obtain Fig. 4.10(c).
3) After the elimination of self-loop transition t_{12} by Rule f, we obtain Fig. 4.10(d).

At this stage, since the net in Fig. 4.10(d) is safe, live and reversible, we can conclude that the original one is bounded, live, and reversible. Please note that one may continue to use Rule b to fuse two transitions t_3 and t_4 in Fig. 4.10(d). Then either reduce it to a marked place by eliminating the self-loop transition or to a single transition by eliminating a self-loop place. □

In the above example, a strict sequence of applying the above rules has to be observed. Concurrent applications of several rules at a single step are prohibited in it. Next we show how to apply several rules concurrently to reduce a Petri net model.

Example 4.11: The concurrent use of these rules is exemplified in Fig. 4.11(a)-(d).

1) After fusing p_2 and p_3 into p_{23} by Rule a, p_5 and p_6 into p_{56} by Rule c, and transitions t_6 and t_7 into t_{67} By Rule d, we obtain Fig. 4.11(b).
2) After fusing transitions t_1 and t_3 into t_{13}, t_4 and t_5 into t_{45}, and t_{67} and t_8 into t_{678} by Rule b, we obtain Fig. 4.11(c).
3) Continuing to use Rule b to fuse transitions t_{45} and t_6 into t_{45678} and Rule e to eliminate place p_1, we obtain Fig. 4.11(d).

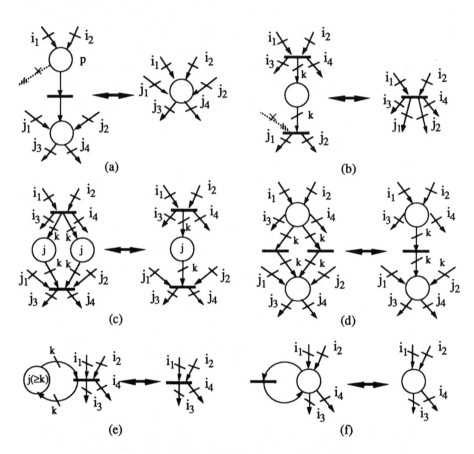

Figure 4.9. A set of reduction rules that preserve boundedness, liveness, and reversibility where k is a positive integer, j, i_j and j_i are non-negative integers and dotted arcs with a cross are prohibited in Rules a and b.

4.6. Reduction Methods

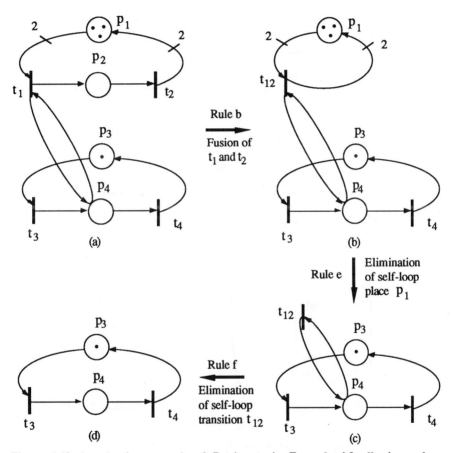

Figure 4.10 A reduction example of Petri nets in Example 4.9 allowing only a strict sequence of applying rules b, e, and f.

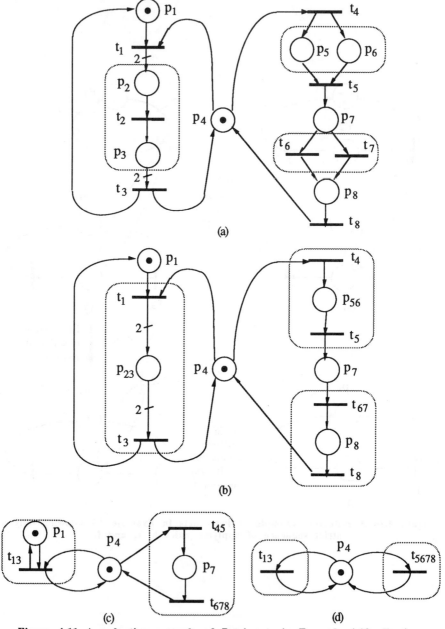

Figure 4.11 A reduction example of Petri nets in Example 4.10 allowing application of several rules at a single step.

4.6. Reduction Methods

We may use Rule f to eliminate both transitions in Fig. 4.11(d) and leave with a marked place. At this stage, we can conclude that this net is bounded, live, and reversible. Thus we can conclude that the original net in Fig. 4.11(a) is bounded, live, and reversible. ◊

Example 4.12: The Petri net shown in Fig. 4.12 is used to illustrate that the wrong use of Rule b guarantees no properties preserved. It is not difficult for a reader to verify that the net in Fig. 4.12(a) is neither live nor reversible. A deadlock results from firing transitions t_1, t_2 and t_1 in Fig. 4.12(a).

The Petri net in Fig. 4.12(b) is live and reversible after transitions t_1 and t_2 in a shaded box are wrongly fused. Note that the input arc from p_2 to t_2 is not allowed for the fusion of t_1 and t_2 according to Rule b. However, transitions t_3 and t_4 in a non-shaded box can be fused into t_{34} by Rule b. This fusion apparently preserves the non-liveness and non-reversibiliy in the original net. In other words, a reader can easily find that the reduced Petri net in Fig. 4.12(c) is neither live nor reversible. The same sequence t_1, t_2 and t_1 for Fig. 4.12(a) will result in a deadlock in Fig. 4.12(c).

The nets in Fig. 4.12(a) and (c) are structurally non-live regardless of the initial markings. If only p_1 and p_2 are initially marked, the deadlock is reached by firing t_1-t_2 as many times as $m_0(p_2)$ and then firing t_1 as many times as $m(p_1)$. If $m(p_1)=a$, $m_0(p_2)=b$, $m_0(p_3)=c$, and $m_0(p_4)=d$, it is easy to verify that firing the sequence

$$(t_3t_4)^d(t_1t_2)^{b+d}t_1^{d+a}$$

leads to a dealock in the net in Fig. 4.12(a). ◊

Example 4.13: This example shows that the fusion of two places cannot preserve the properties if the multiplicity of the concerned arcs is not single. The Petri net with initial marking $m_0=(1\ 1\ 0\ 0\ 0\ 0)^\tau$ in Fig. 4.13(a) is neither live nor reversible since firing transitions t_1 and t_5 results in a dead marking from m_0. Note that once t_1 fires, the net will never go back to its initial marking m_0 that enables both t_4 and t_5 simultaneously. Hence, firing either t_4 or t_5 allows p_5 to trap the token and it remains to have a single token, implying its output transition t_6 is not enabled.

However, after places p_5 and p_6 are fused into p_{56}, then the net as shown in Fig. 4.13(b) is live but not reversible. In this net, p_5 can no longer trap the single token. This is also an example that liveness does not imply the reversibility. A reader can also construct a reversible net that is not live. ◊

Reduction/augmentation plays an important role in complex system design and analysis. Unfortunately, it is very difficult, if not impossible,to develop a set of complete rules that can be used to analyze a property or a set of properties of a Petri net in an automatic way. Other complicated reduction rules and their applications are presented in [Berthelot, 1986 and 1987; Lee and Favrel, 1985; Lee et al., 1987; Silva, 1985].

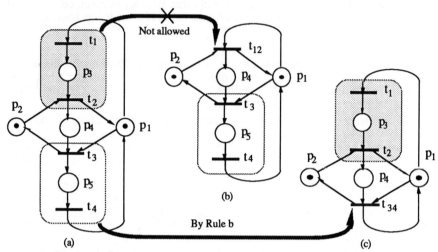

Figure 4.12 An example to illustrate Rule b in Example 4.12.

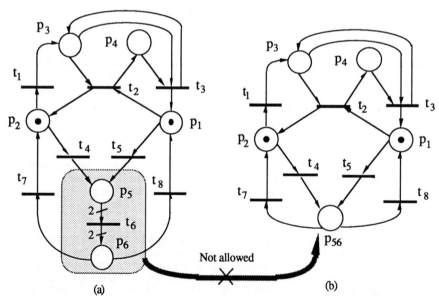

Figure 4.13 (a) A non-live non-reversible net and (b) A live but non-reversible net.

4.7 Other Analysis Methods

In addition to the above discussed reachability analysis, invariant analysis, and reduction methods, we may use siphon and trap-based analysis methods to analyze some special classes of Petri nets, and simulation method [Zhou and Jeng, 1998]. The concept of siphon and trap, as well as the related analysis methods will be discussed in the next chapter. Its usage to certain subclass of Petri nets can avoid the state explosion problems encountered in using the reachability analysis method.

For complex Petri net models, discrete-event simulation is another way to check the system properties. The idea is simple, i.e., using the execution algorithm to run the net. Simulation is an expensive and time-consuming technique. It can reveal the presence of undesirable properties but cannot prove the correctness of the model in general cases. Despite this, Petri net simulation is indeed a convenient and straightforward yet effective approach for engineers to validate the desired properties of a manufacturing system to a certain degree.

A simulation procedure based on Petri nets is outlined as follows:

Algorithm 4.2:
1. Initialization: decide the initial marking and the set of all enabled transitions at the marking;
2. If the number of preset simulation steps or certain stopping criteria are met, stop. Otherwise, if there is no transition enabled, report a dead marking and either stop or go to Step 1.
3. Randomly pick a transition to fire. Remove the same number of tokens from each of its input places as the number of arcs from that place to the transition and deposit the same number of tokens to each of its output places as the number of arcs from the transition to that place.
4. Remove enabled transitions that are modified at the new marking by checking the output transitions of the input places used in Step 3. If the output transitions of the output places in Step 3 become enabled, add those enabled ones. Go to Step 2.

The above algorithm can be modified to simulate extended Petri nets such as timed PNs. Two important modifications are
1. Policies are needed to pick up a transition to fire among conflicting ones and
2. The firing time delays need to be determined, which may be deterministic or sampled from a given random time delay distribution.

Simulation of timed Petri nets allows one to derive the temporal performance for a system under very realistic assumptions in addition to the detection of presence of the behavioral problems. The performance measures include equipment utilization, system throughput, and work-in-process. The simulation method will be used as a performance analysis method for Petri nets augmented with time delays and breakdown handling to be discussed in Chapters 7-9.

4.8 Summary

This chapter starts with the discussions of concepts of conditions, events, and state machines. It proceeds with a formal definition and examples of Petri nets. It introduces PN structures, markings, enabling and firing rules. Important extensions to PNs are also discussed toward the convenience in their use and the increase in their representation power. The chapter then discusses Petri net properties including reachability, boundedness, safeness, conservativeness, liveness, reversibility and home state, repetitiveness and consistence. It presents their implications in flexible manufacturing systems. They are related to system functional behavior, resource capacity overflow, stability, resource conservation, freedom from deadlock, scheduling, and cyclic behavior. Finally, the chapter introduces the analysis methods for these properties. Among them are reachability analysis, invariant analysis, and reduction methods. Other methods include siphon and trap-based analysis methods and simulation method.

1. Reachability analysis method is the most powerful because it can check all properties of bounded PNs that are usually sufficient to model flexible manufacturing systems. However, it is also the most computation-intensive since all system states have to be enumerated.
2. Invariant analysis based on the incidence matrix and state equations are mainly used for analysis of structural properties.
3. Reduction method is useful in performing analysis of many well-structured systems. Its limitation is that the conditions for the reduction rules may not exist or are difficult to find given PN models. The analysis of a reduced net that cannot not be further reduced needs to be performed using the reachability analysis method.
4. Siphon and trap-based methods to be discussed in the next chapter are efficient for some special classes of PNs compared with a reachability analysis method.
5. Simulation is another exhaustive enumeration approach that can report only the situations that are simulated. Hence, it is used as a tool to identify the errors in the design instead a proof of the design correctness.

CHAPTER 5

MODELING FMS WITH PETRI NETS

Modeling and synthesis methods using Petri nets are discussed in this chapter. Special classes of Petri nets such as state machines, marked graphs, free-choice Petri nets, assembly Petri nets, augmented marked graphs, and production-process nets are discussed and their application scope in manufacturing automation is presented. Modeling procedures and methodologies including ad hoc, top-down, bottom-up, hybrid, and modular methods are introduced. Examples are given to illustrate how to use Petri nets to model a variety of FMS. Among them are AT&T's flexible workstation, two-machine production cell, job-shop production system with multiple parts, and a flexible manufacturing cell.

Some common notation needs be introduced to facilitate the discussions of special classes of Petri nets. The word "ordinary" is given a special meaning in this chapter, i.e., a Petri net $Z=(P, T, I, O, m_0)$ is called an ordinary one if $\forall\ p \in P$, $t \in T$, $I(p, t) \leq 1$ and $O(p, t) \leq 1$. In other words, all arcs in an ordinary net have their single multiplicity.

A node is referred to as either a place or a transition. In other words, x is *a node* if $x \in P \cup T$. The *preset* of transition t is the set of all input places to t, i.e., $^\bullet t = \{p: p \in P$ and $I(p, t) \neq 0\}$. The *postset* of t is the set of all output places from t, i.e., $t^\bullet = \{p: p \in P$ and $O(p, t) \neq 0\}$. Similarly, the set of input transitions of place p is its preset $^\bullet p = \{t \in T: O(p, t) \neq 0\}$ and the set of output transitions of p is its postset $p^\bullet = \{t \in T: I(p, t) \neq 0\}$. These concepts can be extended to a subset. For example, if S is a set of places, then the preset of S is the union of the presets of all places in S, i.e., $^\bullet S = \underset{p \in S}{\cup}\ ^\bullet p$. $|S|$ denote the cardinality of set S, i.e., the number of distinct members in S. It is assumed that a Petri net Z has no isolated node, i.e., no node x exists such that $^\bullet x = x^\bullet = \varnothing$.

Given $Z=(P, T, I, O, m_0)$, *an elementary path* is a sequence of nodes: $x_1 x_2 ... x_n$, $n \geq 1$, such that $\exists$ an arc (x_i, x_{i+1}) for $i \in N_{n-1}$ if $n>1$, and $x_i = x_j$ implies that $i=j$, $\forall\ i, j \in N_n$ where $N_n = \{1, 2, ..., n\}$. *An elementary loop or circuit* is

$x_1 x_2 ... x_n$, n>2 such that $\exists$ an arc (x_i, x_{i+1}) for $i \in N_{n-1}$ if n>1, and $x_i = x_j$, $1 \leq i < j \leq n$, implies that i=1 and j=n. Note that a simplest elementary loop is a self-loop with the format of ptp or tpt. In the following discussion, a loop implies an elementary loop.

5.1 State Machine Petri Nets

A state machine can be easily converted into a Petri net by introducing a transition between two states and treating each state as a place. In such net, each transition links two and only two states or places. Its firing represents the transition from one state to another. Formally, it can be defined as follows:

A *state machine Petri net* is a Petri net $Z=(P, T, I, O, m_0)$ satisfying
1) $\forall\ p \in P, t \in T, I(p, t) \leq 1$ and $O(p, t) \leq 1$;
2) $\forall\ t \in T, |\bullet t| = |t \bullet| = 1$, i.e., $|\{p \in P: O(p, t)=1\}| = |\{p \in P: I(p, t)=1\}| = 1$; and
3) There is one and only one place $p \in P$ such that $m_0(p)=1$.

In the above definition, Condition 1 states that Z is an ordinary Petri net in which there is none or only a single arc from a node to another. Condition 2 states that a transition has exactly one input place and one output place. The last condition states that there is only one token initially signifying that a system can be in a state only at any time instant. A four-tuple (P, T, I, O) satisfying the conditions 1 and 2 is called a state machine structure.

A direct graph is *strongly connected* if and only if given any two nodes, there is a direct path connecting one to another.

Property 5.1: A state machine Petri net is safe, live, and reversible if it is strongly connected.

Modeling a system with a state machine was discussed in the previous chapter and is summarized as follows:

Procedure 5.1: Modeling a system with a state machine Petri net
Step 1. Decide all the possible states and tag them "new";
Step 2. Draw and label a circle for each state as a place;
Step 3. While there is a "new" state left,
select a state or place p, decide all events that can take place at the state; for each event, draw and label a transition t, decide the next

5.1. State Machine Petri Nets

state p' resulting from the occurrence of the event; and link an arc from p to t, and another from t to p'. Tag the state "old".

Step 4. Mark with one token the place representing the initial state of the system.

Several examples were presented in the last chapter as finite state machines. They can be readily converted into a state machine Petri net by replacing an arc between two states with a transition and two arcs. Then all states and transitions can be labelled. The following presents another manufacturing system example.

Example 5.1: A manufacturing system is designed to produce a final part from raw material through milling and drilling operations as shown in Fig. 5.1(a). Three inspection stations are used to detect if incoming raw material is acceptable, and a milled, and drilled part is within specifications, respectively. If not, it is scrapped. To model it as a state machine Petri net, we need to decide all the possible states for the system first. They are raw material, mill, drill, final part and scrap. Second, draw a circle for each state as a place as shown in Fig. 5.1(b).

Third, for the above states, we decide all events that can take place at each state. Then we make a transition for each event and link the places. Take the example of state "raw material." At this state, two events that can occur are "start milling" as a result of acceptable material, and "start scrapping". Two transitions t_1 and t_2 are created and the arcs from p_1 (raw material) to t_1 and t_2 and then t_1 to p_2 (mill) and t_2 to p_5 (scrap) are added. Continue this process till all the states are considered. We finally obtain the state machine Petri net as shown in Fig. 5.1(c). Table 5.1 shows the interpretation of all places and transitions.

Fourth, we mark p_1 only with one token to indicate the system's initial state.

This net is safe but not live since this state machine Petri net is not strongly connected. The two deadlock states are "final part" and "scrap". ◊

The structure of a place with two or more output transitions is called a choice, conflict, or decision structure. Such structures exhibit non-determinism. For example, in Fig. 5.1(c), each of places p_{1-3} has two output transitions. When there is a token in any of them, the token can flow through either of its two output transitions to an output place. State machine Petri nets allow clear representations of choices. They, however, cannot model synchronization of concurrent or parallel activities explicitly.

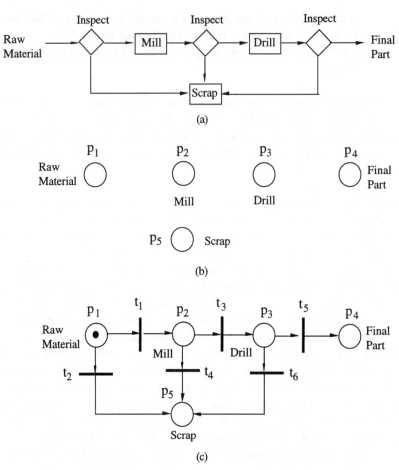

Figure 5.1 (a) A manufacturing system (b) All places and (c) Final state machine PN model.

Table 5.1. Interpretation of all places and transitions of Fig. 5.1(c)

Places	Interpretation	Transitions	Interpretation
p_1	Raw material	t_1	Start milling
p_2	Mill	t_2	Start scrapping an unacceptable raw part
p_3	Drill	t_3	Start drilling
p_4	Final part	t_4	Start scrapping an ill-milled part
p_5	Scrap	t_5	Complete the manufacturing process
		t_6	Start scrapping an ill-drilled part

5.2 Marked Graphs

Marked graphs are also named event graphs. They are used to represent concurrent activities or events in a system. Such systems exhibit deterministic and repetitive or cyclic behavior.

A *marked graph* is a Petri net $Z=(P, T, I, O, m_0)$ satisfying:
1) $\forall\ p \in P,\ t \in T,\ I(p, t) \leq 1$ and $O(p, t) \leq 1$; and
2) $\forall\ p \in P,\ |\bullet p| = |p \bullet| = 1$, i.e., $|\{t \in T: O(p, t)=1\}| = |\{t \in T: I(p, t)=1\}| = 1$.

The first condition states that the net is ordinary. The second one states that each place has exactly one input and one output transition. Marked graphs are a class of well-studied Petri nets. Recall that a loop is $x_1 x_2 \ldots x_n$, $n>2$ such that $\exists$ an arc (x_i, x_{i+1}) for $i \in N_{n-1}$ if $n>1$, and $x_i = x_j$, $1 \leq i < j \leq n$, implies that $i=1$ and $j=n$. Several of the most important properties can be summarized in the following Property [Commoner et al., 1971; Peterson 1981; Murata 1989].

Property 5.2: Given a marked graph $Z=(P, T, I, O, m_0)$,
1. In a live marked graph Z, a marking m is reachable from m_0 iff the token count in each loop of Z is the same in both m and m_0 where the token count in a loop is the total number of tokens in all places in the loop.
2. A marking m is reachable from m_0 iff (a) the token count in each loop of Z is the same in both m and m_0, and (b) for the minimal nonnegative solution x for $Cx = m - m_0$, no transition t such such $x(t) > 0$ is on any token-free directed loop in Z where $C = O - I$ is Z's incidence matrix.
3. The token count in a loop of Z does not change as a result of transition firings.
4. Z is live and reversible iff the token count in each loop of Z is at least one.
5. A live marking is safe iff every place in Z is in a loop with a token count of one.
6. For a strongly connected Z, a firing sequence leads back to m_0 iff it fires every transition for an equal number of times.

The proof of Properties 5.2.1-2 is shown in [Murara, 1977]. Property 5.2.3 can be seen since each place in Z has one input and one output transition, thus firing a transition in a loop removes and deposits a token in the loop while firing a transition outside of the loop cannot affect the token distribution in the loop. The proof of Property 5.2.4 was given in [Commoner et al., 1971].

Property 5.2.5 can be easily seen since a place can obtain at most one token if it is in a loop containing only one token according to Property 5.2.3.

Property 5.2.6's proof can be sketched as follows [Murata, 1989]. Consider a firing sequence f which starts and ends at m_0. Then its firing count vector is a solution of x of the homogeneous equation: $Cx=0$. A connected marked graph Z with n transitions has its incidence matrix C of rank n-1. Thus $Cx=0$ has only one independent solution $x=(k, k, ...,k)^\tau$ where integer $k>0$. This implies a firing sequence f that fires every transition in the marked graph k times. Therefore, Property 5.2.6 holds.

Many repetitive concurrent systems can be modeled with a marked graph. The modeling procedure is as follows:

Procedure 5.2: Modeling with a marked graph
    Step 1. Identify all resources in a system;
    Step 2. For each resource, identify a set of sequentially and cyclically executed activities/events. Take Choices 1 and/or 2 described bellow:
        Choice 1: Represent and label each activity as a place in an order of their execution from the beginning to the end, insert a transition between two consecutive places, and link places to transitions and transitions to places to form a circuit or loop.
        Choice 2: Represent each activity as a transition in an order of their execution from the beginning to the end, insert a place between two consecutive transitions, and link transitions to places and places to transitions to form a circuit or loop.
    Step 3. Merge all the common transitions and transition paths representing the same activity.

Note that a transition path is a direct path starting and ending with two different transitions. They are initially token-free. This procedure is a bottom-up procedure since it begins with each individual resource's activities and then merge them through common transitions and paths. Conventionally, to reduce the net size in modeling a manufacturing system, an operation with relatively short duration is modeled by a single transition. Such operations include robot loading and unloading, and AGV transportation. An operation with relatively long duration is modeled by a transition-place-transition module where the first transition represents the start, the place represents the operation, and the second transition represents the end of the operation. The following manufacturing cell example should help one understand Procedure 5.2.

5.2. Marked Graphs

Example 5.2: A production cell consists of two machines (M1 and M2), each of which is served by a dedicated robot (R1 and R2) for loading and unloading, as shown in Fig. 5.2(a). An incoming conveyor carries pallets with raw materials one by one, from which Robot R1 loads M1. An outgoing conveyor takes the finished product, to which Robot R2 unloads M2. There is a buffer with capacity of two intermediate parts between two machines. The system produces a specific type of final parts. Each raw workpiece fixtured with one of three available pallets is processed by M1 and then M2. A pallet with a finished product is automatically defixtured, then fixtured with raw material, and finally returns to the incoming conveyor.

Step 1: Identify all resources in a system. This system has the following resources: three pallets with raw materials, Machines M1 and M2, Robots R1 and R2, and buffer slots.

Step 2. We identify the following sequentially and cyclically executed activities for all the above resources as shown in Fig. 5.2(b). For example, a pallet with raw material goes through the following activities and status:

                R1 loading M1,
                M1 processing,
                R1 unloading M1,
                intermediate part ready in buffer,
                R2 loading M2,
                M2 processing, and
                R2 unloading.

We use p_1 to represent the availability of pallets with raw materials. Transitions t_1-t_4 are used to to model the relatively short activities of R1 and R2, and places p_2 and p_4 for relatively long processes at M1 and M2. Place p_3 is used to model the status of a intermediate part being ready. We finally link all these places and transitions to form a cycle.

Take M1's modeling as another example. M1 goes through the activities:

                R1 loading M1 (modeled with transition t_1),
                M1 processing a part (modeled with place p_2), and
                R1 unloading (modeled with transition t_2).

Step 3. We merge all the common paths among all the sub-net models. For example, the three models of M1, Raw-material, and R1 share a common transition path t_1-p_2-t_2 and then they merge together as shown in Fig. 5.2(c). Continue this process to obtain the final Petri net model in Fig. 5.2(c). The places and transitions are explained in Table 5.2. Its initial marking is determined by the individual subnet models in Fig. 5.2(b) and is

$$(3\ 0\ 0\ 0\ 1\ 2\ 1\ 1\ 1)^\tau. \Diamond$$

98   Chapter 5. Modeling FMS with Petri Nets

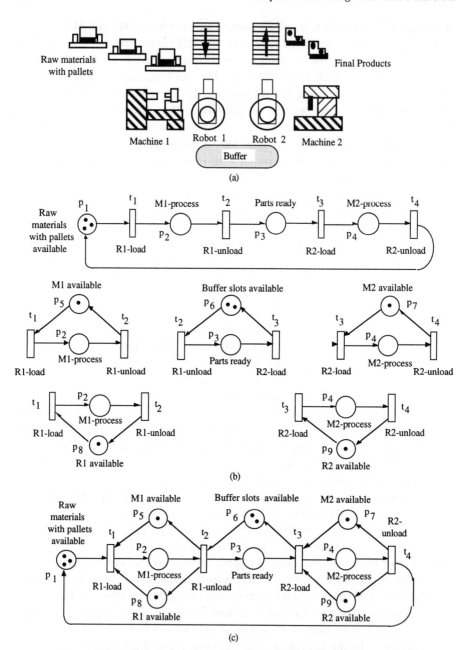

Figure 5.2 (a) A production cell, (b) All sub-net modules, and (c) A complete Petri net model.

5.3. Free-Choice and Asymmetric Choice Petri Nets

Table 5.2. Interpretation of all places and transitions of the marked graph

Places	Interpretation	Places	Interpretation
p_1	Raw materials with pallets available	p_8	R1 available
p_2	M1 processing a part	p_9	R2 available
p_3	Intermediate parts ready	Transitions	
p_4	M2 processing a part	t_1	R1 loading
p_5	M1 available	t_2	R1 unloading
p_6	Buffer slots available	t_3	R2 loading
p_7	M2 available	t_4	R2 unloading

Procedure 5.2 is tedious and the modeling of such a system can be simplified by using an ad hoc approach to be presented in Section 5.6. Marked graphs will be further explored when deterministic time delays are introduced into places, transitions and/or arcs. Their cycle times can conveniently be computed, which will be discussed in the next chapter.

5.3 Free-Choice and Asymmetric Choice Petri Nets

Allowing the conflicts of state machines and concurrency of marked graphs in a net leads to a class of Petri nets called free-choice Petri nets, FC nets for short.

A *free-choice Petri net* is a Petri net $Z=(P, T, I, O, m_0)$ satisfying:
1) $\forall\ p \in P, t \in T$, $I(p, t) \leq 1$ and $O(p, t) \leq 1$; and
2) $\forall\ p \in P$, $|p^\bullet| \leq 1$ or $^\bullet(p^\bullet) = \{p\}$; or equivalently, $\forall\ p_1, p_2 \in P$, if $p_1^\bullet \cap p_2^\bullet \neq \varnothing$, $|p_1^\bullet| = |p_2^\bullet| = 1$.

The first condition states that the net is ordinary. Condition 2 states that every arc from a place is either a unique outgoing arc or a unique incoming one to a transition. In other words, there is no choice or a choice to fire a transition can be made freely since all the output transitions of a place are enabled or disabled at the same time. Two important concepts, siphon and trap, are discussed next, which are useful to analyze FC nets and other classes of Petri nets.

An *asymmetric choice Petri net* is a Petri net $Z=(P, T, I, O, m_0)$ satisfying:
1) $\forall\ p \in P, t \in T$, $I(p, t) \leq 1$ and $O(p, t) \leq 1$; and
2) $\forall\ p_1, p_2 \in P$, if $p_1^\bullet \cap p_2^\bullet \neq \varnothing$ then either $p_1^\bullet \subseteq p_2^\bullet$ or $p_2^\bullet \subseteq p_1^\bullet$.

Condition 2 states that if two places share a non-empty set of output transitions, then one place's output transition set must be the other's subset. In other words, we cannot allow two places that share some output transitions but both have their own non-shared output transitions.

A non-empty subset of places S in an ordinary PN is called *a siphon* if $^\bullet S$ is a subset of $S^\bullet$, i.e., every transition having an output place in S has an input place in S. It is called *a trap* if $S^\bullet$ is a subset of $^\bullet S$, i.e., every transition having an input place in S has an output place in S. Note that the union of two siphons (traps) is a siphon (trap). A siphon is called a basis siphon (trap) if it cannot be expressed as a union of other siphons (traps). A siphon (trap) is minimal if it does not contain other siphons (traps). A minimal siphon (trap) is a basis siphon (trap) but the reverse is not true.

Example 5.3 (Siphon and trap examples):

1. Consider $S_1=\{p_1, p_2, p_3\}$ and $S_2=\{p_2, p_3, p_4\}$ in Fig. 5.3. Then $^\bullet S_1=\{t_2\}$, $S_1^\bullet = \{t_1, t_2\}$, $^\bullet S_2=\{t_2, t_3\}$, and $S_2^\bullet = \{t_2\}$. $^\bullet S_1$ is a subset of $S_1^\bullet$ and thus S_1 is a siphon. $S_2^\bullet$ is a subset of $^\bullet S_2$ and thus S_2 is a trap. Let $S_3=\{p_2, p_3\}$. Then $^\bullet S_3= S_3^\bullet =\{t_2\}$. Thus S_3 is both a siphon and a trap.

2. Consider Fig. 5.4(a). Let $S_1 = \{p_1, p_2, p_5\}$ and $S_2 = \{p_1, p_3, p_5\}$. Then $^\bullet S_1=\{t_5, t_1, t_2, t_3\}$, $S_1^\bullet = \{t_1, t_2, t_3, t_5\}$, $^\bullet S_2=\{t_5, t_1, t_2, t_4\}$, and $S_2^\bullet = \{t_1, t_2, t_4, t_5\}$. $^\bullet S_1=S_1^\bullet$ and $^\bullet S_2=S_2^\bullet$ imply that S_1 and S_2 are both a siphon and a trap. ◊

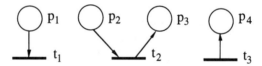

Figure 5.3 Illustration of a siphon and trap.

There are three classes of transitions with respect to a set of places. A transition's firing may decrease, keep constant, or increase the number of tokens in the set. For example, firing t_1 in Fig. 5.3 takes away one token from S_1, firing t_2 does not change the number of tokens in S_1 or S_2, and firing t_3 adds one token to S_2. Therefore, three transitions in Fig. 5.3 belong to three different classes with respect to set $\{p_1, p_2, p_3, p_4\}$. It can be easily observed that once S_1 loses all tokens, it will never obtain them back. Once S_2 obtains tokens, it will have them for good. It can be easily proved that:

1. If a siphon is token-free under some marking, it stays token-free under each successor marking;
2. If a trap is marked, it stays marked under each successor marking; and
3. Given a Petri net Z and a marking at which no transition is enabled, all the empty places under the marking form a siphon.

5.3. Free-Choice and Asymmetric Choice Petri Nets

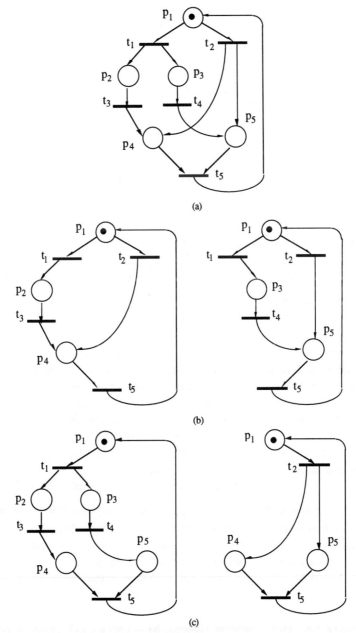

Figure 5.4 (a) A live safe FC net, (b) SM-components, and (c) MG-components.

In the below property, SM means state machine and MG means marked graph.

Property 5.3: Given a free-choice Petri net $Z=(P, T, I, O, m_0)$,
1. Z is live iff every siphon contains a marked trap.
2. A live Z is K-bounded iff Z is covered by strongly-connected SM-components each of which has at most K tokens at m_0. Z is safe if K=1 in the above statement.
3. A live and bounded Z is covered by strongly-connected MG components. Moreover, there is a marking $m \in R(Z)$ such that any of its MG components, Z', is a live and bounded MG under m' where m' is m restricted to Z', i.e., m'(p)=m(p) if $p \in P'$ which is the set of places of Z'.
4. An asymmetric choice net Z is live if (but not only if) every siphon in Z contains a marked trap.
5. An asymmetric choice net Z is live if for each minimal siphon S, either S contains a marked trap or $F(S)>0$ where $F(S)=\text{Min} \sum_{m \in S} m(p)$

    subject to

    $m=m_0+Cy$, $m \geq 0$ and $y \geq 0$.

In Properties 5.3.2-3, an SM-component (MG-component) Z' of a net Z is defined as a subnet generated by places (transitions) in Z' satisfying 1) each transition (place) in Z' has at most one incoming and at most one outgoing arc; and 2) a subnet generated by places (transitions) is the net consisting of these places (transitions), all of their input and output transitions (places), and their connecting arcs. Properties 5.3.1-4 are proved in [Best, 1987]. Property 5.3.5 is due to [Chu and Xie, 1997]. F(S) can be obtained by solving a linear programming problem in polynomial time:

$F(S)=\text{Min} \sum_{m \in S} m(p)$ subject to $m=m_0+Cy$, $m \geq 0$ and $y \geq 0$.

Note that at any marking m, if S loses all the tokens, then F(S)=0. In other words, the condition guarantees that S is always marked. Note that an FC net belongs to an asymmetric choice Petri nets. Since the number of siphons or minimal siphons grows exponentially with the net size in general, the use of Properties 5.3.1, 4, and 5 for liveness check has limitation in their applications to large size Petri nets.

Example 5.4: Two minimum siphons, $S_1 = \{p_1, p_2, p_5\}$ and $S_2 = \{p_1, p_3, p_5\}$, found in Fig. 5.4(a) are themselves marked traps. $S=\{p_1, p_2, p_3, p_4, p_5\}$ contains

5.3. Free-Choice and Asymmetric Choice Petri Nets

two marked traps, i.e., S_1 and S_2. Thus all the siphons contain at least a marked trap. Therefore, the FC net in Fig. 5.4(a) is live. Further, we can find two SM-components that cover the net, as shown in Fig. 5.4(b), and each is a state machine containing only one token. Thus, the net in Fig. 5.4(a) is also safe. This net is covered by two MG-components as shown in Fig. 5.4(c). Each MG-component is indeed safe under the initial marking. ◊

The synthesis of live and bounded FC nets can start with a single loop. Then rules to augment sequential and concurrent activities, and choices can be applied to the single loop net to obtain all live and bounded FC nets. The methods with sufficient details can be found in [Esparza and Silva, 1990] and [Chao, 1987; Chao et al., 1994]. In [Esparza and Silva, 1990], PP-, PT-, TP-, and TT-handles (on a path) or bridges (between two paths) are formulated. In [Chao, 1987; Chao, et al., 1994], PP, PT, TP, and TT-rules are used to formulate a so-called knitting technique to synthesize a wider class of Petri nets. In the above abbreviations, P means place and T means transition.

In general, if a manufacturing system satisfies the following two conditions, then it can be modeled with an asymmetric choice net [Chu and Xie, 1997]:

1. A process step for a product type requires only one resource at a time;
2. If a process step involves selecting one among several possible manufacturing processes, either no shared resource is involved in this choice, or the same set of shared resources is needed for any choice.

The following example discusses such a manufacturing system.

Example 5.5: A manufacturing system has four resources A, B, C and D. Resources A, B, and C are of single capacity and D is of capacity two. Two product types q_1 and q_2 can be processed, which follow the following steps:
Process steps for q_1 are either
1) O_{11}, O_{12}, and O_{13} or
2) O_{11}, O_{12}, O_{14}, and O_{15}
where O_{11}, O_{13-15} requires resources A, B, C, and D, respectively and O_{12} needs no resource.
Process steps for q_2: O_{21} and O_{22} where O_{21} requires resource D only and O_{22} requires both resources A and D.
The Petri net model is built according to the process steps through which each product type has to go. Model each process step as a place. Insert a

transition between two places, which means the end of a preceding operation and start of the current one. We also need a transition to model the start of product types q_1 and q_2, respectively, and a transition to model the completion of product types q_1 and q_2, respectively. Link all these transitions and places according to the sequential process steps for q_1 and q_2. Label all these places and transitions. As a result, we find paths and for q_1 and q_2, respectively as shown in Fig. 5.5.

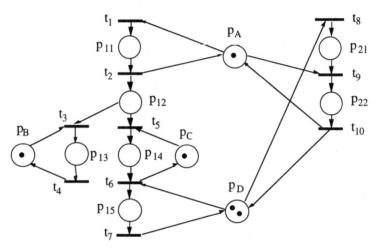

Figure 5.5 An asymmetric choice Petri net model for an FMS in Example 5.5.

Table 5.3. Interpretation of all places and transitions of Fig. 5.5(b)

Places	Interpretation of places	Transitions	Interpretation of transitions
p_{11}	Process step O_{11}	t_1	Start process step O_{11}
p_{12}	Process step O_{12}	t_2	End O_{11} and start O_{12}
p_{13}	Process step O_{13}	t_3	End O_{12} and start O_{13}
p_{14}	Process step O_{14}	t_4	End product q_1's final step O_{13}
p_{15}	Process step O_{15}	t_5	End O_{12} and start O_{14}
p_{21}	Process step O_{21}	t_6	End O_{14} and start O_{15}
p_{22}	Process step O_{22}	t_7	End product q_1's final step O_{15}
p_A	Resource A available	t_8	Start process step O_{21}
p_B	Resource B available	t_9	End O_{21} and start O_{22}
p_C	Resource C available	t_{10}	End product q_2's final step O_{22}
p_D	Resource D available		

Next, model each manufacturing resource as a place. If a process step requires a manufacturing resource, then link the corresponding resource place to its start transition with an input arc. If a process step's completion releases the resource, then link its completion transition to the corresponding resource place with an output arc. Take step process O_{14} modeled with place p_{14} as an example. Its start transition is t_5 and completion transition t_6. Since O_{14} needs resource C, an input arc from place p_C (modeling resource C) to t_5 is constructed. Its completion releases resource C, and thus an output arc from t_6 to p_C is drawn. The complete model is shown in Fig. 5.5, which is indeed an asymmetric choice net. Table 5.3 lists all the places and transitions and their interpretation.

This Petri net has four minimal siphons $\{p_B, p_{13}\}$, $\{p_C, p_{14}\}$, $\{p_{11}, p_A, p_{22}\}$, and $\{p_{15}, p_D, p_{21}, p_{22}\}$. All these four siphons are also traps. Therefore, when each of resource places p_{A-D} contains at least one token, the net is live according to Property 5.3.5. The net is unbounded due to the fact that we can fire transitions t_1-t_2 for an infinite number of times and thus place p_{12} is unbounded. ◊

Other examples requiring the use of Property 5.3.5 are discussed later in Section 5.5.

5.4 Acyclic Petri Nets and Assembly Petri Nets

This section discusses those nets in which there exist no loops. They can be used to represent some acyclic manufacturing, assembly, and disassembly processes.

A Petri net containing no directed circuit is called *an acyclic Petri net*.

A *tree marked graph* is a marked graph of a tree structure.

A *disassembly Petri net* $Z=(P, T, I, O, m_0)$ is an acyclic ordinary Petri net such that

1) There is a place called a product or root denoted by p_1 with no input arc, a set of places called subassemblies, a set of places called parts or leaves denoted by P', each of which has no output arcs,
2) Each transition has at most one input arc and at least two output arcs,
3) $\forall\ p \in$ P-P', and $\forall\ t,\ t' \in p^\bullet$, $\Omega(t)=\Omega(t')$ where $\Omega(t) = \{q \in$ P': there is a direct path from t to q$\}$,
4) $m_0(p_1)=1$ and $m_0(p)=0$, $\forall\ p \in$ P-$\{p_1\}$. When no transition is enabled, the final marking $m_f(p)=1$ if $p \in$ P' and $m_f(p)=0$ $p \in$ P-P'.

Condition 2 states that each transition is a disassembly process that separates an assembly into two or more components or subassemblies. Condition

3 states that all output transitions of a place should share a common set of leaves. A place with multiple output transitions represents a subassembly which have multiple ways to be disassembled. Then these different disassembly choices should lead to a common set of parts contained in this subassembly. Multiple output transitions from a place form a Logic-OR relation and multiple output places from a transition form a Logic-AND relation. That is why disassembly Petri net is commonly represented by an AND/OR graph. It is clear that a disassembly Petri net belongs to the class of FC nets.

$Z=(P, T, I, O, m_0)$ is *an assembly Petri net* if 1) (P, T, O, I) satisfies Conditions 1-3 of a disassembly Petri net and 2) $m_0(p)=1$ if $p \in P'$ and $m_0(p)=0$ $p \in P-P'$. When no transition is enabled, the final marking $m_f(p_1)=1$ and $m_0(p)=0$, $\forall\ p \in P-\{p_1\}$.

Note that net $Z'=(P, T, O, I, m_0)$ is graphically obtained from $Z=(P, T, I, O, m_0)$ by reverting the direction of all the arcs. Z' is called the dual of Z mathematically. The assembly net, however, does not belong to the class of FC nets. An example can be seen after the following property.

Property 5.4 (Acyclic Petri Nets):
1. In an acyclic Petri net $Z=(P, T, I, O, m_0)$, m is reachable from m_0 iff there exists a nonnegative integer solution y satisfying: $m = m_0 + Cy$ where $C = O - I$ is the incidence matrix.
2. Any two markings of Z are mutually reachable iff Z is a tree marked graph.
3. Z is unbounded, live and reversible if Z is a tree marked graph.
4. Both assembly and disassembly Petri nets are safe, but neither live, nor reversible.

Note that the existence of a nonnegative integer solution y satisfying
$$m = m_0 + Cy$$
is a necessary condition for m to be reachable from m_0. Property 5.4.1's sufficiency proof can be seen in [Murata 89]. Property 5.4.2 is a special case as a result of Property 5.2.1 since a tree marked graph contains no loop and thus the condition in Property 5.2.1 is automatically satisfied. Property 5.4.3 is obvious by noting the existence of a source transition that has no input places and thus is always enabled, and by using Property 5.4.2. Property 5.4.4 is also obvious based on the definition of assembly and disassembly Petri nets.

5.4. Acyclic Petri Nets and Assembly Petri Nets

Example 5.6: A handlight is used as an example of assembly and disassembly Petri nets. Fig. 5.6(a) shows the components contained in this handlight: Cover, Glass, Head-housing, Bulb, Spring and Main-housing, denoted by C, G, H, B, S and M for short, respectively. The metal head-housing is screwed onto the same material main-housing, and both the cover and bulb are screwed onto the head-housing. The spring is welded onto the main-housing.

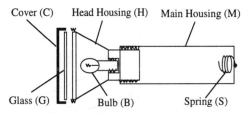
Figure 5.6 A handlight.

Starting with the node of a handlight, we have two choices to disassemble it: unscrewing between CGHB (Cover + Glass + Head-housing + Bulb) and SM (Spring + Main-housing) or unscrewing among Cover, Glass and the remaining. The first choice is represented by transition t_1 and the second by t_2. t_1 has two output places: CGHB and SM, and t_2 three outputs: Cover, Glass, and HBSM. CGHB is further disassembled into Cover, Glass and HB through t_3 and t_3's three output places. HB is disassembled into Head-housing and Bulb through transition t_6 and t_6's two output places. SM is then disassembled into Head-housing and Bulb through transition t_7 and its two output places, H and B. At HBSM, two choices emerge, unscrewing between HB and SM or HSM and Bulb. They are represented by t_4 and t_5. Subassemblies HB and SM are further disassembled as before. HSM is disassembled into SM and Head-housing via unscrewing. Finally we label all the places and mark p_1 with one token. All the possible disassembly paths are clearly shown in Fig. 5.7(a). The graph is a free-choice net. It can be easily verified that the net is safe. The deadlock marking is the final one, i.e., $(0,0,0,0,0,0,1,1,1,1,1,1)^\tau$.

The handlight's assembly Petri net is shown in Fig. 5.7(b) with initial marking
$$(0,0,0,0,0,0,1,1,1,1,1,1)^\tau$$
and final marking
$$(1,0,0,0,0,0,0,0,0,0,0,0)^\tau.$$
The assembly PN is neither a free-choice net nor an asymmetric choice one. It is also safe and has its final marking as a deadlock one. Note that transition t_7 means "welding" and others "screwing" in Fig. 5.7(b). ◊

108   Chapter 5. Modeling FMS with Petri Nets

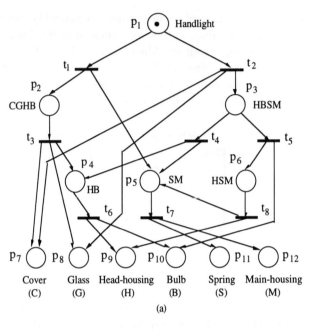

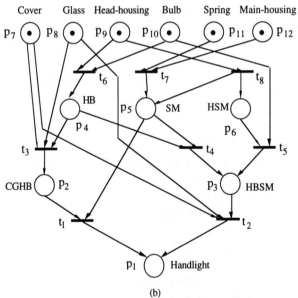

Figure 5.7. (a) Disassembly Petri net, and (b) Assembly Petri net.

In [Suzuki et al., 1993], it is proved and illustrated that the problem to search for the optimal task sequence for a class of assembly Petri nets in which each transition has only two output places can be formulated and solved by a linear programming problem as follows:

Minimize: $c^\tau x$

subject to: $Cx = m - m_0$ and $x \geq 0$

where c is a cost vector corresponding to each (dis)assembly task or a transition, C is the incidence matrix of a (dis)assembly Petri net, m is a final marking and m_0 is an initial marking, both of which are known according to the model. Compared with a heuristic search method for an optimal task sequence, the solution of the above linear programming problem is more efficient. Note that once x is found, the sequence of transition firings corresponding to (dis)assembly is determined. Furthermore, the solution is proved a (0, 1) solution.

5.5 Production-Process Nets and Augmented Marked Graphs

In many manufacturing systems, part flows can be readily recognized. From a raw material piece to a final product, each step involves in using manufacturing resources that may be machines, robots, transportation devices, and buffer slots. If the routing of a product type is fixed and each step takes one and only one manufacturing resource, then we can obtain so-called Production-Process Nets (PPN). The modeling of such a system takes two steps:

Step 1. Model the routing of each product type as a sequentially-connected transitions and places with a transition as a starting node and a place as a final node. Each place represents the status of raw material, final product, or operations or process steps; and

Step 2. Model the usage of resources by using a place for each type of resources and marking initially the place with the same number of tokens as the quantity of the resources of the type.

This class of nets present a framework for designers to study deadlock avoidance policies and algorithms [Banazak and Krogh, 1990; Xing et al., 1995]. A key research issue is to identify a policy that avoids deadlock and allows to maximize the utilization of the resources and system productivity.

Example 5.7. Consider an FMS which consists of two machine centers, M1 and M2 with capacity C_{M1} and C_{M2} respectively. There are two types of products q_1 and q_2 to be produced. Their process routes are as follows:

Routing for q_1: M1, M2, and M1

Routing for q_2: M1, and M2

Assume that the transportation between two cells is performed automatically with dedicated devices.

As a consequence of Step 1, we have the Petri net model of product routes shown in Fig. 5.8(a). Two start transitions t_{10} and t_{20} model "start to obtain raw materials for q_1 and q_2", respectively. Tokens in p_{10} mean the ready raw materials for q_1 and those in p_{20} mean so for q_2. For i=1 and 2,

Transition t_{i1} starts the first step for product q_i, and

Transition t_{i2} ends the first step and starts the second step of q_i.

Transition t_{13} ends the second step and starts the last step for q_1,

Transition t_{14} ends the process for q_1; and

Transition t_{23} ends the process for q_2.

Places p_{11} to p_{13} represent the process steps of q_1, and p_{21} and p_{22} represent the process steps of q_2. The number of tokens in places p_{14} and p_{23} represents the number of final products produced for q_1 and q_2, respectively.

Next step, two resource places representing availability of M1 and M2 are drawn, labeled as p_{M1} and p_{M2}, and then the arcs are connected according to the process routes. For example, the first process step of q_1 requires M1. Thus an arc from place p_{M1} links to transition t_{11}. When q_1 ends its use of M1 at the end of the first step, an arc from t_{21} links back to p_{M1}. After employing the similar method, we finally obtain the net as shown in Fig. 5.8(b). Then an appropriate number of tokens are used to mark the corresponding places that represent the capacity of all machining centers.

It is clear that the resulting PPN is unbounded. Without any restriction on the firing of transitions, the net may be trapped into a (partial) deadlock state. For example, if $C_{M1}=C_{M2}=1$ for the net in Fig. 5.8(b), firing t_{10}, t_{11}, t_{12}, t_{20}, and t_{21} leads to a partial deadlock state at which only t_{10} and t_{20} can continue to fire, but none of the other transitions. Thus the system cannot produce any products once the system enters the above partial deadlock. Different algorithms and strategies are derived to prevent the system from such states in [Banazak and Krogh, 1990; Xing et al., 1995]. ◊

5.5. Production-Process Nets & Augmented Marked Graphs

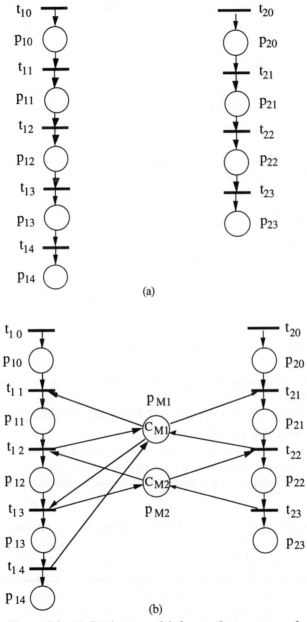

Figure 5.8. (a) Petri net model for product routes and (b) A complete Petri net model for the FMS

A closely related class of nets is introduced in [Proth and Minis, 1995], called *controllable nets*. They are suitable for acyclic manufacturing systems in which each product part has one or multiple routes, and each process requires one and only one manufacturing resource. In such nets, the acquisition, use and release of a resource is modeled by a transition and thus only a self-loop relation exists between a resource place and a transition. A resource place may associate itself with several transitions, thus modeling resource sharing situations. This class of nets has been used to schedule acyclic jobshop systems [Proth and Minis, 1995].

An augmented marked graph (AMG) is a Petri net $Z=(P \cup R, T, I, O, m_0)$ satisfying the following conditions:

1. Z' obtained from Z by removing the places in R and their related arcs is a marked graph;
2. $\forall \, r \in R$, there is k_r (>0) pairs of transitions in T denoted by $D^r = \{(t_{aj}^r, t_{bj}^r), j \in N_{k_r}\}$ such that

    2.1 $r^\bullet = \{t_{aj}^r, j \in N_{k_r}\}$ and $^\bullet r = \{t_{bj}^r, j \in N_{k_r}\}$,

    2.2 $t_{ai}^r \neq t_{aj}^r$, and $t_{bi}^r \neq t_{bj}^r$ when $i \neq j$, $\forall i, j \in N_{k_r}$, and

    2.3   If $t_{ai}^r \neq t_{bi}^r$, there exists in Z' an elementary path $EP(t_{ai}^r, t_{bi}^r)$ denoting a path from t_{ai}^r to t_{bi}^r.

3. Z' under its restricted m_0 is live and bounded, $m_0(r) > 0$, $\forall \, r \in R$, and m_0 marks no places in $EP(t_{ai}^r, t_{bi}^r)$, $\forall \, r \in R$, $i \in N_{k_r}$.

For the above defined augmented marked graph Z, it can be proved that [Chu and Xie, 1997]:

1. Place $r \in R$ and all the places in all $EP(t_{ai}^r, t_{bi}^r)$, $i \in N_{k_r}$ form a p-invariant.
2. Z is reversible if it is live.
3. It is live and reversible if each minimal siphon S containing at least one place from R contains either one trap or $F(S) > 0$ where

$$F(S) = \min_{m \in S} \sum m(p) \text{ subject to } m = m_0 + Cy, m \geq 0 \text{ and } y \geq 0.$$

In other words, the conditions derived for asymmetric choice nets (Property 5.3.5) are also applicable to an augmented marked graph. Compared with PPNs, augmented marked graphs allow one to model a system with cyclic production systems. In such production systems, product routes are all unique although the resources can be shared among the concurrent operations. This class of the systems represent a significant portion of applications. Their further extension is desired to handle flexible routes in many FMS.

5.5. Production-Process Nets & Augmented Marked Graphs

The following example is adopted from [Zhou and DiCesare, 1993; Chu and Xie, 1997].

Example 5.8. The manufacturing system is composed of three workstations, W_1, W_2, and W_3, and a robot as shown in Fig. 5.9(a). A part needs to be processed by W_1 first, then W_2, and finally W_3 to have its final shape. Workstations W_{1-2} have their unit processing capacity while W_3 has capacity b. The robot is shared by W_1 and W_3 and is used for loading and unloading these two workstations. Once the robot starts loading either workstation, it cannot be interrupted until it finishes unloading the same one. Fixtured parts are awaiting processing in the input storage area. Final products will be automatically transported to the output storage area, whereas fixtures will be released to the input storage area by the robot as soon as the robot completes the unloading of W_3. The total number of fixtures is a. First, we can identify the one condition and three major operations that are modeled as four places in this system's Petri net model in Fig. 5.9(b):

Condition (place p_1): Fixtured parts in storage are available to W_3;

Operation 1 (place p_2): The robot loads a part from the storage area; Workstation W_1 processes the part; the robot unloads W_1; the fixtured part is automatically moved to Workstation W_2.

Operation 2 (place p_3): Workstation W_2 processes fixtured parts.

Operation 3 (place p_4): The robot loads a part to W_3; W_3 processes the part; the robot unloads W_3, and finally releases the fixture to storage.

Next, we add transition t_1 to start Operation 1; t_2 to end Operation 1 and start Operation 2; t_3 to end Operation 2 and start Operation 3; t_4 to end Operation 3. Connect place p1 to t_1, t_1 to p_2, ..., t_4 to p1 to obtain a loop representing the flow of fixtures with raw material.

Finally, we can add resource places p_5 and p_6 to represent the availability of W_2, and robot, respectively. Add directed arcs according to if the operations need W_2 and robot to obtain the Petri net model in Fig. 5.9(b). The initial marking is

$$m_0=(a\ 0\ 0\ 0\ b\ 1)^\tau.$$

We have excluded the modeling of W_1 and W_3 for Operations 1 and 3 assuming their availability whenever Robot is available.

The obtained model belongs to AC net as well as augmented marked graphs. There are four four minimal siphons:

$S_1=\{p_1, p_2, p_3, p_4\}$,
$S_2=\{p_3, p_5\}$,
$S_3=\{p_2, p_4, p_6\}$, and
$S_4=\{p_4, p_5, p_6\}$.

114   Chapter 5. Modeling FMS with Petri Nets

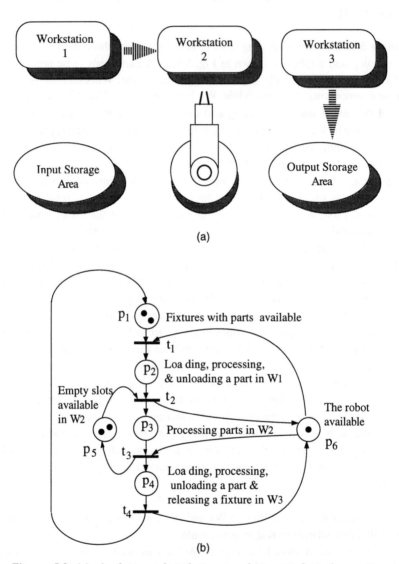

Figure 5.9 (a) A three-workstation, one-robot manufacturing system and (b) its Petri net model

5.5. Production-Process Nets & Augmented Marked Graphs

The first three siphons are also traps. By expressing $F(S_4)=0$ in terms of the initial marking, it can be derived that $F(S_4)>0$ iff $m0(p_5)+m_0(p_6)>m_0(p_1)$, i.e., $b+1>a$. If $a>0$, and $b>0$, then S_{1-3} are all initially marked. The net is live and reversible iff $a>0$, $b>0$, and $b+1>a$. The necessity can be derived as follows. Suppose the net is live but $b+1\leq a$. Firing the sequence $(t_1t_2)^b t_1$ leads to a system deadlock. This contradiction proves the necessity of the above condition. ◊

Example 5.9. Choice-synchronization structure is very useful in modeling certain manufacturing systems [Zhou et al., 1993; Zhou and DiCesare, 1993]. It is used when raw parts are released from a same entry, and processed parts have to be assembled following a fixed part ratio after their processing. One such FMS example is shown if Fig. 5.10(a). It consists of AGV, two machines, and a robotic assembly station. It machines two types of intermediate parts and produces a final assembly containing one from each type. The Petri net model as shown in Fig. 5.10 can be used to capture the behavior of such an FMS. The number of initial tokens in p1 indicate the number of AGVs. The availability of machines and robotic is not explicitly modeled. The places and transitions can be explained when p_4 and p_6 are initially marked with a single token, respectively and p_5 and p_7 have none as below:

p_1: AGVs with raw material-loaded pallets available
p_2: Processing a block at M1
p_3: Processing a peg at M2
p_4: Block processed first
p_5: Peg processed second
p_6: Assembly cell available
p_7: Block available at the assembly cell
t_1: Start processing a block
t_2: Start processing a peg
t_3: Send a block to the robotic assembly cell
t_4: Load and assemble a peg to a block and unload the final product

By treating p_1 as a resource place, we find that the remaining net structure after removing p_1 and related arcs from the net is indeed a marked graph. Suppose that initially p_1, p_4 or p_5, and p_6 or p_7 are marked and $m_0(p_2)=m_0(p_3)=0$, then the model is an augmented marked graph. Note that this net is also an AC net. Its minimal siphons are found as follows:

$S_1=\{p_1, p_2, p_3\}$,
$S_2=\{p_4, p_5\}$,
$S_3=\{p_6, p_7\}$,

$S_4=\{p_1, p_3, p_6\}$, and
$S_5=\{p_1, p_2, p_7\}$.

The first three siphons are also traps. S_4 and S_5 do not contain any trap. Let $\mu_i=m_0(p_i)$.

$F(S_4)=0$ implies that $m(p)=0$, $\forall\ p \in S_4$ and $m(p)\geq 0$, $\forall\ p \notin S_4$ for marking m satisfying the state equation. Solving $m(p)=0$, for all $p \in S_4$ and noticing $\mu_2=\mu_3=0$, we have:

$$y_1=\mu_1+\mu_6+y_2,$$
$$y_3=\mu_6+y_2, \text{ and}$$
$$y_4=y_2.$$

Substituting these results into the remaining state equations leads to:

$$m(p_2)=\mu_1,$$
$$m(p_4)=\mu_4-\mu_1-\mu_6,$$
$$m(p_5)=\mu_5+\mu_1+\mu_6, \text{ and}$$
$$m(p_7)=\mu_7+\mu_6$$

By noting that $\mu_i \geq 0$, we have $m(p_2)\geq 0$, $m(p_5)\geq 0$, and $m(p_7)\geq 0$. Hence, $F(S_4)=0$ iff $m(p_4)\geq 0$, i.e., $\mu_4-\mu_1-\mu_6\geq 0$. Equivalently, $F(S_4)>0$ iff $\mu_4-\mu_1-\mu_6<0$.

Similarly, we can derive that $F(S_5)>0$ iff $\mu_5-\mu_1-\mu_7<0$. Combining with constraints related to S_{1-3}, the Petri net is live and reversible if

$$\mu_1>0,$$
$$\mu_4+\mu_5>0,$$
$$\mu_6+\mu_7>0,$$
$$\mu_1+\mu_6>\mu_4, \text{ and}$$
$$\mu_1+\mu_7>\mu_5.$$

These conditions are also necessary for this net [Chu and Xie, 1997] and the proof is left as an exercise for the reader.

If we initially mark places p_4 an p_6 with one token and none with p_5 and p_7 to indicate that we fire t_1 and t_2, t_3 and t_4 alternatively with t_1 and t_3 first, respectively. In the context of manufacturing, this implies that we allow one type of raw parts to enter than the other into the system for processing and assembly in a strictly order. According to the above equation, $\mu_1=1$ will suffice to make the model live and reversible. This is easily validated by noting that only the repeated sequence is $t_1t_3t_2t_4$. In practice, to maximize the productivity, one will design $\mu_1=2$ such that concurrent operations can be performed in p_2 and p_3.

◊

5.5. Production-Process Nets & Augmented Marked Graphs 117

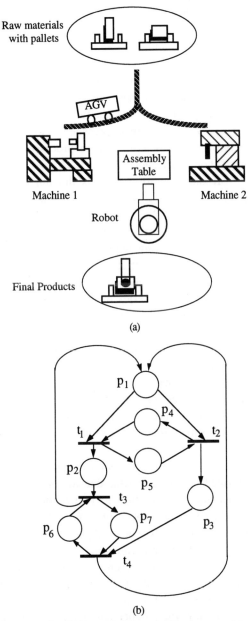

Figure 5.10 (a) An example FMS containing AGV, two machines, and a robotic assembly station and (b) Its PN model represented as a choice-synchronization structure

5.6 General Modeling Method

Before a general or ad hoc modeling method is discussed, basic relations among events, operations, and processes are introduced in the context of manufacturing. An automated manufacturing system consists of a variety of components ranging from robots, machines, raw materials, sensors, actuators, computers and accessories related to a specific process. At the modeling stages, one needs to focus on the major operations and their precedent, concurrent, and conflicting relationships. The basic relations among these processes or operations can be classified as follows [Zhou and Robbi, 1994]:

Sequential: If one operation follows the other, then the places and transitions representing them should form a cascade or sequential relation in Petri nets. Such an example is shown in Fig. 5.11(a).

Concurrent: If two operations are initiated by an event, they form a parallel structure starting with a transition, i.e., two places are two outputs of the same transition. An example is shown in Fig. 5.11(b). The pipeline concurrent operations can be represented with a sequentially-connected series of places/transitions in which multiple places can be marked simultaneously or multiple transitions are enabled at certain markings.

Conflicting: If either of two operations can follow an operation, then two transitions form two outputs from a same place. An example is shown in Fig. 5.11(c).

Cyclic: If a sequence of operations follow one after another and completion of the last one initiates the first one, then the cyclic structure is formed among these operations. An example is shown in Fig. 5.11(d).

Mutually Exclusive: Two processes are mutually exclusive if they cannot be performed at the same time due to constraints on the usage of a shared resource. A structure to realize this is through a common place marked with one token plus multiple output and input arcs to activate these processes. For example, a robot may be shared by two machines for loading and unloading. Two such structures are parallel mutual exclusion and sequential mutual exclusion discussed in [Zhou and DiCesare, 1991]. An example is shown in Fig. 5.11(e). Roughly speaking, if t_1-p_2-t_2 and t_3-p_3-p_4 have no sequential relation, or they are independent without place p_1 and the related arcs, the structure is a parallel mutual exclusion. Examples can be

5.6. General Modeling Method

found in Fig. 5.5. If they form a sequential relation, then a sequential mutual exclusion results. One example is shown in Fig. 5.9(b).

The resources can be classified into dedicated and shared ones. The dedicated one features with a place with single input and single output arcs only; the shared resource features a place with multiple-input and/or multiple-output arcs. A condition representing the status of a sensor or an actuator is also modeled with a place whose holding a token implies the truth of the modeled condition. The same kind of resources may be represented by a place with the number of tokens corresponding to the number of resources. Initiation of an operation requires often several kinds of conditions and resources available - modeled as a transition with several input places. Completion of an operation may release some resources and change the status of the conditions - modeled as a transition with several output places.

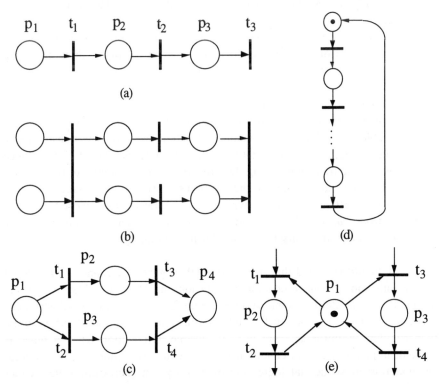

Figure 5.11. Examples of Basic Relations (a) Sequential (b) Concurrent (c) Conflicting (d) Cyclic (e) Mutually exclusive.

A general modeling method can be summarized into the following steps:

Identification of Operations and Resources: Given an industrial automated system description, identify the major events, operations, processes, resources, conditions, routing information, etc.

Identification of Relations: The relationships among the above identified events, operations, and/or processes have to be determined. The resources should be divided into different classes and major allocation policies for shared ones and routing information on products should be determined. Then the initial net structure can be roughly decided.

Petri Net Design: Design and label the places and/or transitions which represent events, operations, and/or processes. Arrange them according to the relationship identified in Step 2. Designate and label places which model status of resource(s) and the condition(s). Insert necessary places and transitions such that no two places can link to each other and neither two transitions. For each transition, draw an input arc to it from a place if enabling it requires the resource(s), truth of the condition, or the completion of the operation(s) represented in the place; draw an output arc from it to a place if firing it releases resource(s), changes the condition(s), or signals the initiation of the operation in the place. The number of input arcs from a place to the transition should equal the required quantity of the tokens (often implying resources) in the place to enable the transition. The number of output arcs from the transition to a place should equal the quantity of the tokens to the place to be produced due to the transition's firing. Determine the initial number of tokens over all places according to the system's initial state. Associate other characteristics, e.g., timing, with places and transitions if needed.

Petri Net Modification: Check the net model to see whether it reflects the system operation and modify the net until it models the system.

It should be noted that the above method generally cannot lead to a net model with known behavioral properties [Zhou and DiCesare, 1993; Zhou and Robbi, 1994]. In addition, some systems cannot be modeled using ordinary Petri nets unless some extensions such as inhibitor arcs or priority transitions are used. In particular, inhibitor arcs are used to disable a transition when a place is marked, implying a certain condition met. Their use is explored in the bottom-up modeling of sequential logic controller design in [Ferrarini, 1992]. Other useful modeling extensions are also discussed in Chapter 9 for error-prone manufacturing systems.

5.6. General Modeling Method

Example 5.8: An AT&T FWS-200 flexible workstation is developed for high-volume, low-cost production of printed circuit boards. The front view of FWS-200 is shown in Fig. 5.12. It consists of the following major components:
1. The frame structure supports the control cabinet, robot arms, platen, work table, etc.
2. The control cabinet houses hardware including an AT&T personal computer, Industrial I/O modules and STD bus, control electronics, display, pneumatic and ventilation systems, and power supplies.
3. Dual robot arms are able to accomplish pick-and-place operations, with their X, Y, Z, and theta motions in high accuracy (± 0.001 inch or $\pm 2.5 \times 10^{-4}$ meter along X, Y, and Z axes and $\pm 0.15°$ about Z axis).
4. The AT&T PC 6386 computer serves as a supervisory controller for the system.
5. The drive electronics contains the electronics for controlling the various axes, with two servo motor amplifiers for separate Z and theta motions and a stepper motor amplifier for both X and Y motions.
6. The control panel contains power buttons, touch screen display, pressure and vacuum gauges, and emergency stop buttons.
7. The vision system includes IRI SV512 and ICOS M10000 modules.
8. The peripheral equipment includes part feeders and others.

The system can be used to accomplish tasks such as assembly of surface mount components, assembly of through-hole components, wire harnessing, etc. The common work area is of size 24 inch × 32 inch (60.96 cm × 81.28 cm).

With focus on the activities of the two robot arms in the system, we design a Petri net model in the following steps:
1. Each robot performs repeatedly the following operations: *picking*, *moving*, *inserting*, and *moving*. To use the common workspace, each robot has to request first, then gets access to, and leaves the common workspace. The resources include the shared feeder area, shared PCB area, and the two robots. However, since the two arms work on the same circuit board and obtain components from the same feeder area, avoidance of arm collision has to be considered. Collision avoidance can be achieved by mutual exclusion structure implemented using semaphores in the Modular Manufacturing Language (M^2L) program [AT&T, 1989]. Thus, when one robot is doing a job above the PCB area, the other is prohibited from moving into the same area. Possible collision may also occur in the feeder area. Based on the similar reasoning, the use of a mutual exclusion structure is also needed for this area.

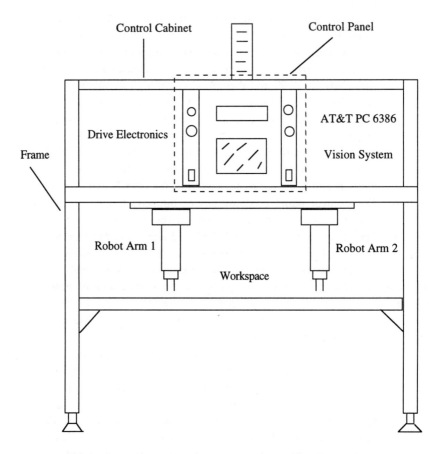

Figure 5.12 The front view of FWS-200 developed by AT&T

5.6. General Modeling Method

2. The relations among these operations and events for each robot are sequential. R1 and R2's starting pick-up as well moving to PCB area for insertion are mutually exclusive, respectively and thus the robot collision can be avoided.
3. For each robot, design the four places for R1 and R2's operations, and five transitions for their start and/or end and arrange them in a sequential order, as shown in Fig. 5.13(a). Designate six places to represent the availability of two robots, two kinds of components, feeder area, and PCB area, respectively. Label these places and transitions as shown in Fig. 5.13(a). Next, add the arcs. Since the precedence relation between picking, moving, inserting, and moving, we add the arcs from t_{i2} to p_{i2}, p_{i2} to t_{i1}, ..., t_{i4} to p_{i5}, and p_{i4} to t_{i5} for i=1, 2. Since enabling t_{i1} requires the availability of the feeder area, robot arms Ri, and Ri's component, input arcs from p_3, p_i, and p_{i1} are added to t_{i1}. Firing t_{i2} makes the feeder area available, and thus an output arc from t_{i2} to p_3 is formed. Enabling t_{i2} requires the availability of the PCB area, and thus an input arc from p_4 to t_{i2} is created. Continue this process till all the relations are represented. Finally, we add a token to all the resource places to signify the initial system state. The use of one token in p_{i1} and the arc from t_{i5} back to p_{i1} models actually the case that components to be inserted will be always available over the concerned period. The final model is shown in Fig. 5.13(b). The initial marking is:

$$(1, 1, 1, 1, 1, 0, 0, 0, 0, 1, 0, 0, 0, 0)^\tau.$$

The net has in total 14 places, and 10 transitions.
4. Check it and we find that the model represents the system operations.

124   Chapter 5. Modeling FMS with Petri Nets

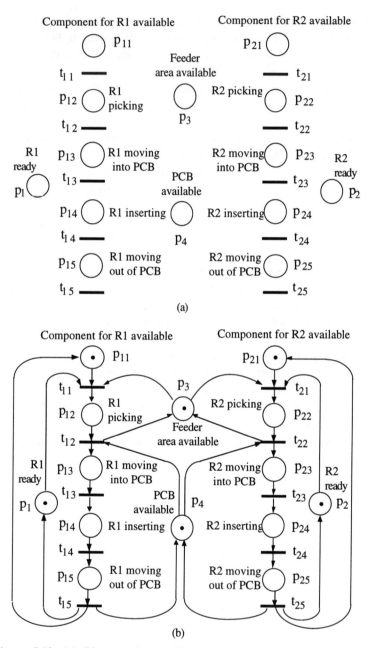

Figure 5.13. (a) Places and transitions and (b) Complete net model.

5.7 Systematic Modeling Methods

Petri net modeling and synthesis is a very important research area which attracts much attention. Top-down, bottom-up, hybrid methods are proposed to handle Petri net modeling of flexible manufacturing systems. The current focus is on the synthesis of a Petri net model satisfying both the system specifications and certain behavioral properties such as freedom from deadlock, no capacity overflow and cyclic behavior. An in-depth treatment of this topic can be found in [Zhou and DiCesare, 1993; Proth and Xie, 1996] and an overview of these techniques can be referred to [Jeng and DiCesare, 1993 and 1995].

5.7.1 Bottom-up Methods

The idea behind bottom-up modular approaches can be summarized as two words: decomposition and composition. Decomposition involves dividing a system into several subsystems. In manufacturing, this division can be performed based on the job types and/or physical plant layout. Depending upon the system's complexity, hierarchical decomposition may be needed. Then all the subsystems are modeled as Petri nets. Composition involves integrating these submodels into a complete model. In general, the composition process does not guarantee the properties of the resulting model.

The following are the Petri net models used to model each individual subsystem, resource's activities, product routes, or tasks:

1. Elementary loops or circuits that are safe and live. They can be used to model a single resource to perform repetitive activities, and single capacity buffer. Safeness is guaranteed by marking the loop with a single token initially and allowing only single arcs in the loop. Some examples are shown in Fig. 5.2(b).
2. Elementary loops or circuits that are bounded and live. They can be used to model multiple resources to perform repetitive activities, buffer with more than one slot, and processes that require multiple parts as a batch. One example in shown in Fig. 5.2(b). The multiplicities of arcs are not limited to single.
3. Finite state machines that are safe and live. They can be used to define a task's sequences and alternative routes.
4. Marked graphs that are bounded and live. They can model concurrent repetitive systems.
5. General Petri nets.

In particular, we present the following commonly-used modules as shown in Fig. 5.14 in the context of FMS [Zhou and Jeng, 1998]:
- (a) **Resource/operation module:** It describes a single operation stage that requires a dedicated resource. The module as shown in Fig. 5.14(a) evolves from Fig. 5.11(a) by associating a dedicated resource place to t_1-p_2-t_2. Place p_1 and p_3 often represent the buffer places that stock intermediate parts. Place p_4 represents the availability of a single or multiple capacity resource. Three interpretations may be associated with t_1, t_2, and p_2:
    1) Transitions t_1 and t_2 describe the start and end of an operation, respectively, and place p_2 represents the operation;
    2) t_1 and t_2 represent "loading" and "unloading," respectively, and p_2 represents the operation; and
    3) t_1 and t_2 represent "setup" and "operation" or "process," respectively, and p_2 represents the status of "the resource being ready to operate".
- (b) **Periodically-maintained resource/operation module:** The above module can be extended to a resource that requires periodic maintenance. Suppose that it needs maintenance service every k processes. Then Figure 5.14(b) shows such a module. Place p_4 is initially marked with k tokens, implying that the resource can be used for k consecutive times. Place p_6 is needed to guarantee that a single operation is ongoing at a time. Note that transition t_3 (maintenance service) is enabled only after place p_5 accumulates k tokens. A delay can be associated with t_3 to indicate the time needed by the service in a timed Petri net to be explored next chapters.
- (c) **Fault-prone resource/operation module:** The module in Fig. 5.14(a) can also be extended to a fault-prone resource by associating a loop with it, resulting in a module in Fig. 5.14(c). Transition t_3 means that the machine breaks down and t_4 means that it is repaired. Place p_5 stands for the status of "the resource being in repair."
- (d) **Priority Module:** It describes different routes, each of which requires a dedicated resource. The module as shown in Fig. 5.14(b) evolves from Fig. 4(c) by associating a dedicated resource place to each route. These resources often exhibit different performances. Thus when a part is ready to be processed by either of these resources, it can take the one with the highest priority among the available resources. The priorities are determined by their processing times, quality functions, etc.

5.7. Systematic Modeling Methods

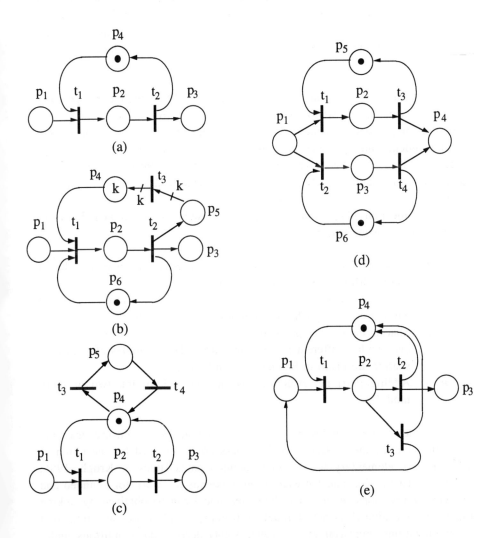

Figure 5.14. Commonly used Petri net modules for flexible manufacturing (a) Resource/operation, (b) Periodically-maintained resource/operation, (c) Fault-prone resource/operation, (d) Priority, and (e) Rework

(e) **Rework module:** If a process fails, a part may need rework although in many cases such a part may be scraped in practice. The module as shown in Fig. 5.14(e) evolves from Fig. 5.14(a) by introducing a new transition that reroutes the part back to the place p_1 (input buffer).

All the above modules can be extended to obtain their corresponding resource sharing cases when the dedicated resource becomes a shared one among either parallel or sequential processes. These modules can be used as a building block. Linking these modules together with merging of some shared places can lead to a PN model of a complex manufacturing system. Their applications to a semiconductor manufacturing system can be found in [Jeng et al., 1996].

All the individual models can be composed into a complete Petri net model in the following ways:

1. Sharing places. These places are often resources that are shared among different operations and processes.
2. Sharing transitions. They serve as synchronization points that enforce certain synchronization among the subsystems.
3. Sharing elementary paths that can start with a place or transition and ends with a different place or transition. The examples of such path sharing are given in Example 5.2 and Fig. 5.2.
4. Adding Petri net structures as an interface to link two or more modules.

Under certain conditions, bottom-up approaches can ensure desirable properties. For example safeness and liveness are guaranteed for the union of safe and live elementary loops along common paths [Beck & Krogh, 1986; Valavanis, 1990]. Koh and DiCesare (1991) extends the work for bounded and live elementary loops. A complete Petri net model can be obtained by linking finite state machine Petri nets through self-loops, inhibitor arcs and transition synchronizations. Ferrarini et al. (1992, 1994) derives the conditions under which the safeness, liveness and reversibility of the resulting model are guaranteed. The results are very useful in implementing logic control units.

The following example is adopted as a portion of of a complex system [Jeng et al., 1996]. To avoid the technical details in semiconductor manufacturing, we use the generic terms to describe it.

5.7. Systematic Modeling Methods

Example 5.10. A batch of wafers in a semiconductor manufacturing process needs go through the following steps: q_{1-5} using resources M1-5 and M2. Step q_1 needs M1. Step q_2 needs M2 or M2'. M2 tends to break down from time to time. There is an alternative M2' that is available if M2 is out of service. M3 and M5 that are required by q_3 and q_5 respectively process the batch and inspect the results. If it is not satisfied at M3, it needs be rerouted to the beginning of q_2. If it is not satisfied at M5, it needs just M5 to reprocess. M4 required in Step q_4 needs regular maintenance, once every k uses. Buffer slots are always available between these stages. Then we have:

1. Resource/operation module (Fig. 5.14(c)) for q_1 or M1;
2. Priority module (Fig. 5.14(d)) for q_2 where the upper half uses fault-prone resource/operation module (Fig. 5.14(c)) for M2 and the lower half uses Resource/ operation module (Fig. 5.14(c)) for M2'. Transition t_3 has priority over transition t7.
3. Rework module (Fig. 5.14(e)) for q_3 or M3. If the batch fails, the system needs reroute it to the beginning of q_2, i.e., p_3 (output buffer place of q_1, input buffer place of q_2);
4. Periodically-maintained resource/operation module for q_4 or M4 (Fig. 5.14(b)); and
5. Rework module (Fig. 5.14(e)) for q_5 or M5;

The final Petri net model is shown in Fig. 5.15 by sharing the output buffer place of Process i and input buffer place of Process i+1 for i=1, 2, 3, and 4.

Transition t is designated at the end of Step q_5, signifying that a new batch is allowed to release into the system after a batch is finished. The number of batches allowed in the system can be represented by the number of tokens that marks place p_0. The initial marking of each module is fixed already as shown in Fig. 5.15. ◊

Note that it usually takes a batch of wafer to go through hundreds of steps on different or same semiconductor manufacturing resources before it is completed. Design and implementation of these technologically and procedurally complex manufacturing systems present formidable challenges to their designers and analysts. Petri nets have emerged as a powerful tool and modeling method to tackle the research and development issues encountered in the design and implementation of these systems. Modeling and analysis of some realistic semiconductor manufacturing systems are referred to [Jeng et al., 1996, 1997; Kim and Desrochers, 1997; Zhou and Jeng, 1998].

130   Chapter 5. Modeling FMS with Petri Nets

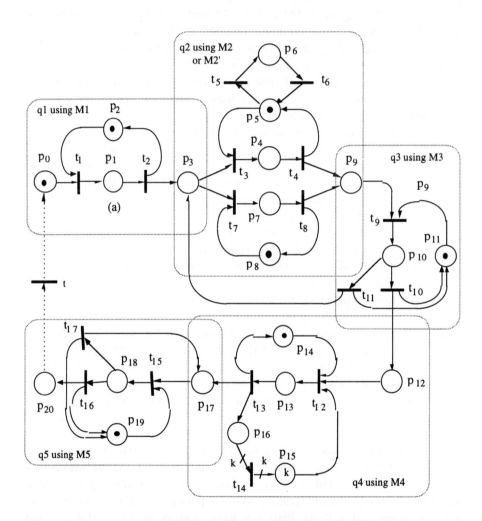

Figure 5.15. An example of linking different modules for a complete Petri net model of a semiconductor manufacturing process

5.7.2 Top-Down Methods

Top-down design of Petri nets refers to a design process for Petri nets using stepwise refinement of places, transitions, or substructures. This is accomplished by replacing them with more complicated subnets or pre-designed modules. Each successive step contains increasing detail until the desired level is reached. The desired level depends on the purpose of the model, which can be relatively simple for initial analysis or early planning, or complex for detailed analysis or control implementation. Refinement theory is used to guarantee that each step of refinement preserves the desirable system properties of Petri nets. Since those refinements are often local and required to have no side-effects, interaction among subnets is very difficult to handle using such a strategy.

The first step to initiate top-down design is to find an appropriate Petri net as a first-level model for the system. At this point, overall analysis of a system to be modeled and its specification are often necessary. The decomposition of a large system into independent subsystems is also completed at this stage. The investigation into Petri net modeling of many manufacturing systems has led to two kinds of widely used first-level Petri nets [Zhou and DiCesare, 1993]:

1. A Petri net with a single loop as shown in 5.11(d). It fits to many concurrent and repetitive manufacturing systems. The choice structures are required only in subnet models. It is also used as a basis for the knitting technique to synthesize a Petri net [Chao et al., 1995]. Its extended versions can be any live and bounded marked graphs. This has been a case for the FMS example discussed later in Section 5.8.

2. A Petri net based on a choice-synchronization structure. Some manufacturing systems have only one raw material or part entry. Different raw pieces are fixtured at the entry. Then they are sent for further machine processing. Finally they are assembled into a product. Since an assembly can be done only when the required two or more pieces are all finished, a control system has to enforce the synchronization among these pieces. A choice is encountered when raw pieces need go through different routes at the beginning, and a synchronization occurs when the parts are assembled. The first level Petri net model has to capture such choice-synchronization structures. One such example is illustrated in Fig. 5.10(b) in Example 5.9. Their detailed discussions and usage in FMS can be seen in [Zhou and DiCesare, 1993].

After the first level Petri net is decided, those transitions, places or simple structures represent aggregate activities need to be refined to represent more details. Each refinement introduces new details and it can be made to satisfy certain conditions and thus the overall net's desired properties are guaranteed.

The top-down approaches are appropriate to describe a hierarchical task structure. Each task may consist of subtasks. They are also fit to the description of assembly/disassembly Petri nets as discussed in Section 5.4.

5.7.3 Hybrid Methods

Combination of top-down and bottom-up approaches results in a hybrid approach. The hybrid synthesis approach formulated in [Zhou and DiCesare, 1993] has two stages: top-down design and bottom-up design. Top-down design aims at refining the net step by step so as to include enough system operation details for the purpose of implementation. Bottom-up design focuses on a correct construction of interactions among subsystems or detailed operation processes. All shared resources are considered at this stage. Their appropriate use and release becomes a key to constructing a Petri net with desirable qualitative properties, i.e., boundedness, liveness, and reversibility. Two main classes of places are operation and resource places. The former is used to represent an operation or process and the latter to represent availability of one or more resources. Resource places are further divided into two categories. The first category contains those resources whose capacity or quantity, thus the initial number of tokens is fixed, e.g., machines and robots. The second one has those resources whose number may vary, e.g., the number of jobs, fixtures, and pallets. Some places are also introduced for the control purpose such that the desired properties are guaranteed. The method is briefly described in the following steps:

1. Choose a bounded, live, and reversible Petri net as the first-level model of a system that is working when all major resources are available, and determine operation and resource places.
2. Decompose this system into several subsystems expressed as operation places using the defined basic design modules, some of them are given in Fig. 11(a)-(c). Replace these operation places by more detailed basic design models until no operations can be further divided or until one reaches a point at which the further division is not necessary. Basic modules include sequence, choice, concurrent, mutually exclusive and choice-free structures [Zhou et al., 1989].

3. Properly add non-shared resource places at each stage when operations require certain resources. The arcs are linked to the resource places.
4. Add shared resource places and their related arcs such that certain conditions are met. These resources may include a shared buffer resource, and a manufacturing resource shared by different concurrent and sequential processes. Initial markings which enable the system to be live and reversible have to be derived based on such structures. An interested reader is referred to [Zhou and DiCesare, 1993] for additional information and applications. Alternative way to determine the proper initial markings is based on siphons and traps as discussed in Section 5.5 and more examples are given in [Chu and Xie, 1997].

5.8 Top-down Modeling of an NJIT's FMS Cell: A Case Study

5.8.1 Description of A Flexible Manufacturing Cell

The top view of the FMS cell developed at the Center for Manufacturing Systems at New Jersey Institute Technology is shown in Figure 5.16. It consists of the following major components:

1. *SI Handling Conveyor System*: It consists of four carts, A, B, C and D, with fixture mounted on each, two transfer tables, TT1 and TT2, and dual conveyors which transport materials to each workstation.
2. *NASA II CNC Milling Machine*: The milling machine accepts rectangular solid blanks and machines "NJIT FMS" on the part or any other impressions that in turn can be generated from a CAD system.
3. *The GE P50 Robot* is a shared resource used to load and unload the material between the CNC milling machine and the conveyor system, and between the parts presentation station and the conveyor system.
4. *Parts Presentation Station*: This includes a gravity-chute that supplies rectangular solid blanks as raw materials. This station also contains two bins; one for accepted parts and one for rejected parts.
5. *Computer Vision System*: The vision system provides the visual automated inspection of the parts.

6. *Drilling Workstation*: It includes an IBM 7535 Industrial Robot with a 1/4" drill as an end-effector and drills holes at the four corners of the part.

The system is currently controlled by a Programmable Logic Controller (PLC) that coordinates all the movements of the individual components.

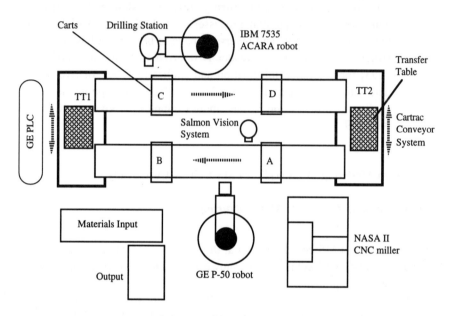

Figure 5.16. Top-view of an FMC developed at New Jersey Institute of Technology

The working cycle for this system proceeds in the following manner, where Steps 1 and 2 constitute the Loading Stage for this FMS cell, and Steps 3 through 13 constitute the Working Stage for this FMS cell:

1. Initially, all four carts on the conveyor system are empty and available for the raw materials to be loaded into them from the parts presentation station.
2. The GE P50 robot loads four parts, one by one, into the four carts on the conveyor system. The carts move clock-wise as they are being loaded.
3. Once the four parts are loaded, the positions acquired by the four carts are as shown in Figure 5.17.

5.8. Top-Down Modeling of an NJIT's Cell: A Case Study

4. The IBM 7535 robot drills four holes (one at each corner) on each blank part as the cart stops at the drilling workstation.
5. The GE P50 robot goes to the conveyor, removes the blank part from the cart at position x_1 (Figure 5.17) and loads it into the fixture located on the CNC machine tool table.
6. Once the part is loaded onto the CNC milling machine, the robot backs off and the milling machine mills "NJIT FMS" on the rectangular part.
7. After the milling operation, the robot arm goes to the milling machine to remove the piece machined from the fixture.
8. The robot returns the finished part to the same cart on the conveyor.
9. A signal is sent to the vision camera to inspect the part.
10. The vision system outputs a signal that directs the robot to accept or reject the part.
11. The robot runs either an *accept program* to place the part on the accepted pile or runs *a reject program* to place it on the rejected pile.
12. The GE robot goes to the parts presentation station and loads a new blank part into the cart.
13. The cart is released to the system and the next cycle is started.

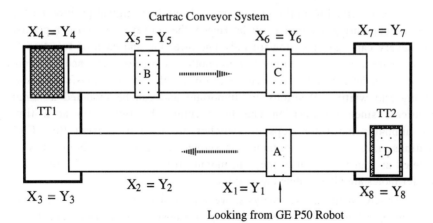

Figure 5.17. Fully-loaded stage of the material handling system.

5.8.2 Top-Down Design Process: First Level Net and Modules

This section applies top-down approaches, i.e., the system decomposition, refinement, and modular composition to the modeling process of Petri nets for the described FMS cell. The system has two stages: initialization or loading state and working state. Also, the system can be naturally decomposed into the following parts:
1. Conveyor System,
2. NASA II CNC Milling Machine, GE Robot, Computer Vision System, and Parts Presentation Station, and
3. Drilling Workstation.

These parts are related to each other through the material handling system. The initial positions for the carts with raw materials in the loading state are shown in Fig. 5.17.

MODELING OF CONVEYOR SYSTEM

To facilitate the presentation, we first concentrate on the material handling system. After its model is constructed, we will discuss
1. How it is further developed to model the FMS at the initialization or loading state, and
2. How it is used as the first-level model during the course of the top-down synthesis of the entire model of the FMS at the working state.

It is observed that all four carts move from their original position back to the same one during a complete cycle. Hence, the conveyor system is repetitive. Since every cart can move toward only one direction and each of two transfer tables carries a cart from one position to another, and goes back once it is done, the system is decision-free. The basic operations with carts and transfer tables are moving and waiting. Positions are important and can be modeled as single resources, since no position can be occupied by two carts at a time. Programmable hardware stops or switches are used to prevent this. Eight positions, i.e., x_1-x_8, are obtained as shown in Figure 5.17. Four carts are modeled as four tokens since they are moving from a position to another. Two transfer tables are also modeled as two tokens.

Based on the above analysis, we represent each moving between two positions as a transition, and each waiting at a position as a place, and display transitions t_1-t_{10} and places, px_1-px_8 as shown in Fig. 5.18. For example, t_1 means a cart's moving from Position x_1 to x_2 and t_3 means a transfer table's moving from Position x_3 to x_4. px_1 means a cart's waiting at Position x_1. The

5.8. Top-Down Modeling of an NJIT's Cell: A Case Study

availability of the positions x_1-x_8 is modeled as places py_{1-8}. For example, place py_1 represents availability of position x_1 and py_3 represents the availability of transfer table TT1 at position x_3.

Next, we add arcs to specify the relations among the transitions and places. Since firing t_1, i.e., a cart's moving from position x_1 to x_2, requires that a cart occupy position x_1 and position x_2 be available, arcs (px_1, t_1) and (py_2, t_1) are thus added. Also, arcs (t_1, py_1) and (t_1, px_2) are added since firing t_1 means that position x_1 becomes available, or py_1 should be marked and a cart moves from position x_1 to x_2. Based on the same reason, we add all other arcs as shown in Fig. 5.18. The meaning of all places and transitions is shown in Table 5.4.

The initial marking can be designed as

$$m_0 = (1, 0, 0, 0, 1, 1, 0, 1, 0, 1, 0, 1, 0, 0, 0, 0)^\tau$$

whose components are associated with the series of places px_1, px_2, ..., px_8, py_1, ...py_8. $m_0(px_1)=1$ implies that there is a cart at position x_1 initially and $m_0(px_8)=1$ means that TT2 is available at position x_8 and ready to transfer a cart to Position x_1. It is noted that place px_1 and px_5 are macro-places that represent more detailed activities to be discussed later.

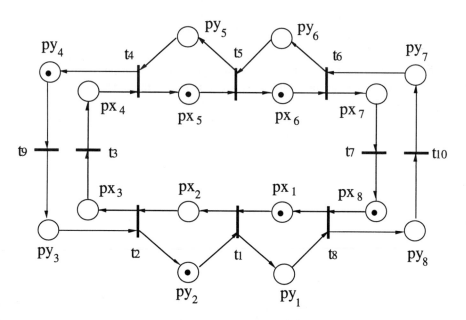

Figure 5.18. A net model for the material handling system

Table 5.4. Description of Places and Transitions in Figure 5.18 where TT means transfer table.

px_1	Cart available at loading station at X_1	py_1	Availability of cart halting position at X_1
px_2	Cart halting position at X_2	py_2	Availability of cart halting position at X_2
px_3	Cart loaded on transfer table TT_1 at X_3	py_3	Availability of TT_1 at X_3
px_4	Cart transported to position X_4 by TT_1	py_4	Availability of TT_1 at X_4
px_5	Cart available at drilling station at X_5	py_5	Availability of cart halting position at X_5
px_6	Cart halting position at X_6	py_6	Availability of cart halting position at X_6
px_7	Cart loaded on transfer table TT_2 at X_7	py_7	Availability of TT_2 at X_7
px_8	Cart transported to position X_8 by TT_2	py_8	Availability of TT_2 at X_8
t_1	Cart moves from loading station to X_2	t_6	Cart is loaded on transfer table TT_2
t_2	Cart is loaded on transfer table TT_1	t_7	TT_2 and loaded cart move together to X_8
t_3	TT_1 and loaded cart move together to X_4	t_8	Cart unloads from TT_2 and moves to X_1
t_4	Cart unloads from TT_1 and moves to X_5	t_9	Cart moves from position X_8 to X_7
t_5	Cart moves from drilling station to X_6	t_{10}	Cart moves from position X_4 to X_3

Since every place has exactly one input and one output transition, this subnet for this material handling system is a marked graph. The properties can be easily verified based on Property 5.2. Since every loop in this net has at least one token, the net is live and reversible. Since each place belongs to a loop whose token count is one, this net is also safe.

5.8. Top-Down Modeling of an NJIT's Cell: A Case Study

Initialization or Loading Stage

To bring the system from unloaded state to the loaded state shown in Fig. 5.17, the system has to load all four carts. The loading is accomplished when the cart is in Position x_1 via GE P50 Robot. Thus the substructure px_1 - t_1 has to be augmented as shown in Fig. 5.19. After the cart moves into position x_1, if both the robot (represented by p_{01}) and a raw piece (represented by the place p_{02}) are available, firing t (loading) loads a raw piece to the cart. When t_1 fires, it releases the robot and position x_1 becomes available. It is clearly seen that the net in Fig. 5.19 modeling the initialization state derives from the first level net in Fig. 5.18 by replacing the sub-structure px_1 - t_1 with the detailed structure px_1, p_{01}, p_{02}, p_{03} and t with their related arcs). They are explained below:

p_{01}: Availability of Robot GE P50
p_{02}: Availability of raw pieces
p_{03}: Availability of a loaded cart
t: loading a raw piece to the cart at position x_1.

The places in this net except place p_{02} are safe. After all the tokens in p_{02} are exhausted, the net eventually reaches a deadlock state due to the exhaustion of all four tokens in p_{02}. This deadlock state is, however, a desired state since it serves as the initial one for the next repetitive working state of this FMS. Note that the Petri net after the removal of p_{02} in Fig. 5.19 is a safe and live marked graph.

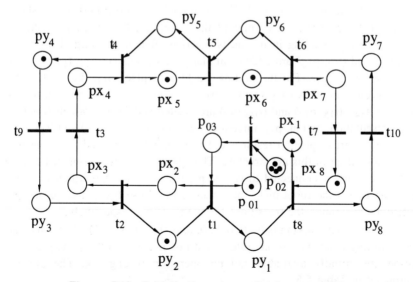

Figure 5.19. Initialization stage: Petri net model

MODELING OF MILLING MACHINE, GE ROBOT, COMPUTER VISION SYSTEM, AND PARTS PRESENTATION STATION

We model all these components together because they are all tied to the GE P50 robot. For this set of components, we can list the sequential activities as {loading, milling, unloading, inspecting, moving to the accepted area or rejected area, loading a new raw-material to the cart}. The places used to represent these activities are pictured as p_{2-8}. Two places, i.e., p_6 and p_7, moving to the accepted area, and moving to the rejected area, respectively, are pictured in parallel as shown in Fig. 5.20(a). Before loading, a place, i.e., p_1, representing the availability of a raw material in a cart needs to be added at the top. After the robot loads a new raw material to the cart (p_8), a place p_9 needs to be added to represent the availability of the cart with a new part to be released. Insert transitions t_{11-19} between the places as shown in Fig. 5.20(a). They represent the start or termination of operations.

The places to model the availability of the GE robot (p_{10}), milling machine (p_{11}), vision system (p_{12}), and raw material in the part presentation station (p_{13}) are drawn in parallel with the related places. Now consider what arcs should be created to model the relations among places and transitions. In order to start the loading operations for t_{11}, the robot (p_{10}) must be present and a part (p_1) must be available. This leads to two arcs (p_1, t_{11}) and (p_{10}, t_{11}). Arc (t_{11}, p_2) is added since firing t_{11} (start loading) leads to an operation (loading) in p_2. For the other transitions, we can apply the same method to obtain other arcs shown in Fig. 5.20(a). From the view point of the resources, the robot, p_{10}, should be available to start loading in p_2 and is released after the loading. Thus two arcs (p_{10}, t_{11}) and (t_{12}, p_{10}) are added. Similarly transitions t_{13}, t_{14}, t_{15}, t_{16}, and t_{19} are connected to p_{10}. For the milling machine or place p_{11}, the relationship is shown through the arcs (p_{11}, t_{12}) and (t_{13}, p_{11}). For the vision system, p_{12}, three arcs (p_{12}, t_{14}), (t_{15}, p_{12}) and (t_{16}, p_{12}) are created and for the raw material in the parts presentation station, p_{13}, three arcs are (p_{13}, t_{17}), (p_{13}, t_{18}) and (t_{19}, p_{13}).

The part inspection result is either accepted or rejected. This means that there is a conflict in this net structure. Thus p_5 has two arcs connected to t_{15} and t_{16} that are further connected to p_6 and p_7, respectively. Nevertheless, the next operation is that the robot picks up a blank part. These two places p_6 and p_7 are unified again through the transitions t_{17} and t_{18} to p_8, respectively.

The initial marking is $(0, 0, 0, 0, 0, 0, 0, 0, 0, 1, 1, 1, 1)^\tau$. It means that the robot, milling machine, vision system, and raw material in the parts presentation station are initially available and no operation is ongoing. The places are interpreted in Table 5.5.

5.8. Top-Down Modeling of an NJIT's Cell: A Case Study

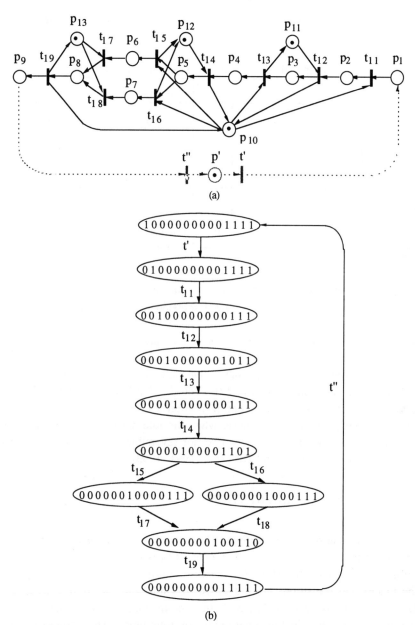

Figure 5.20. (a) Module for milling machine, GE robot, vision system, and parts presentation station, where t', p', and t" are added to check for its properties; and (b) Reachability graph

Table 5.5. Description of Places and Transitions in Fig. 5.20(a).

p_1	the cart with a raw piece	p_9	Cart with raw piece to enter the system
p_2	Miller loaded by GE P50 robot	p_{10}	Availability of GE P50 robot
p_3	Milling raw piece	p_{11}	Availability of milling machine
p_4	Finished part returned to the cart	p_{12}	Availability of the vision system
p_5	Part inspected by vision camera	p_{13}	Raw pieces in the parts presentation station
p_6	GE P50 robot places accepted part	t_{11-19}	End of the preceding operation and start of the next operation whenever applicable
p_7	GE P50 robot places rejected part	t', t''	Connecting transitions
p_8	GE P50 loads the cart with a new raw piece	p'	ideal place marked with one token initially

Next, we need to prove properties such as liveness, safeness, and reversibility. To achieve this objective, an idle place p' is initially marked with one token, two transitions t' and t" are associated with place p_1 and p_9 [Valette, 1979; Zhou et al., 1989]. The resulting net is called associated net [Valette, 1979]. The module plus p', t' and t" and related arcs in the dotted lines leads to its associated net in Fig. 5.20(a). Its reachability graph is constructed as shown in Fig. 5.20(b). Since the associated net is live, safe, and reversible, we say that the module is live, safe, and reversible [Zhou et al., 1989]. This conclusion can also be derived based on the sequential mutual exclusion concept [Zhou and DiCesare, 1991, 1993]. Furthermore, when the idle place is initially allowed to have two tokens, this net is neither live nor reversible but 2-bounded using the theory developed in [Zhou and DiCesare, 1991, 1993].

MODELING OF DRILLING STATION

There is only one operation: drilling. Three more places are needed to represent the availability of a raw material with a cart, a finished part with the cart, and the drilling station. Thus the structure is easily constructed as shown in Fig. 5.21. The places and transitions are explained in Table 5.6. The initial marking for this substructure is $(0, 0, 0, 1)^\tau$. Note that the drilling operation may be further decomposed as four drilling operations for four holes. This structure can be proved to be live, safe, and reversible using the same technique introduced in the last section. Note that the subnet in Fig. 5.21 is identical to the module presented in Fig. 5.14(a).

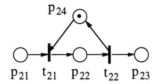

Figure 5.21. A net model for the drilling workstation

Table 5.6. Description of Places and Transitions in Figure 5.21

p_{21}	Raw piece with the cart	p_{24}	Availability of the drilling machine
p_{22}	Drilling the raw piece	t_{21}	Initiation of drilling operation
p_{23}	Drilled part with the cart available	t_{22}	Completion of of drilling operation

5.8.3 Final Petri Net Model

The final Petri net model as shown in Fig. 5.22 is obtained by refining the place px_1 and px_5 in Fig. 5.18 by the structures shown in Figs. 5.20(a) and 5.21, respectively. Note that both detailed structures in Figs. 5.20(a) and 5.21 are live, safe, reversible. These two refinements then guarantee the resulting net to be live, safe, and reversible [Valette 1979; Suzuki and Murata, 1983; Zhou and DiCesare, 1993]. Therefore, we have shown the procedure to get this Petri net for the FMS cell. More importantly, we have shown that the final net has the desired system properties: liveness, safeness, and reversibility.

The final Petri net model contains 31 places and 21 transitions with the initial marking as below:

$$(1\ 0\ 0\ 0\ 0\ 0\ 0\ 0\ 1\ 1\ 1\ 1\ 1\ 0\ 0\ 1\ 0\ 0\ 0\ 1\ 0\ 1\ 0\ 1\ 0\ 1\ 0\ 0\ 0\ 0)^\tau$$

based on the order of p_{1-13}, p_{31-24}, px_{3-4}, px_{6-8}, and py_{1-8}. Note that $m_0(p_1)=m_0(px_1)=1$ and $m_0(p_{21})=m_0(px_5)=1$.

The modeling of time delays into the above Petri net model will be discussed next chapter so that one can determine the cycle time of this FMS.

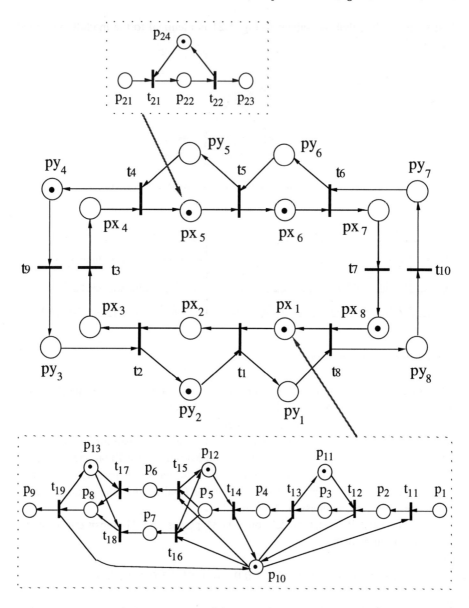

Figure 5.22. The final Petri net model for the FMS cell

5.9 Summary

This chapter discusses those common and useful special classes of Petri nets and their particular applications in manufacturing. Their properties are introduced and discussed. They include:
- State machine Petri nets that admit choices but not concurrency,
- Marked graphs that admit concurrency but not choice,
- Free-choice (FC) Petri nets that allow the choices and concurrent structures existing in the previous two classes of nets and only free or symmetric choices,
- Asymmetric choice (AC) Petri nets that extends FC nets by allowing asymmetric choices,
- Acyclic Petri nets and assembly/disassembly Petri nets with the latter aiming to represent assembly and disassembly of a product.
- Production-process nets and augmented marked graphs that allow the production processes modeled together with resource-sharing structures.

We can summarize their application scope in Table 5.7. Many manufacturing examples are provided to explain these nets, their synthesis, and applications.

Table 5.7. Special Petri net classes and their modeling scope in manufacturing

SM	A part may choose one among different operation sequences. No part (dis)assembly is allowed. Resources and concurrent operations are implied and not explicitly modeled.
MG	A part has a choice-free operation sequence. Parts can be disassembled and assembled. Each operation may request dedicated resources. No resource sharing is allowed.
FC	A part may choose one among different operation sequences. In a choice, no resource is involved. Each choice-free operation may request dedicated resources. Parts can be disassembled and assembled.
AC	A part may choose one among different operation sequences. In a choice, either no resource or the same set of resources is involved. Part (dis)assembly is allowed.
APN/DPN	Assembly (disassembly) of components/subassemblies (a product) into a product (components) is modeled. Choices and limited concurrency are allowed.
PPN/AMG	A part has a choice-free operation sequence (for AMG, or may have limited choices, e.g., in choice-synchronization structure). An operation may request dedicated/shared resources.

Basic relations among events are described and an ad hoc method and its application to modeling of AT&T flexible station are presented. Systematic synthesis methodologies are also briefly discussed. Among them are: bottom-up design method, top-down design method, and hybrid design method. Useful modules in PN modeling of automated manufacturing systems are presented. Bottom-up design of a partial semiconductor manufacturing system and top-down design of an FMS are performed. These presentations should help designers and analysts develop Petri net models for automated manufacturing systems.

CHAPTER 6

FMS PERFORMANCE ANALYSIS

The development of human-made systems requires that both functional and performance requirements be met. The ordinary Petri nets do not include any concept of time. With this class of nets, it is possible only to describe the logical structure and behavior of the modeled system, but not its evolution over time. Responding to the need for the temporal performance analysis of discrete-event systems, time has been introduced into Petri nets in a variety of ways. In this chapter two fundamental types of timed Petri nets are discussed in the context of the performance evaluation. They are deterministic timed Petri nets, and stochastic Petri nets. An example of a simple production system is used to illustrate the basic solution techniques for these two classes of nets. The focus of the discussion is on the analytical performance evaluation of Petri net models. This choice was motivated mainly by the recognition of the importance of using the analytical techniques for quick performance evaluation.

This chapter presents a Petri net (PN) approach to performance modeling and analysis of automated manufacturing systems. One case study on timed Petri net analysis of NJIT's flexible manufacturing cell is performed. Deterministic timed Petri net models are constructed and analyzed. Another case study focuses on the performance evaluation of two possible control structures for the AT&T Flexible Workstation using stochastic Petri nets when the robot arms in the system are subject to failure.

6.1 Deterministic Timed Petri Nets

When time delays for operations or activities in a concurrent conflict- or choice-free system are fixed, we can model the system as a deterministic timed Petri net. When a choice is involved in such a system and the system is allowed to make a choice freely, then its behavior becomes non-deterministic. For the class of choice-free Petri nets, marked graphs or event graphs are suitable and sufficient, which can represent concurrent activities but not choices. The performance index for such a model is cycle time. Transitions, places, and arcs all can be associated with time delays in a marked graph, resulting in a timed marked graph. These elements with time delays can be converted into each other.

1. If a transition is used to model an event or operation that takes certain time τ, then it is associated with τ. It takes τ for the transition to fire after its being enabled;
2. If a place is used to model an operation/activity/process that takes certain time τ, then it is associated with τ. After the place receives a token, the token needs to take τ for it to become available to enable their output transitions;
3. If an arc from a place to a transition is used to model the transportation process or material flow process that takes certain time τ, then it is associated with τ. The token in the place needs to wait for τ to be able to enable the transition; and
4. If an arc from a transition to a place is used to model the transportation process or material flow process that takes certain time τ, then it is associated with τ. The token generated by the transition firing needs to wait for τ to be able to reach the place.

Given a loop in a timed marked graph, the summation of all the delays in the loop remains constant regardless of the ways to associate time delays to the model. The number of tokens in in a loop of a marked graph also remains constant.

A *deterministic timed Petri net* is (DTPN) defined as a marked graph $Z = (P, T, I, O, m_0)$ and either zero or positive time delays are associated with places, transitions, and/or arcs. If non-zero time delays are associated with transitions (places, arcs) only, Z is also call timed transition (place, arc) Petri net.

In a DTPN or timed marked graph, deterministic time delays can be associated with its places, transitions and/or arcs. A token in a place with delay τ can be available or not-available to enable its output transition. After a place receives a token(s), it takes τ for the toke(s) to become available from non-available token(s). If an arc from a place to a transition has a delay time τ, it takes τ for an available token in the place to become ready to enable its output transition. A token that is ready to enable a transition is called a ready one. If the arc from a place to a transition is associated with zero time delay, an available token in the place is a ready one. If an arc is from a transition to a place with delay τ', the token generated due to the firing of the transition has to spend τ' in reaching the place. After the delay associated with the place, the token becomes available in the place. The previous firing rules for ordinary PNs with single arcs only are expanded as below:

6.1. Deterministic Timed Petri Nets

1. At any time instance, transition t becomes enabled at m if each input place p of t contains at least one ready token, and
2. Once transition t is enabled at m, it starts firing by removing one token from each input place. It completes firing after the time delay associated with t, deposits one token to each output place p after the delay associated with the arc from t to p. The newly deposited token becomes ready after the time associated with p plus the time associated with the p's output arc.

It is assumed that as soon as transitions are enabled by the ready tokens, they fire. A transition t is k-enabled at marking m if $m(p) \geq kI(p, t)$, $\forall\, p \in P$ where k is a positive integer. If t is k-enabled by the ready tokens, it fires k times simultaneously. From the viewpoint of queueing theory, we adopt the infinite server policy [Campos, 1992].

Theorem 6.1. A deterministic timed Petri net can be converted into a timed transition or place or arc Petri net.

Proof: We need to prove the case that any deterministic timed Petri net can be converted into a timed transition Petri net. To prove this, we present the following transformations from :

(a) a timed place,
(b) a timed arc from a transition to a place, and
(c) a timed arc from a place to a transition

into their corresponding equivalent structures that have delays associated with transitions only but none with places and arcs, respectively. Such transformations are shown in Fig. 6.1(a)-(c). It can be easily shown that all these transformations preserve the behavioral properties. Second the time duration for the pairs of Petri net structures is also preserved. Therefore, a DTPN can be converted into a timed transition Petri net without changing its temporal behavior.

Similarly, we can construct transformations that can convert any DTPN into a timed place or arc Petri net. ◊

Given $Z = (P, T, I, O, m_0)$, x is *a node* if $x \in P \cup T$. *An elementary path* is a sequence of nodes: $x_1 x_2 ... x_n$, $n \geq 1$, such that $\exists$ an arc (x_i, x_{i+1}) for $i \in N_{n-1}$ if $n > 1$, and $x_i = x_j$ implies that $i = j$, $\forall\, i, j \in N_n$ where $N_n = \{1, ..., n\}$. *An elementary loop or circuit* is $x_1 x_2 ... x_n$, $n > 1$ such that $x_i = x_j$, $1 \leq i < j \leq n$, implies that $i = 1$ and $j = n$. In the following discussion, a loop implies an elementary loop. A marked graph is strongly connected if there is an elementary path from any node to any other node.

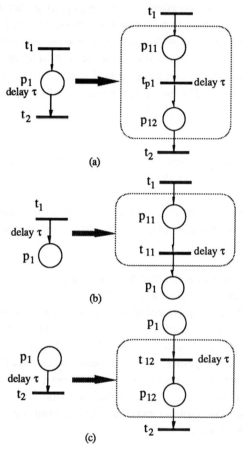

Figure 6.1 (a) a timed place, (b) a timed arc from a transition to a place, and (c) a timed arc from a place to a transition are converted into equivalent timed transition Petri nets.

6.1. Deterministic Timed Petri Nets

A fundamental approach to studying DTPNs is based on two concepts: the total time delay in a loop and total number of tokens in a loop or token count. Both remains constant for a loop in a marked graph. A strongly-connected marked graph consists of only a limited number of elementary loops. The total time delay in a loop is obtained as a sum of delays introduced by all transition, places, and arcs comprising the loop. The total number of tokens is obtained as a sum of tokens present in all the places that belong to the loop. Then, we have the following theorem due to [Ramchandani, 1974; Ramamoorthy and Ho, 1980].

Theorem 6.2. The cycle time of a strongly connected deterministic timed Petri net is determined as follows:
$$\pi = \underset{i}{\text{Max}} \{D_i / N_i\}$$
where D_i is the total time delay of loop i, and N_i is the token count of loop i. The ratio D_i / N_i is called the cycle time of loop i.

If a time delay is given as a time interval, i.e., $[\tau_{Min}, \tau_{Max}]$, then the upper and lower cycle times are given as follows:
$$\pi_{Min} = \underset{i}{\text{Max}}\{D_{Min,i}/N_i\} \text{ and } \pi_{Max} = (\underset{i}{\text{Max}}\{D_{Max,i}/N_i\}$$
where $D_{Min,i}$ ($D_{Max,i}$) is the total time delay in the i-th loop based on τ_{Min} (τ_{Max}) for all delays. Note that a transition with delay $[\tau_{Min}, \tau_{Max}]$ takes at least τ_{Min} and at most τ_{Max} to fire after it is enabled. The similar explanation is applicable to the places and arcs of such interval time delays.

The upper and lower cycle times can also be derived for marked graphs in which time delays are random and their first moments or means are known [Campos 92]. More accurate bounds can be obtained if their standard deviations are also known [Xie, 1994]. In general, the number of loops increases exponentially as the size of a marked graph measured by the number of its places and transitions. Thus the enumeration of all loops may not be feasible for large size nets. Fortunately, the problem to find the cycle time of a marked graph can be converted into a linear programming problem [Campos et al., 1992; Magott, 1984; Morioka and Yamada, 1991; Yamada and Kataoka, 1994].

Suppose that a marked graph is a timed place Petri net. In other words, a time delay function f: $P \rightarrow R^+$ is defined and there is no delay on the arcs and transitions. Then, according to [Campos et al., 1992], we can formulate the following linear programming (LP) problem to find the cycle time:
$$\pi = \text{Max } x^\tau f$$
Subject to $x^\tau C \leq 0$, $x^\tau m_0 \leq 1$, $x \geq 0$

where x is an n-dimensional vector (n is the number of places), C is the incidence matrix, and m_0 is the initial marking. This problem has n (the number of places) decision variables and n+s+1 constraints where s is the number of transitions. The above LP problem can be reduced to one with n-s+1 decision variables and n constraints by a transformation discussed in [Yamada and Kataoka, 1994], or one with s+1 decision variables and n+1 constraints [Magott, 1984]. Please note that a linear program problem can be solved using the simplex method that has exponential worst-case complexity or the algorithms of polynomial worst-case complexity [Nemhauser et al., 1989].

For a timed transition marked graph with function $\tau: T \rightarrow R^+$ defined and no delays associated with places and arcs, then the above f in the LP problem can be selected as follows:

$$f\tau = (f_1, f_2, ..., f_n) = (\tau(\cdot p_1), \tau(\cdot p_2), ..., \tau(\cdot p_n))$$

where $\cdot p_i$ is the input transition of p_i, which is unique for a place in a marked graph. Equivalently, f can be selected as $(\tau(p_1\cdot), \tau(p_2\cdot), ..., \tau(p_n\cdot))$.

For a timed marked graph with functions:

$\tau: P \rightarrow R^+$

$\tau': T \rightarrow R^+$ and

$\tau'': W \rightarrow R^+$ where W = {(p, t): I(p, t)≠0)} ∪ {(t, p): O(p, t)≠0)}, a set of all arcs in the net, the i-th component of the above f is as follows:

$$f_i = \tau(p_i) + \tau'(\cdot p_i) + \tau''(\cdot p_i, p_i)) + \tau''(p_i, p_i\cdot).$$

Example 6.1 (Deterministic Timed PN). The production cell consists of two machines, two robots, and an intermediate part buffer space with capacity two as modeled in Example 5.2. It has the following time characteristics: each of loading and unloading by Robots R_1 or R_2 takes equal time, 1 time unit, M1 takes 10 time units, and M2 takes 16 time units. Return of a pallet takes 3 time units. What is the cycle time of this cell?

First, we can associate the time delays with the following transitions and places shown in Fig. 5.2(c) according to the system's timing characteristics. The results are presented in Table 6.1. Other places, transitions and arcs have no time delay.

Next, we enumerate all the loops in the net model. In fact, all the loops are found in Fig. 5.2(b). Then we compute token count and total time delay in each loop. Loop cycle time is calculated based on the formula in Theorem 6.2. The results are summarized in Table 6.2.

6.2. Analysis of an FMS Cell: A Case Study

Table 6.1 The transitions, places, arcs and their associated time delays.

Transition, place, or arc	t_1	p_2	t_2	t_3	p_4	t_4	(t_4, p_1)
Time delay	1	10	1	1	16	1	3

Table 6.2 Loops, total time delays, token counts, and loop cycle times.

Loops	D_i	N_i	D_i/N_i
$p_5 t_1 p_2 t_2 p_5$	12	1	12
$p_6 t_2 p_3 t_3 p_6$	2	2	1
$p_7 t_3 p_4 t_4 p_7$	18	1	18
$p_1 t_1 p_2 t_2 p_3 t_3 p_4 t_4 p_1$	33	3	11
$p_8 t_1 p_2 t_2 p_8$	12	1	12
$p_9 t_3 p_4 t_4 p_9$	18	1	18

Finally, the cycle time is 18 time units as decided by the loops $p_7 t_3 p_4 t_4 p_7$ and $p_9 t_3 p_4 t_4 p_9$. If robots and machines need setup times after each operation, they could again be added to the corresponding places. The similar analysis can be conducted for these cases. It should also be noted that increasing the pallet number from three (modeled by tokens in p_1) will not shorten this system's cycle time due to the bottleneck machine M2. Thus, deterministic time Petri net analysis can help decide the appropriate number of pallets and/or fixtures for repetitive and concurrent systems. ◊

6.2 Analysis of an FMS Cell: A Case Study

In Section 5.8, an FMS cell was synthesized using a top-down method. This section discusses its temporal analysis by finding its cycle time. By observing the operations of this FMS cell, we have measured the timing delays for all individual operations in the FMS cell, as summarized in Table 6.3. All the arcs have no delay. This final net is not a marked graph since the module for the GE robot, vision system, and parts presentation station as shown in Fig. 5.20(a) is not a marked graph. Thus before the formula in Theorem 6.2 to calculate cycle time for timed marked graphs can be used, the net model in Fig. 5.22 needs to be converted into a marked graph.

The information we obtained in the top-down synthesis can help us achieve this reduction. The synthesized non-marked graph module related to GE robot, etc. is reduced to a single place with its timing characteristics preserved as discussed below.

First, the overall net model is safe and so is the module under consideration. Thus each time only one token goes through p_1 and none enters p_1 until p_9's token leaves the module in Fig. 5.20(a), which deals with GE robot, vision system, and parts presentation station. According to the module's reachability graph as shown in Fig. 5.20(b), each time p_1 receives a token, it goes through sequentially $t_{11}p_2t_{12}p_3t_{13}p_4t_{14}p_5$-$t_{15}p_6t_{17}$(or $t_{16}p_7t_{18}$)-$p_8t_{19}p_9$. At place p_5, the token goes through $t_{15}p_6t_{17}$ or $t_{16}p_7t_{18}$. Fortunately, both paths take the same time delays, i.e., 0+3+0 = 3 Sec. In other words, from the viewpoint of the behavior, we can treat the module as one of a strictly sequential activities even though place p_5 has a choice structure and p_{10} has multiple input and output transitions. The total time duration for a token to pass from p_1 to p_9 is thus $\tau(p_1) + \tau(p_2) + \tau(p_3) + \tau(p_4) + \tau(p_5) + \tau(p_6) + \tau(p_8) + \tau(p_9)$ since all transitions have zero time delay as shown in Table 6.3. Thus, the substructure in Fig. 5.20(a) is equivalent to place px_1 with $\tau(px_1) = 74$ (sec).

Similarly, the substructure in Fig. 5.21 is equivalent to px_5 with $\tau(px_5) = 73$ (sec). Therefore, we can use px_1 and px_5 instead of the two modules to evaluate this system's cycle time. In other words, we can use Fig. 6.2 that is identical to Fig. 5.19 when we enumerate all loops and determine the cycle time. The results are summarized in Table 6.4. The maximum cycle time is 82 Seconds and the bottleneck of the system occurs at loops γ_1, γ_2, and γ_7. This concludes the analysis of this flexible manufacturing cell.

Table 6.3 Time Delays (Sec.) for Places and Transitions.

Places or Transitions	Delay	Places or Transitions	Delay
p_9, p_{13}, p_{23-24}, py_i, t_{11-22}	0	px_1	74
p_2, p_4, p_8, px_2, t_4, t_8	5	px_3, px_7	2
p_3	45	px_5	73
p_5	10	px_6	75
p_{6-7}, px_4, t_{1-3}, t_{5-7}, t_{9-10}	3	px_8	66
p_1, p_{21}	1	p_{22}	72

6.2. Analysis of an FMS Cell: A Case Study

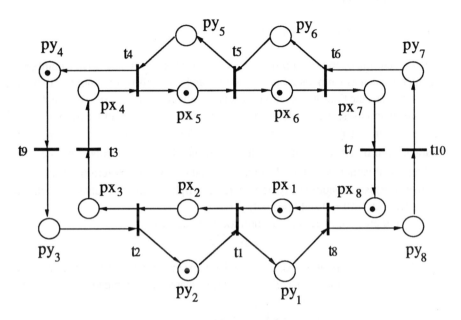

Figure 6.2 The equivalent marked graph of the Petri net model in Fig. 5.22.

Table 6.4. Elementary loops or circuits and their cycle times

Elementary loops or circuits	$D_i(s)$	N_i	D_i/N_i (s)
$\gamma_1 = px_1 t_1 px_2 t_2 px_3 t_3 px_4 t_4 px_5 t_5 px_6 t_6 px_7 t_7 px_8 t_8 px_1$	328	4	82
$\gamma_2 = px_1 t_1 py_1 t_8 px_1$	82	1	82
$\gamma_3 = px_2 t_2 py_2 t_1 px_2$	11	1	11
$\gamma_4 = px_5 t_5 py_5 t_4 px_5$	81	1	81
$\gamma_5 = px_6 t_6 py_6 t_5 px_6$	81	1	81
$\gamma_6 = px_3 t_3 px_4 t_4 py_4 t_9 py_3 t_2 px_3$	19	1	19
$\gamma_7 = px_7 t_7 px_8 t_8 py_8 t_{10} py_7 t_6 px_7$	82	1	82
$\gamma_8 = py_1 t_8 py_8 t_{10} py_7 t_6 py_6 t_5 py_5 t_4 py_4 t_9 py_3 t_2 py_2 t_1 py_1$	28	2	14

The maximum cycle time for bottleneck loops γ_1, γ_2, and $\gamma_7 = \text{Max}(D_i/N_i) = 82$ Sec.

6.3 Stochastic Petri Nets

6.3.1 Exponential Distribution

A manufacturing operation or process may be complex, depending upon a number of parameters such as temperature and pressure, and the skill of operators if they are involved. Thus the time an operation takes is not always deterministic. For such non-deterministic cases, a time delay of random nature is used and assumed to be continuous in the following discussion. A random time variable τ is completely characterized by its (cumulative) distribution function $F(x) = \text{Prob}(\tau \leq x)$, i.e., the probability of random variable τ less than or equal to x where x is zero or positive real number. $F(x)$ is a monotonically non-decreasing right-continuous function such that $0 \leq F(x) \leq 1$ with $F(0) = 0$ and $F(\infty) = 1$.

The probability density function $f(x)$ is defined as
$$f(x) = \frac{dF(x)}{dx}$$
provided that this derivative exists everywhere, except possibly at a finite number of points. In particular, the exponential distribution characterizes a random time variable τ that can only take nonnegative real variable values, and it is defined as
$$f(x) = \lambda e^{-\lambda x}, \ x \geq 0 \text{ and } f(x) = 0, \ x < 0.$$
where $\lambda > 0$. Its corresponding distribution function $F(x)$ is obtained by integrating $f(x)$ over the interval $[0, x]$ to yield
$$F(x) = 1 - e^{-\lambda x}, \ x \geq 0.$$
λ is the sole parameter that characterizes the exponential distribution, called a rate. Both $f(x)$ and $F(x)$ are continuous. A transition in a Petri net can be associated with an exponentially distributed random time delay expressing the delay from the enabling to the firing of the transition. In this case, this parameter λ is also called its firing rate. If a manufacturing operation takes exponential random time with rate λ, λ is also called the operational speed or rate. When this operation is associated with a machine, it is called the machine's rate. Failure rate and repair rate are used to describe the timing features associated with a machine's failure and repair time. The physical meaning of this rate is reflected by the fact that the mean or expected time of the exponential random variable τ is $E(\tau) = 1/\lambda$. Thus if the average machining time is known, then its reciprocal is the machine's rate assuming that the machining time is an exponential random time.

Note that uniform and normal exponential distributions are two other widely used ones in engineering practice.

6.3.2 Definition and Solution of Stochastic Petri Nets

When time delays are modeled as random variables, or probabilistic distributions are added to the deterministic timed Petri net models for the conflict resolution, stochastic timed Petri net models are yielded. In such models, it has become a convention to associate time delays with the transitions only. When the random variables are of general distribution or both deterministic and random variables are involved, the resulting net models cannot be solved analytically for general cases. Thus simulation or approximation methods are required. The stochastic timed Petri nets in which the time delay for each transition is assumed to be random and exponentially distributed are called stochastic Petri nets (SPN) [Florin and Natkin, 1982; Molloy, 1982]. The SPN models that allow for immediate transitions, i.e., with zero time delay, are called generalized SPN (GSPN) [Ajmone Marsan, Balbo, and Conte, 1984; Ajmone Marsan et al., 1995]. Both models, including extensions such as priority transitions, inhibitor arcs, and probabilistic arcs can be converted into their equivalent Markov process representations. Thus their analysis can be conducted by solving a set of mathematical equations.

A Stochastic Petri net is $(P, T, I, O, m_0, \Lambda)$ where $\Lambda:T \to R^+$ is a firing function associating positive firing rates to all transitions. Normally, we use λ_i to denote the firing rate of t_i.

Note that each firing rate of a transition could be marking-dependent. The following theorem is due to [Molloy, 1982].

Theorem 6.3: Any finite place, finite transition, marked stochastic Petri net is isomorphic to a one-dimensional discrete space Markov process. A bounded stochastic Petri net is isomorphic to a finite Markov process or chain.

Note that a one-dimensional Markov process is one that has only one random variable. In an SPN equivalent Markov process, time is the only random variable.

Theorem 6.3 opens the way to performance analysis of stochastic manufacturing systems using Petri nets. Each transition that takes an exponentially distributed random time delay is associated with a rate called firing rate. The procedure to use SPN for such analysis is as follows:

Procedure 6.1 (Performance analysis with a bounded SPN)
Step 1: Model a system using a Petri net and associate exponential time delays with transitions;

Step 2: Generate the reachability graph $R(m_0)$. The Markov process is obtained by assigning each arc with the rate of the corresponding transition (possibly marking dependent). Label all states or markings with $m_0, m_1, m_2, ..., $ and m_{q-1} where q is the total number of states, i.e., $q = |R(m_0)|$.

Step 3: Analyze the Markov process. The steady-state probabilities, denoted by $\Pi = (\pi_0, \pi_1, ..., \pi_q)$, are obtained by solving the following linear matrix equation:

$$\Pi A = 0$$
$$\sum_{i=0}^{q-1} \pi_i = 1$$

where $A = (a_{ij})_{q \times q}$ is the transition rate matrix. For $i = 0, 1, ...,$ and q-1, A's ith row elements, i.e. a_{ij}, $j = 0, 1, ...,$ and q-1 are determined as follows:

   3.1. If $j \neq i$, a_{ij} is the sum of the rates of all the outgoing arcs from state m_i to m_j.

   3.2. Since any row of elements in A satisfies $\sum_{j=0}^{q-1} a_{ij} = 0$, after Step 3.1,

$$a_{ii} = -\sum_{j \neq i}^{q-1} a_{ij}.$$

   Note that $-a_{ii}$ represents the sum of firing rates of transitions enabled at m_i, i.e., transition rates leaving state m_i.

Step 4. Find the required performance estimates of a system modeled by the SPN from the obtained stead-state probabilities Π and transition firing rates Λ. The common ones include:

   a. The probability of a particular condition. Let G be the subset of $R(m_0)$ satisfying a particular condition. Then required probability is:
$$\text{Prob}(G) = \sum_{m_i \in G} \pi_i$$
   For example, if the particular condition is that a resource performs useful work, then Prob(G) is the resource utilization.

   b. The expected value of the number of tokens. Let G_{ij} be the subset of $R(m_0)$ such that at each marking in G_{ij}, the number of tokens in place p_i is j. Then the expected value of the number of tokens in a k-bounded place p_i is:
$$E(m(p_i)) = \sum_{j=1}^{k} j \text{Prob}(G_{ij})$$
   For example, if p_i represents the availability of type-A finished products, then $E(m(p_i))$ is the average inventory of type-A products.

   c. The mean number of firings per time unit. Let G_j be the subset of $R(m_0)$ in which transition t_j is enabled. Since t_j may fire only when it is enabled, the mean number of t_j's firings per unit of time is
$$f_j = \sum_{m_i \in G_j} \pi_i \lambda_{ji}$$

6.3. Stochastic Petri Nets

where λ_{ji} is the firing rate of t_j at marking m_i. For the marking-independent case, since $\lambda_{ji} = \lambda_j$, the above formula is simplified as:

$$f_j = \sum_{m_i \in G_j} \pi_i \lambda_j = \lambda_j \sum_{m_i \in G_j} \pi_i$$

d. The mean system throughput or production rate. Suppose that a product is produced each time a transition in a subset T' fires. Then

$$g = \sum_{t_j \in T'} f_j$$

Example 6.2: To demonstrate the above procedure, we consider the example of the system shown in Fig. 5.2(a). Suppose that

1. Machine M1 performs faster than M2 does, however subject to failures when it is processing a part. On the average, M1 takes two time units to break down, and a quarter time unit to be repaired. Thus, its average failure and repair rates (1/time unit) are 0.5 and 4, respectively. M2 and the two robots are failure-free.
2. R1's loading speed is 40 per unit time. The average rate for M1's processing plus R1's unloading is 5 per unit time.
3. The average rate for M2's processing plus the related R2's loading and unloading is 4 per unit time.
4. All the time delays associated with the above operations are exponential.

We are interested in obtaining the average utilization of M1, and the production rate of the system (throughput), assuming that

1. Only one pallet is available, and
2. Two pallets are available.

First, model the above production system using a Petri net and associate exponential time delays with transitions. Following the general modeling method in Section 5.6, we identify three activities in sequence to produce a part,

1. R1's loading,
2. M1's processing and R1's unloading a part, and
3. R2's loading and unloading and M2's processing.

Model them as transitions (to fit to the SPN evaluation) and insert places to represent the condition or status. Link them through arcs. The solid arcs show the portion of the model as shown in Fig. 6.3 up to this stage.

Then we add a resource place to represent its availability for each resource. According to whether an activity in a transition needs or releases a resource represented as a token in a resource place, we draw input and output arcs. The dotted arcs show these in Fig. 6.3. Finally, to model the failure and repair loop of M1 when it is processing, we add a transition to represent the machine's failure transition and another for its repair. Insert a place between them to represent that the status of "M1 is in repair". The shaded arcs represent this portion as shown in Fig. 6.3. Label all the transitions and places as shown in Fig. 6.3. Associate the firing rates λ_{1-5} to all the five transitions. The initial marking is $(k, 0, 0, 0, 1, 2, 1, 1)^T$ where $k = 1$ and 2 for the first and second cases, respectively. The places and transitions are all interpreted in Table 6.5.

Next, generate its reachability graphs as shown in Fig. 6.4 for $k = 1$ and 2. Note that we have eliminated the redundant information in each marking on the token values in places p_{7-9} since p_7 holds the same number of tokens as p_5 does and each of the other two always has one token. Label the transitions and their rates for each arc in the graph. The total numbers of states in two Markov chains are 4 and 7, respectively.

Third, to analyze the Markov process, find the following transition rate matrices as follows.

Case 1: $k = 1$. The total number of states $q=4$.
When $i = 0$, a_{0j}, $j = 1, 2$, and 3 are first obtained:
$$a_{01} = \lambda_1, a_{02} = a_{03} = 0.$$
Then $a_{00} = -\sum_{j=1}^{q-1} a_{0j} = -\lambda_1$. When $i = 1$, a_{1j}, $j = 0, 2, 3$ are first obtained:
$$a_{10} = 0, a_{12} = \lambda_2, a_{13} = \lambda_4.$$
Then $a_{11} = -\sum_{j \neq 1}^{q-1} a_{0j} = -\lambda_2 - \lambda_4$. Continue this process and obtain:

$$A = \begin{pmatrix} -\lambda_1 & \lambda_1 & 0 & 0 \\ 0 & -\lambda_2 - \lambda_4 & \lambda_2 & \lambda_4 \\ \lambda_3 & 0 & -\lambda_3 & 0 \\ 0 & \lambda_5 & 0 & -\lambda_5 \end{pmatrix}$$

The steady-state probabilities can be obtained by solving the following equations:
$$(\pi_0, \pi_1, \pi_2, \pi_3)A = 0$$
$$\pi_0 + \pi_1 + \pi_2 + \pi_3 = 1$$

This yields

$$\Pi^T = \begin{pmatrix} \pi_0 \\ \pi_1 \\ \pi_2 \\ \pi_3 \end{pmatrix} = \begin{pmatrix} \lambda_2 \lambda_3 \lambda_5 / \lambda \\ \lambda_1 \lambda_3 \lambda_5 / \lambda \\ \lambda_1 \lambda_2 \lambda_5 / \lambda \\ \lambda_1 \lambda_3 \lambda_4 / \lambda \end{pmatrix}$$

where $\lambda = \lambda_2 \lambda_3 \lambda_5 + \lambda_1 \lambda_3 \lambda_5 + \lambda_1 \lambda_2 \lambda_5 + \lambda_1 \lambda_3 \lambda_4$.

6.3. Stochastic Petri Nets

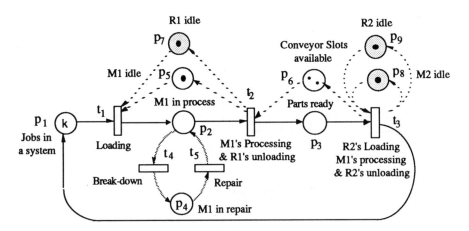

Figure 6.3 A Stochastic PN model (shaded places p_{7-9} are redundant from the analysis viewpoint).

Table 6.5 Places, transition, and their firing rates in Fig. 6.3.

Places	Interpretation	
p_1	Pallets with workpieces available	
p_2	Machine M1 in process	
p_3	Intermediate parts available for processing at M2	
p_4	M1 in repair	
p_5	M1 available	
p_6	Conveyor slots available	
p_7	R1 available (redundant from the analysis viewpoint)	
p_8	M2 available (redundant from the analysis viewpoint)	
p_9	R2 available (redundant from the analysis viewpoint)	
Transitions	Interpretation	Firing rates
t_1	R1 loads a part to M1	$\lambda_1 = 40$
t_2	M1 machines and R1 unloads a part	$\lambda_2 = 5$
t_3	R1 loads/unloads and M1 machines a part	$\lambda_3 = 4$
t_4	M1 breaks down	$\lambda_4 = 0.5$
t_5	M1 is repaired	$\lambda_5 = 0.5$

162 Chapter 6. FMS Performance Analysis

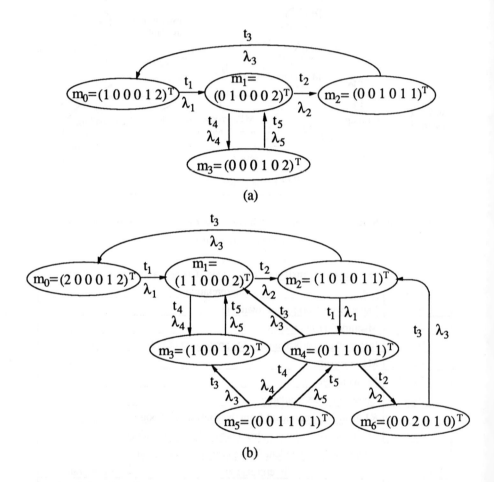

Figure 6.4 Reachability graphs and Markov chains for (a) k = 1 and (b) k = 2.

Case 2: k = 2. The total number of states q=7.

Using the same method, we obtain the below transition rate matrix for the Markov chain in Fig. 6.4(b):

6.3. Stochastic Petri Nets

$$A = \begin{pmatrix} -\lambda_1 & \lambda_1 & 0 & 0 & 0 & 0 & 0 \\ 0 & -\lambda_2-\lambda_4 & \lambda_2 & \lambda_4 & 0 & 0 & 0 \\ \lambda_3 & 0 & -\lambda_3-\lambda_1 & 0 & \lambda_1 & 0 & 0 \\ 0 & \lambda_5 & 0 & -\lambda_5 & 0 & 0 & 0 \\ 0 & \lambda_3 & 0 & 0 & -\lambda_2-\lambda_3-\lambda_4 & \lambda_4 & \lambda_2 \\ 0 & 0 & 0 & \lambda_3 & \lambda_5 & -\lambda_3-\lambda_5 & 0 \\ 0 & 0 & \lambda_3 & 0 & 0 & 0 & -\lambda_3 \end{pmatrix}$$

By solving the following equations:

$$(\pi_0, \pi_1, \pi_2, \pi_3, \pi_4, \pi_5, \pi_6)A = 0$$
$$\pi_0 + \pi_1 + \pi_2 + \pi_3 + \pi_4 + \pi_5 + \pi_6 = 1$$

we obtain the steady-state probabilities as follows:

$\pi_0 = \lambda_2\lambda_3^2\lambda_5\beta/\lambda$

$\pi_1 = \lambda_1\lambda_3^2\lambda_5(\beta+\lambda_1\lambda_3+\lambda_1\lambda_4+\lambda_1\lambda_5)/\lambda$

$\pi_2 = \lambda_1\lambda_2\lambda_3\lambda_5\beta/\lambda$

$\pi_3 = \lambda_1\lambda_3^2\lambda_4(\beta+\lambda_1\lambda_2+\lambda_1\lambda_3+\lambda_1\lambda_4+\lambda_1\lambda_5)/\lambda$

$\pi_4 = \lambda_1^2\lambda_2\lambda_3\lambda_5(\lambda_3+\lambda_5)/\lambda,$

$\pi_5 = \lambda_1^2\lambda_2\lambda_3\lambda_4\lambda_5)/\lambda$

$\pi_6 = \lambda_1^2\lambda_2^2\lambda_5(\lambda_3+\lambda_5)/\lambda))$

where $\lambda = [\lambda_2\lambda_3^2\lambda_5\beta + \lambda_1\lambda_3\lambda_5(\lambda_1+\lambda_3) + \lambda_1\lambda_2\lambda_3\lambda_5 + \lambda_1\lambda_3\lambda_4(\lambda_1+\lambda_3)]\beta + \lambda_1^2\lambda_2^2\lambda_5(\lambda_3+\lambda_5)$ and $\beta = \lambda_3\lambda_4 + (\lambda_2+\lambda_3)(\lambda_3+\lambda_5)$

Finally, we derive the performance measures as follows.

Case 1: k = 1.

Machine M1's utilization is determined by the probability that M1 is machining a raw workpiece. This corresponds to the marking m_1 at which p_2 is marked, or state probability π_1. Therefore, the expected M1's utilization is π_1. Using the rates in Table 6.5, we obtain Π = (0.05, 0.4, 0.5, 0.05). Therefore, the M1's utilization is 40%.

The completion of a product is modeled by firing transition t_3. This transition is enabled in marking m_2. Thus, the system production average rate is given by the product of the probability that t_3 is enabled and the firing rate of t_3, i.e., $\pi_2 \lambda_3$ workpieces/time unit. Since $\pi_2 = 0.5$, and $\lambda_3 = 4.0$, the throughput is 2 workpieces/time unit.

Case 2: k = 2.

Machine M1's utilization is determined by the probability that M1 is machining a raw workpiece. This corresponds to the markings m_1 and m_4 at which p_2 is marked, or state probabilities π_1 and π_4. Therefore, the expected M1's utilization is $\pi_1 + \pi_4$. Given the rates in Table 6.5, we obtain

$$\Pi = (0.006, 0.273, 0.061, 0.051, 0.264, 0.017, 0.330).$$

Since $\pi_1 = 0.273$ and $\pi_4 = 0.264$, M1's utilization is 53.7%.

The completion of a product is modeled by firing transition t_3. This transition is enabled in markings m_2, m_4, m_5, and m_6. Thus, $G_3 = \{m_3, m_4, m_5, m_6\}$, the mean number of t_3's firings in unit time and the system throughput in this case is

$$f_3 = \sum_{m_i \in G_3} \pi_i \lambda_3 = (\pi_2 + \pi_4 + \pi_5 + \pi_6)\lambda_3$$

Using the above results, we have $\pi_2 = 0.061$, $\pi_4 = 0.26$, $\pi_5 = 0.07$, $\pi_6 = 0.33$, and $\lambda_3 = 4.0$; and thus the throughput is 2.69 workpieces/time unit. ◊

The analysis of a Markov model may involve the steady-state (as shown in the example) or transient analysis (the study of the system evolution from its initial state to the steady state), or both depending on the requirements and the Markov process itself. For example, if the net is prone to deadlock, i.e., the underlying Markov process has at least an absorbing state, then the steady-state analysis makes no sense.

The manual derivation of the results for a stochastic Petri net is very tedious. Fortunately, the above process has already been automated and implemented by many research groups. Two most notable ones are GreatSPN [Chiola, 1985 and 1987] and SPNP [Dugan et al., 1985; Ciardo, 1989], and more example applications on stochastic timed Petri nets can be referred to [Ajmone Marsan et al., 1995]. The application of SPNP to an AT&T Flexible Workstation will be illustrated in Section 6.4.

6.3.3 Generalized Stochastic Petri Nets

Given a non-timed Petri net model, modeling each transition firing time with an exponentially distributed random time delay often leads to a huge Markov chain that could soon become unsolvable due to the limited memory resources in a computer as well as numerical difficulty for large size problems. Considering some transitions take relative long time and others short, Generalized Stochastic Petri Nets (GSPN) are introduced in [Ajmone Marsan, Balbo, and Conte, 1984; Ajmone Marsan et al., 1995]. A GSPN has two types of transitions, timed and immediate. A timed one has an exponentially distributed firing rate, and an immediate one has no firing delay and is used to model a logical control or activity whose delay is negligible compared with those associated with timed transitions. If only one of the enabled transitions is immediate, the immediate one is always fired before the others. If multiple immediate transitions are enabled at a marking and their firings do not disable each other, they can fire simultaneously. If they are in conflict, probability density functions on them have to be specified based on the firing frequency of these transitions. A set of immediate transitions together with its associated probability density function is called a random switch. This distribution is called a switching probability. Different markings may result in a single random switch if they always enable the same set of immediate transitions on which a single marking-dependent or independent switching distribution is defined. If the probability for an enabled immediate transition is zero, it cannot fire, and thus behaves as if it were not enabled. If the marking links to another through one or more immediate transitions in a reachability graph, it is called vanishing marking or state since the system stays at this marking for zero time delay. Such a vanishing marking can be eliminated.

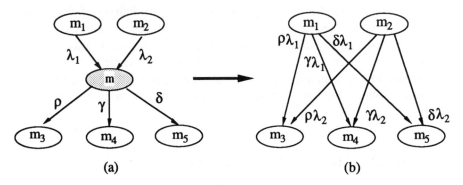

Figure 6.5 An example of eliminating a vanishing state (a) Markov chain with m and (b) its equivalent Markov chain without m.

An example shown in Fig. 6.5 is used to illustrate the elimination of a vanishing marking. Marking m is a vanishing marking in Fig. 6.5 since it has three immediate transitions linking it to other markings. The probability distribution shown in Fig. 6.5 shows that t_{1-3} have their firing probabilities ρ, γ, and δ, respectively with $\rho+\gamma+\delta=1$ and $\rho>0$, $\gamma>0$ and $\delta>0$. The exponential rates from m_1 and m_2 to m are λ_1 and λ_2. Then the rates from m_1 to m_{3-5} are $\rho\lambda_1$, $\gamma\lambda_1$, and $\delta\lambda_1$; from m_2 to m_{3-5} are $\rho\lambda_2$, $\gamma\lambda_2$, and $\delta\lambda_2$, respectively as shown in Fig. 6.5(b). All the vanishing states can be eliminated using the described fashion. Procedure 6.1 with the addition of the step to eliminate vanishing states after Step 2 can be readily applicable to analysis of a GSPN model.

6.3.4 Other Extensions to Stochastic Petri Nets

Extended stochastic Petri nets (ESPNs) are introduced in [Dugan, 1984]. In ESPN, timed transitions are classified into three types: exclusive transitions - for all markings enabling an exclusive transition t, t is the only enabled transition; competitive transitions - for all markings enabling a competitive but non-exclusive transition t, t is in conflict with all other enabled transitions; and concurrent transitions - for some marking enabling a concurrent but non-exclusive transition, there exists one or more other enabled transitions that are not in conflict with it. If 1) all concurrent transitions are associated with exponentially distributed firing times and 2) the firing policy of all competitive transitions is such that each time they are enabled, new firing delays are resampled from their corresponding distribution, then the ESPN can be converted into a semi-Markov chain regardless of the distributions on exclusive and competitive transitions. They can be solved with the analysis technique for semi-Markov chains.

Deterministic stochastic Petri nets (DSPNs) are introduced in [Ajmone Marsan and Chiola, 1986], which allow firing delays of timed transitions to be either constant or exponentially distributed random variables. A strict restriction on the deterministic transitions has to be introduced to result in amenable analysis, i.e., for concurrent transitions, all markings can enable at most one deterministic transition. The related solution methods and software packages are reported in [Linderman, 1993].

6.4 Performance Analysis of a Flexible Workstation

Before the SPN analysis on the AT&T flexible assembly system modeled in Chapter 4, we introduce the Petri net tool called SPNP developed at Duke University [Dugan

6.4. Performance Analysis of a Flexible Workstation

et al., 1985, Ciardo, 1989] and then present the evaluation results for the production cell when the number of pallets varies. SPNP's other applications in automated manufacturing systems can be seen in [Al-Jaar and Desrochers, 1990; Desrochers, 1994; Watson and Desrochers, 1991; Zhou et al., 1990; Zhou and Ma, 1993; Zhou and Leu, 1994; Zhou, and Thorniley, 1994; Zhou, 1995].

6.4.1 Overview of SPNP

Developed in the early 80's, Stochastic Petri Net Package (SPNP) has become one of the most commonly used Petri net tools for performance evaluation. The package can deal with Petri net text file inputs that define places, transitions, their input and output relations, and transitions' random firing delay times with exponential distributions. In SPNP, Petri nets with immediate transitions and priorities of enabled transitions are allowed.

First, a C file is created to describe a PN. To declare places, transitions and arcs, the following statements are used, taking p_{1-2} and t_{1-2} in Fig. 6.3 for example:

```
int k;
float lamda1;
k = input("The number of initial tokens (pallets) in p1:");
lamda1=input("The firing rate (Robot 1's loading speed) of t1:");
place("p1"); init("p1", k);
place("p2");
transition("t1"); rateval("t1", lamda1);
transition("t1"); rateval("t1", 5.0);
iarc("p1", "t1");
oarc("p2", "t1");
...
```

Suppose that t_a and t_b are two immediate transitions in conflict with fixed probabilities c and d such that $c + d = 1$, c>0 and d>0. They are declared as follows:

```
transition("ta"); probval("ta", c);
transition("tb"); probval("tb", d);
```

The results can be obtained by executing *SPNP filename*. It involves automatically generating the reachability graph of a net model. The performance indices include the probability for a place to be marked, the average number of tokens in it, probability for a transition to be enabled, and throughput of a transition. Based on such data, we can obtain resource utilization rate, average in-process inventory and production rate or throughput. For example, the system throughput in the above production cell example is the throughput of t_3 and its reciprocal is the average cycle time for a product. The utilization rate of Machine M1 is the probability of place p_2 being marked since a token in p_2 implies that M1 is processing a part. The average in-process inventory is the sum of average numbers of tokens in places p_{2-4}. The package can also be used to derive the transient solution for a stochastic Petri net. SPNP runs under several operating systems on a wide array of platforms such as VAX and SUN [Ciardo 90].

```
# include "user.h"
/* 12-20-93 This is the SPNP file called ex.c for the production line*/
int m01;
parameters() {
iopt(IOP_PR_FULL_MARK, VAL_YES);
iopt(IOP_PR_MC,VAL_YES);
iopt(IOP_PR_RGRAPH,VAL_YES);
iopt(IOP_PR_PROB,VAL_YES);
    m01 = input("Number of pallets in the system:"); }
net() {
    place("p1"); init("p1",m01); place("p2"); place("p3"); place("p4");
    place("p5"); init("p5",1); place("p6"); init("p6",4);
    trans("t1");     rateval("t1",40.0);   trans("t2");     rateval("t2",5.0);
    trans("t3");     rateval("t3",4.0);    trans("t4");     rateval("t4",0.5);
    trans("t5");     rateval("t5",0.5);
    iarc("t1","p1"); iarc("t1","p5"); oarc("t1","p2");
    iarc("t2","p2"); iarc("t2","p6"); oarc("t2","p5"); oarc("t2","p3");
    iarc("t3","p3"); oarc("t2","p6"); oarc("t3","p1");
    iarc("t4","p2"); oarc("t4","p4"); iarc("t5","p4"); oarc("t5","p2");
/*This defines the sochastic Petri net*/ }

/* the following lines should appear in all programs */
assert() {return(RES_NOERR);}
ac_init() {}
ac_reach() {fprintf(stderr,"/nThe reachability graph has been generated/n/n");}

/* User-defined output functions */
reward_type ef0() {return(rate("t3"));} /* System production rate = */
reward_type ef1() {return(mark("p2"));} /* Machine 1 utilization */

/*Output*/
ac_final() {pr_expected("System production rate = ",ef0);
        pr_expected("Machine 1 utilization = ",ef1);
        pr_std_average(); }
```

Figure 6.6 The C-file to describe the net shown in Fig. 6.3.

6.4. Performance Analysis of a Flexible Workstation

Example 6.3 (Application of SPNP to a Production Cell). First, we convert the net model shown in Fig. 6.3 of Example 6.2 into a C-file as shown in Fig. 6.6. The file is named ex.c. Note that the description of places p_{7-9} and related arcs are not shown because, again, markings of these places are redundant. First time when we issue command *spnp ex*, the computer (Sun workstation) takes relatively more time to generate its execution file. Then it requests our input for the number of pallets allowed in the system. Once we input the number, it generates its output file called ex.out. After we record the data from ex.out, we can again issue *spnp ex*. It will immediately request the number of pallets allowed in the system and generate the results in ex.out. Continuing this process, we can obtain all the data and results as shown in Fig. 6.7 when the number of pallets varies from 1 to 7. ◊

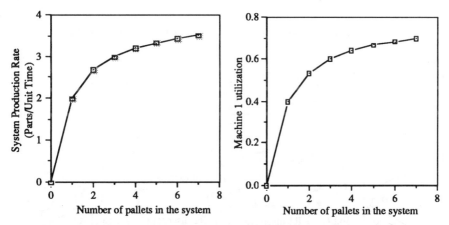

Figure 6.7 The results when the number of fixtures changes from 1 to 7.

6.4.2 Performance Evaluation of a Flexible Workstation

In this section we use the SPNP software to evaluate two control structures for the flexible assembly operation. Based on the net model of this AT&T flexible workstation as shown in Fig. 5.12, a temporal net model is obtained as shown in Fig. 6.8 in which two robot failure loops are included. For i = 1 and 2, t_i represents that Robot i malfunctions or needs maintenance, t'_i represents that R_i is being repaired or serviced, and p'_i represents that R_i is in repair or service. This model presents the mutual exclusion structures implemented by the semaphores in Manufacturing Modular Langauge - M^2L.

Another model is based on the fixed sequence on two robots' movements, i.e., Robot 1 first picks up a component and then goes to insert. Meanwhile Robot 2 goes to pick up and waits to insert until Robot 1 completes its insertion. Next when Robot 2 is performing insertion, it is again Robot 1's turn to pick up a component. Continue this process. With this fixed sequence, the control is easier to implement. The resulting control structure can be described by a marked graph as shown in Fig. 6.9. Instead of two places p_3 and p_4 in mutual exclusion structures, four places are used to fix the two robots' movements. They are:

p_3: R1 picks first if p_3 is initially marked but p_2 not;
p_4: R2 picks second if p_1 is initially marked but p_2 not;
p_5: R1 inserts first if p_5 is initially marked but p_6 not;
p_6: R2 inserts second if p_5 is initially marked but p_6 not;

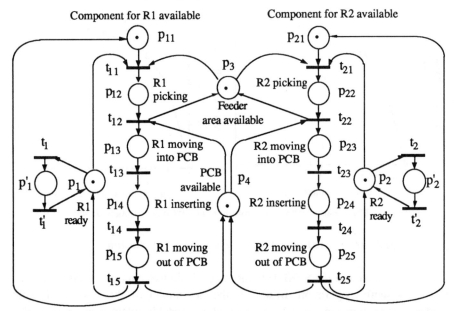

Figure 6.8. The net with mutual exclusion structure plus robot failure loops

Note that the placement of initial tokens can result in different interpretation of the places. Furthermore, their inappropriate placement may lead to incorrect and undesired behavior. For example, if we initially mark p_4 and p_6, then R2 will pick and insert a component first and R1 next, which is desired behavior. If we mark p_3 and p_6

6.4. Performance Analysis of a Flexible Workstation

initially but not p_4 and p_6, then the net contains a deadlock that is at the marking (0 1 0 0 0 1 0 1 0 0 0 1 0 0 0 0)^T right after t_{11}'s firing from the initial marking for the net structure in Fig. 6.9.

The GSPN model is obtained in Fig. 6.10 in which two robot failure loops are included. It should be noted that the net shown in Fig. 6.10 is no longer a marked graph due to the inclusion of the robot failure loops. It is a free-choice Petri net while Fig. 6.8 is an asymmetric Petri net.

We assume that all the time delays are associated with transitions instead of places in order that we can directly use the SPNP package. The firing rates assumed for the transitions in Figs. 6.8 and 6.10 are summarized in Table 6.6. Another simplification is that the time delays are all assumed to have exponentially distributed functions, except transitions t_{15} and t_{25} that are immediate. When an immediate transition is enabled, it can fire instantaneously. Suppose that α denotes the robot failure rate, a variable. Then 1-α is the robot working rate. It should be noted that the two robots in operation are supposed to have the same temporal characteristics.

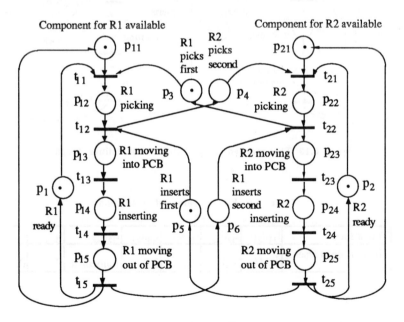

Figure 6.9 A marked graph model which models fixed sequence on two robots' movements.

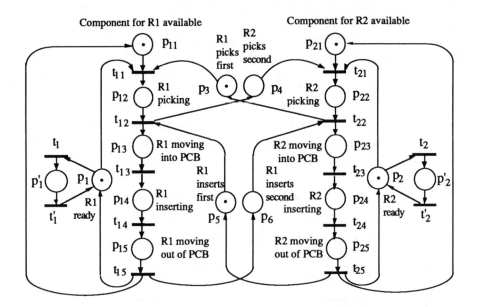

Figure 6.10 The marked graph plus robot failure loops.

Table 6.6 Firing rates of transitions.

Transitions	t_{11},t_{21}	t_{12},t_{22}	t_{13},t_{23}	t_{14},t_{24}	t_{15},t_{25}	t_1,t_2	t'_1,t'_2
Firing Rates (1/sec.)	$1-\alpha$	3.0	0.8	3.0	Immediate	α	0.1

6.4. Performance Analysis of a Flexible Workstation

The results of performance analysis are shown in Figs. 6.11-12 for the two temporal Petri net models shown in Figs. 6.8 (Control Structure 1) and 6.10 (Control Structure 2). They show the changes in system throughout and robot utilization when the robot failure rate increases from 0 to 0.10. Both the system throughput and robot utilization decrease as the robot failure rate increases. The results also show that

1. When there is no robot failure ($\alpha = 0$), the two control structures for this flexible assembly system lead to the smallest differences between their values of performance indices.
2. The net model with mutual exclusion structure (Control Structure 1) has better performance than the marked graph based control model (Control Structure 2).

The performance analysis results show that although the conflict is resolved in the marked graph model, the system performance is degraded.

Using the SPNP package or other similar analysis software, we can also derive the variations in performance measures when other parameters such as insertion time delay change [Zhou and Leu, 1991].

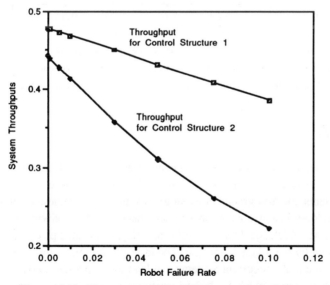

**Figure 6.11 The system throughput vs. robot failure rate.
Control Structure 1: Using semaphore or mutual exclusion (Fig. 6.8)
Control Structure 2: Fixing the sequence of two robots (Fig. 6.10).**

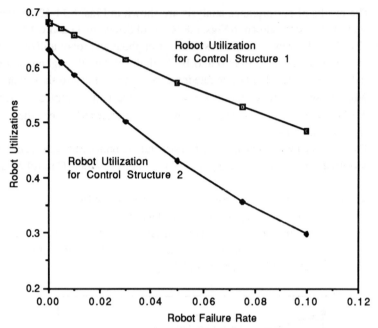

Figure 6.12 The robot utilization vs. robot failure rate
Control Structure 1: Using semaphore or mutual exclusion (Fig. 6.8)
Control Structure 2: Fixing the sequence of two robots (Fig. 6.10).

6.5 Summary

This chapter introduced the fundamental analytical techniques to analyze temporal Petri nets for system performance. For decision-free concurrent and repetitive systems, marked graphs or event graphs are a nice framework to derive a cycle time. All loops in a marked graph are enumerated and then individual loop cycle time is computed, which equals the sum of time delays associated places, transitions, and arcs divided by the token count in it. The largest loop cycle time is the system cycle time. Some real concurrent systems can lead to a Petri net model that is not exactly a marked graph model. Such Petri net models, however, may be converted into a marked graph model for the cycle time analysis. The NJIT's FMS cell is such a system. The detailed analysis of this model is performed with the help of reduction techniques.

6.5. Summary

When there are choices in a system or the time delays are better described with random time delays, stochastic Petri nets (SPN) are used to derive mean production rate, resource utilization and work-in-process. In order to apply existing analytical techniques based on Markov chains, all time delays have to be of exponential distribution. In generalized stochastic Petri nets (GSPN), immediate transitions (with zero time delays) are allowed and probability functions for conflicting immediate transitions are also introduced. Under certain conditions, some time delays can be deterministic in deterministic and stochastic Petri nets (DSPN), and arbitrary distribution in extended stochastic Petri nets (ESPN). The manual derivation of the underlying Markov or semi-Markov processes from a Petri net representation is very tedious. Fortunately, the process has been automated in many software tools including GreatSPN and SPNP. SPNP is introduced and used to study the average utilization of machines and production rate of a production cell subject to change of the number of pallets. It is also used to derive the performance of a robotic station whose robot arms are subject to failures.

Other worth-to-mention analysis technique is based on moment generating function concept. It is applicable to SPN, GSPN, DSPN, and ESPN and leads to symbolic solution [Guo et al., 1993; Zhou et al., 1993; Akella and DiCesare, 1992].

Allowing any distributions and firing transition selection policies in a Petri net lead to arbitrary stochastic Petri nets. They cannot be analytically analyzed. Thus approximation or simulation methods have to be utilized [Jungnitz and Desrochers, 1991]. The techniques to convert a non-exponentially distributed random time delay into a subnet with only exponentially distributed ones are very useful [Chen et al., 1989; Watson and Desrochers, 1991]. For Petri nets with an extremely large state space, reduction and approximation methods are needed [Ma and Zhou, 1990; Zhou and Ma, 1993].

CHAPTER 7

PETRI NET SIMULATION AND TOOLS

7.1 Introduction

Depending on the development stage of a system, the knowledge of either an approximate or exact (or both) performance may be required. For example, at the design stage, the approximate performance of the alternative design models is required in order to eliminate the alternatives which are highly unlikely to meet the performance requirements when fully developed and implemented. Analytical techniques discussed in Chapter 6 play an important role at this stage. They allow a designer to obtain the required performance measures, involving a relatively small time investment for the model construction and its solution. The selected design alternatives are then refined by increasing the level of details in order to include in the model the actual operational policies and time characteristics. As a result, the model complexity, or the presence of heuristic algorithms may prohibit the use of the analytical techniques. The discrete-event simulation is, then the only viable alternative for the performance evaluation, despite the fact that it is an expensive and time-consuming technique.

A discrete event model comprises system building blocks and a schedule or rules that govern the flow of work in an FMS. Discrete event simulation is a process through which a discrete event model mimics the behavior of a discrete-event system such as FMS event by event. Qualitative and quantitative data from this process are obtained to predict the behavior of the system and its level of performance. Simulation has two basic motives, rapid prototyping to determine the correctness of system behavior, and performance prediction. Some performance measures of interest are throughput, resource utilization, buffer capacity, yield, and effects of failures. Simulation studies are conducted using computers. The software model can capture all the dynamics and interactions of a real system. Since real manufacturing systems are expensive to build, simulation is an important means to predict performance accurately, investigate effects of parameter changes, identify bottlenecks, and choose the best design among alternatives.

This chapter focuses on the use of Petri nets as a discrete event model and the basic principles of PN simulation and explores the use of PN simulation methods and tools in FMS design. Section 7.2 introduces a generic simulation procedure. Then various models are briefly discussed in addition to Petri net models. It also discusses discrete-event simulation schemes and tools. Section 7.3 introduces timed Petri nets and token game-based simulation. Section 7.4 describes the software we have developed for Petri net simulation. Section 7.5 indicates other Petri net tools available for Petri net modeling, analysis, and discrete event simulation. The next two chapters present the simulation tools' particular application fields. Chapter 8 presents a PN simulation approach to performance evaluation of two commonly used operating paradigms: push and pull for an FMS. Chapter 9 applies the approach to performance modeling and analysis of error-prone flexible assembly systems.

7.2 Discrete Event Simulation

7.2.1 Discrete Event Simulation Procedure

In discrete event simulation, the model contains entities or objects, attributes, events, activities, and the interrelationships among them. The collection of entities and their statuses define the system state. A system state may change only at discrete points in time. These changes are driven by the event occurrences, generally asynchronous. In manufacturing systems, an entity can be a machine, a robot, material (work in progress), and a controller, with its schedule. The attributes of a machine include its operation rate, the nature of its operation, and its reliability. Event examples are the arrival of raw material, loading, unloading, the change of a tool, and the start of an operation.

The following steps are involved in discrete event system simulation as shown in Fig. 7.1 [Banks 1984, Rembold 1993]:

Goal Definition and Requirements Specification: Determine the requirements specification and simulation goal of the manufacturing system under study or design. The goals are to determine the best system among several alternatives and to investigate the system behavior and performance. An appropriate level of detail is selected to match the modeling goal. For example, it could be to optimize the performance of an individual flexible manufacturing system (FMS) cell, or to optimize the performance of a shop floor containing many cells. Informal requirements have to be converted into formal requirements specifications which are understandable to both managers and designers or analysts. Simulation environments and related languages may be used to formulate requirements although they may differ from the language used to construct the simulation model.

7.2. Discrete Event Simulation

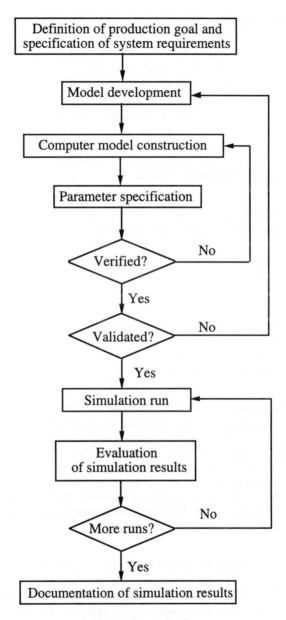

Figure 7.1 Simulation Steps

Model Development: Formulate a model which could be a Petri net or other models, e.g., a queuing model, state machine, and object-oriented model. Construction of a model can be a difficult task, requiring a modeler's understanding of the problem and modeling experience. The model has to capture the essence of the real system without excessive detail. Hierarchical models that could represent different levels of detail are required for large scale flexible manufacturing systems.

Simulation results can be presented to planners and managers for their decision making and serve the basis for design and real-time implementation of FMS. Assumptions and simplifications have to be justified. Ad hoc, bottom-up, top-down, and hybrid methods are used in modeling large manufacturing systems to formulate a hierarchical and modular Petri net model.

Computer Model Construction: Construct a computer representation of the model, e.g., program a mathematical formulation using a simulation language or general-purpose programming language, or build up a graphical model in a computer using a graphic editor and simulator in which simulation algorithms are embedded and not apparent to a user. The latter approach is used in all Petri net based methods.

Data Collection: Analyze all parameters involved in the model and specifications set, and collect data based on accounting data, experience, or from lower-level simulation runs. The quality of data has direct impact on the results obtained from the simulation. Thus care has to be taken during this process.

Simulation Run: Run the computer model or simulator to verify its correctness. The model verification process ensures that the computer simulation models the system properly. If the logic structure, inputs, and outputs are correctly represented in the computer model, verification is completed. Simple cases and common sense are used during this process. Various animation techniques can facilitate this process. Validation ensures that the model is an accurate representation of a real system. Thus the model can be used as a substitute for the actual system for the prediction of performance with a high level of confidence. The use of Petri nets and other graphical tools are very helpful in this step.

Evaluation of Simulation Results: Obtain and evaluate simulation results. In this step statistical data analysis techniques may be needed to analyze the system simulation results and to validate them. Generally, the performances of two or more alternative system designs are compared. The system is simulated over a range of key operational parameters and thus the optimal settings can be determined.

7.2. Discrete Event Simulation

Documentation: Document the input data, methods, simulation tools, computational time, and results. The results should be presented in graphs for patterns and trends over the parameters of interest. Histograms and bar charts are often used to present the simulation data pictorially.

The use of formal description techniques in the construction of simulation models allows for the verification of these models, with respect to their behavioral properties, using strict mathematical techniques. For example, Petri nets allow for the construction of simulation models, as well as their formal verification through analysis methods introduced in Chapter 4. Two advantages of this approach are:

1. The correctness, if present, or absence of behavioral properties of the simulation model can be established at the early stages of the simulation system development. This results in increasing confidence in the validity of the simulation and computer models; and
2. As a result of incorrect or missing behavioral properties identified during the test runs, the cost involved in redevelopment of the simulation and computer models can be reduced with this approach.

The simulation results need be compared with real-world data, if available, for existing systems, or with results produced by theoretical models. For complex systems, obtaining analytical solutions for models involving all facets of the system functionality is often computationally prohibitive. Queuing theory and models are used, in most cases, for obtaining the reference results. The use of queuing theory, as any other technique that yields results representing steady state operation of the modeled systems, poses an additional problem. The effects of the initial bias have to be eliminated. These effects are due to the transient period, which follows the start of the simulation run, and influenced by the nature of the system simulated as well as the simulation environment.

A successful simulation of a factory system design needs close cooperation with the factory personnel to ensure the model's correctness and to make the results acceptable to them [Rembold 93]. With the increasing use of powerful PCs and workstations, graphic simulation and animation of manufacturing systems is possible. Its operation can then be viewed in real time and interactive simulation can be conducted. This type of simulation is most useful for debugging, fine manipulation, and material flow observance [Rembold 93]. A graphic simulation example of a robotic pick-place cell with Silicon Graphic's IGRIP is given in [Zhou et al., 1994]. The tools for highly accurate animation include SLAMSYSTEM, Cinema/SIMAN, and SIMFACTORY [Globle, 1980; Law and Kelton, 1991; Miles et al., 1988; Pritsker 1986].

182   Chapter 7. Petri Net Simulation and Tools

7.2.2 Simulation Models, Schemes and Tools

There are three basic types of discrete event system models: queuing models, state-transition models whose representatives are state machines and Petri nets, and object-oriented models.

Queuing Models

A queuing model of a manufacturing system can be obtained if one treats resources such as machines and robots as servers, storage areas and conveyor systems as buffers (queues), and jobs or parts as customers. When strict, perhaps unrealistic and simplifying, assumptions are made for the model, analytic results can be derived for performance evaluation. When these assumptions cannot hold, simulation or approximation methods [Kamath, 1994] can be used to derive the desired results. Note that a heuristic can be applied to complex queuing models.

State-Transition Models

Typical state-transition models include state machines and Petri nets as discussed in Chapter 4. When a system is complex (contains a multiplicity of machines, AGVs, etc.), the graphical state machine representation may not be practical or possible due to the exponential state-space explosion problem. Note that the number of system states is exponentially related to the number of system components for many real industrial systems. For example, addition of a flexible AGV to a system can lead to a drastic increase in the number of its states since it can move to different locations, each of which represents a state. When multiple such AGVs are introduced, one can quickly see the exponential growth in the number of system states.

When time delays are introduced in the simulation, the event concept can be extended to include time. There are two ways to associate time delays with state machine models, as discussed below using a single machine example as shown in Fig. 7.2. It has three states, *idle*, *busy*, and *down* and four events, **start**, **complete**, **breakdown**, and **finish repair**.

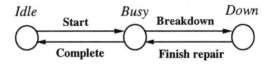

Figure 7.2 A single machine example

7.2. Discrete Event Simulation

Associate time with the events. For example, suppose it takes random time to trigger event **Start** at state *idle*, **Complete** and **Breakdown** at *busy*, and **Repair** at state down. At states *idle* and *down*, there are unique events which will occur, eventually. At state *busy*, depending upon the underlying physical processes, a policy has to be established to select an event and calculate its time of occurrence. An example is given below to illustrate this.

Suppose that at time k, the system enters state *busy* from *idle* or *down*. Based on the sampling of the random variables, **Complete** and **Breakdown** should happen after time intervals x and y (called event lifetime), respectively. If x<y, then **Complete** happens at time k + x and the system enters state idle. If y<x, **Breakdown** takes place at time k + y, implying that the part is processed for y time units and the machine breaks down and needs repair. Once it is repaired after taking time, z, it enters state *busy* again. If the process is interruptible, then the simulator does not need to resample the random variable associated with event **Complete**. Its lifetime becomes x - y. Then x - y and y' are compared where y' is a newly generated value from the random variable associated with event **Breakdown**. The event with the smaller lifetime is selected to take place. If the process is not interruptible, the random variable associated with event **Complete** is resampled. When only exponential random variables are involved, these two policies produce the same results due to the memoryless property of exponential random variables, i.e., the knowledge of event age is irrelevant in determining the event residual lifetime. Otherwise, the policies produce different results.

Associate time and probability with the states. A random or fixed time delay is assigned to each state, implying that the delay starting at the moment the system enters the state has to pass before triggering any further events. Then if there are multiple events possible at the state, probabilities are used to select an event and determine a new state. In the example shown in Fig. 7.2, it may take exponential random time to stay in state *idle*, a uniform random times to stay in state *busy*, and a uniform random time in state *down*. At state *busy*, probabilities Pr and 1−Pr ($1 \geq Pr \geq 0$) are assigned to events **Complete** and **Breakdown**, respectively.

Both timing techniques lead to Markov chains if all the time delays are exponentially distributed random variables. Thus analytical solutions can readily be obtained by solving a set of algebraic equations associated with Markov chains.

As we discussed in Chapter 6, we can associate time delays with transitions, places, and arcs to obtain a timed Petri net model.

High-level Petri Nets

By distinguishing among tokens in an ordinary Petri net and modifying Petri net execution rules, one creates a "colored" Petri net model. Note that color is a euphemism for a set of attributes which may be assigned to a token, and which may change as the token moves through the net. Such a model is useful in describing a large factory with many similar subsystems consisting of machines, robots, AGVs, and parts. It is a more compact model than its ordinary Petri net counterpart; the complexity is hidden in the token attributes, functions associated with arcs, and execution rules. CPN/Design is a commercially available tool for modeling and simulation using colored Petri nets [http://www.daimi.aau/dk/designCPN]. Other high-level Petri nets include Predicate-Transition and object-oriented nets associated with their development environments [Kordon and Kaim, 1994].

Object-oriented Models

A manufacturing system can be viewed as a collection of objects with rules that govern their dynamics and interactions to generate desired objects (product). The objects can be represented graphically as simplified images, icons, and stored in a database as members of a class of similar objects, sharing common properties. Example objects include robots, machines, machine tools, conveyors, etc. A simulation model can be built up by retrieving such objects from a database to represent a system model. Such models can lead to high reuse of simulation components. It should be noted that each object's dynamic behavior in such a model could be represented by other techniques, e.g., state machines, Petri nets, or high-level Petri nets.

Comparisons

Queuing models offer mathematically concise models allowing to develop analytic solutions for first-cut quick decision making under certain restrictions. It can be difficult to map a manufacturing system in terms of only queues, servers, and customers. The development of hierarchical models with different levels of detail is not straightforward.

State-transition models are easier to relate to a manufacturing system, allowing facile validation of a simulation model. When a system is complex, their visualization capability is diminished. Fortunately, they can be used for hierarchical modeling to facilitate system understanding and perform efficient simulation. Within state-transition models, state machines tend to be too complex for realistic industrial systems. They have also limited modeling capability in representing explicitly concurrent manufacturing operations. Petri nets offer a more powerful tool to handle discrete event dynamics and are more compact, in general. Colored Petri nets offer a more compact model than Petri nets at the sacrifice of clarity in some cases. Strict assumptions about the nature of their time delays and operation rules render all these models solvable analytically, i.e. no simulation is needed if computationally feasible.

7.2. Discrete Event Simulation

Object-oriented models arose from the application of object-oriented technology to modeling and simulation of discrete event systems. Their basic elements are objects whose models, as well as their interactions, can be the ones discussed above. By using inheritance, data abstraction, dynamic binding and encapsulation, the objects can be reused for construction of new models. Object-oriented technology will be discussed in more detail when it is combined with Petri nets to develop FMS control software in Chapter 12.

7.2.3 Simulation Schemes and Tools

System simulation can be performed using either an event scheduling or a process-oriented scheme [Banks and Carson, 1984; Cassandras, 1993; Law and Kelton, 1991]. In the former, events are listed and scheduled to take place in temporal order, and the system state advances accordingly. The latter scheme better fits resource-contention environments, where each entity (e.g. a part) undergoes a process as it flows through a discrete event system. A process can be treated as a sequence of functions or events triggered by an entity. Events are a fundamental concept for both schemes.

In the event scheduling scheme, the simulation process works as follows [Cassandras, 1993]:

1. Set the initial values of the related variables including the initial feasible events and their scheduled times of occurrence in a so-called SCHEDULED EVENT LIST. The events are ordered by the system on a smallest-scheduled-time-first basis.
2. Remove the first entry from the SCHEDULED EVENT LIST.
3. Advance the clock time to the removed event's scheduled time.
4. Update the system state due to the occurrence of the event.
5. Delete any entries corresponding to infeasible events in the new state from SCHEDULED EVENT LIST.
6. Add new feasible events and their scheduled occurrence times (current time plus the time obtained from their corresponding random variable generators or set to the fixed delay for deterministic cases);
7. Reorder the updated SCHEDULED EVENT LIST based on smallest-scheduled-time-first and go to Step 2.

The event scheduling scheme is suitable for all the discrete event simulation models. When it is applied to a special model, e.g., Petri nets, the algorithm can be easily modified to fit to it. The algorithm for Petri net simulation is used as an example as follows:

1. Form the set of all newly enabled transitions in the current state. Note that the initial state must contain tokens enabling at least one transition to get started. For each newly enabled transition with a random time delay, generate a delay based on its distribution and associate that delay with the transition. Associate appropriate fixed delays with the other newly enabled transitions.
2. Remove the required number of tokens from the input places of the above set. Add the newly enabled transitions to the to-be-fired list, in temporal order. Transitions on this list are not candidates for Step 1 above.
3. If the to-be-fired list is empty, stop (deadlock).
4. Determine the set of to-be-fired transitions with the minimum delay and advance the clock by that amount. Fire them by depositing the required number of tokens in output places, and removing them from the list. For each transition, log the information regarding its enabled time and increment the number of its firings.
5. Increment the simulation step counter. If the number of preset simulation steps is reached or if a state-dependent stopping criterion is met, stop. Otherwise, repeat Step 1.

The above algorithm, or equivalent, is built into timed Petri net based graphical simulation environments. With a "Step Mode" the execution of a Petri net can be visualized on a workstation screen. Interaction with a user can also be easily conducted.

Almost all the simulators using state-transition models are built based on an event scheduling scheme. General-purpose programming languages such as FORTRAN, C, and C++, and many simulation languages such as GASP, SIMAN, SIMSCRIPT, SLAM and SIMFACTORY can be used to develop event scheduling simulation.

In an FMS, a product is produced after it undergoes many steps. The sequence of these steps is called a manufacturing process. A process may contain conditional logic so that the steps taken by a part may depend on its own attributes and the current state of the system [Askin and Standridge, 1993]. A model may contain multiple processes with a variety of types of entities. Since manufacturing resources may be shared among processes, an entity may have to compete and wait for service. Viewing discrete event systems as a number of entities undergoing processes results in the process-oriented simulation scheme [Cassandras, 1993]. Its detailed discussions and simple examples can be seen in [Cassandras, 1993; Zhou et al., 1997]. Relevant simulation languages and tools include GPSS, SIMAN, SIMSCRIPT, and SLAM [Cassandras, 1993, Law and Kelton, 1991, Pritsker, 1986, Russell, 1983].

7.3 Timed Petri Nets and Token Game Simulation

A PN approach to analyzing a system consists of two parts: modeling with PN and analysis of the PNM by either analytical methods [Narahari and Viswandaham, 1985] or simulation [Dubois, 1989; Valvanis, 1990]. Simulating a PN using the token is one of the several methods to conduct PN simulation. This section first presents the concepts of timed PNs (TPNs) and token game and then discusses the software package developed to simulate the TPN models using the token game.

For the performance evaluation of the system, timed PNs (TPNs) are used in the subsequent chapters. *Instantaneous description (ID)* (Venkatesh et al., 1992) defines the state of a TPN and is a four tuple ID = (m,F,Q,A) where:

1. m is a marking;
2. F is a binary selector function, F: T $\rightarrow$ {0,1}. If F(t) = 1, t is enabled, otherwise disabled;
3. Q: T $\rightarrow$ R$^+$, is remaining firing time function where R$^+$ is the set of all positive integers.
   If Q(t) = q, there is q amount of time to complete firing t. Q is a cumulatively decreasing time function;
4. A: T $\rightarrow$ R$^+$, is active time function. If A(t) = q', t is said to be active for q' amount of time. A is a cumulatively increasing time function.

ID is useful for the quantitative and behavioral analysis of the system. In order to illustrate the timed PN and token game simulation concepts, an assembly cell shown in Figure 7.3 (a) is modeled. This cell contains two part feeders (PF1 and PF2) and a robot (R). Part feeders supply the parts required for assembly and the robot does all assembly operations. PF2 feeds a part to the empty assembly area automatically. The system operates as follows: 1) to start a cycle, robot (R) and parts must be available, 2) R transfers a part from PF1 to assembly area and starts assembly, the timing duration for this is one time unit, and 3) R assembles the parts, and transfers the finished product to the output buffer, the time is two units. Figure 7.3 (b) shows the PN model for this assembly cell where the number on the right hand side of a transition indicates the transition's firing time. Various TPN models (referred to as PNMs) are shown in Figs. 7.3 (c-f) in a chronological order.

At time zero (see Figure 7.3 (c)), initial marking models the initial condition; *F-function* indicates that t_1 is enabled; *Q-function* models that there is one time unit necessary to finish t_1's firing; *A-function* means that there is no transition active. After one time unit (i.e. after assembly starts), the change in marking is shown in Figure 7.3 (d) indicating that R is not ready because it is doing assembly, PF1 and PF2 contains one and two parts respectively, assembly is in progress, and there is no finished product. *F-function* shows that t_2 is enabled, *Q-function* shows that t_2 needs two time units to complete its firing, and *A-function*

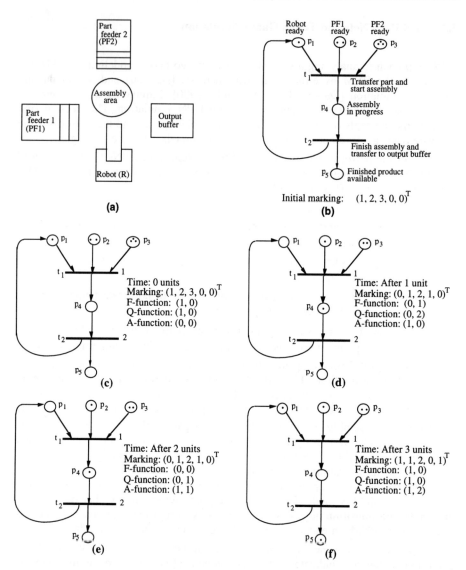

Figure 7.3 Timed Petri net model of the assembly cell.
(a) A simple assembly cell, (b) Petri net model of the assembly cell, (c) Before firing of transition t_1, (d) After firing of transition t_1, (e) During firing of transition t_2, and (f) After firing of transition t_2.

7.4. Description of the Software Package

shows that t_1 is active for one time unit. Similar explanations can be given for PN models in Figs. 7.3. (e-f). The evolution of ID is shown in Table 7.1.

Table 7.1 ID evolution for PN models shown in Figs. 7.3(c) - (f)

Time Units	Marking	F-function	Q-function	A-function
0	$(1, 2, 3, 0, 0)^T$	(1, 0)	(1, 0)	(0, 0)
1	$(0, 1, 2, 1, 0)^T$	(0, 1)	(0, 2)	(1, 0)
2	$(0, 1, 2, 1, 0)^T$	(0, 0)	(0, 1)	(1, 1)
3	$(1, 1, 2, 0, 1)^T$	(1, 0)	(1, 0)	(1, 2)

7.4 Description of the Software Package to Simulate Petri Nets

A software package is developed based on the token game to conduct PN simulation [Venkatesh *et al.*, 1992; Venkatesh and Ilyas 1995]. This package has been used in simulating PN models presented in the subsequent chapters. Figure 7.4 describes the algorithm used in developing the software package. The algorithm is essentially identical to the one presented in Section 7.2.3. The algorithm presented in Fig. 7.4 is implemented using a software package. The software package runs on IBM compatible PCs. The biggest PN model that can be simulated using this package contains one hundred places and one hundred transitions. However, larger nets can be controlled when this software package is installed in a workstation with more memory. At the beginning of the simulation run, the user is prompted to enter for the simulation time in time units (which can be seconds, minutes, etc.). The software package has five major modules whose functionality is briefly described next.

Read_Petri_net

This module reads an input file that specifies the structure of the PNM. The structure of the PNM means the input and output connectivity between places and transitions including the weights on the connecting arcs, initial marking, and transition timing duration. The timing duration for transitions is given in time units. An input file may also contain information that can be used to give priorities that are used to resolve conflicts during simulation.

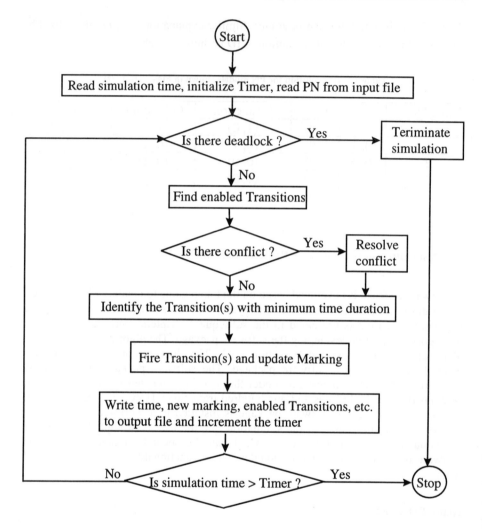

Figure 7.4 Algorithm used for developing software package using Token Game.

Enabled_transitions

This module scans the whole PNM at each unit of time and finds the transitions that are ready to fire. Thus, this module generates a list of enabled transitions that are enabled to fire with respect to real-time.

7.4. Description of the Software Package

This list of enabled transitions is defined as an F-function in the terminology of Timed PNs as explained in the earlier section. The F-function is the input for modules "Conflict_resolver", "New_marking", and "Main".

Conflict_resolver

A conflict in a PN model results when an element is shared by two other elements of the system (e.g. a single robot serving two machines that demand service at the same time). In such cases, this module detects the conflicts and resolves it based on the criteria given by a user. During simulation this module determines the transitions that are in conflict and stops the program execution until the conflict is resolved. Once the conflict is resolved, the program execution is resumed. Conflicts can be resolved by giving priorities to transitions.

Priorities can either be interactively given during simulation or be entered in the input file. This is because in some PN models conflicts can be determined even before starting simulation. In such cases by giving priorities to resolve conflicts in the input file, user intervention during simulation can be avoided. However, in complex PN models conflicts can be identified only during simulation time. In such cases, this module displays the time of the conflict and the transitions in conflict on the screen. It then prompts the user to enter priorities to resolve conflicts. Then, based on the information given by the user it resolves the conflict and resumes simulation.

Minimum_time

This module scans the whole PN model and detects the transition(s) with minimum timing duration to fire. As there can be more than one transition with minimum time, the outputs from this module are both the number of transitions with minimum time and their identity.

New_marking

Whenever a transition fires, this module updates the marking of PN model accordingly by using the PN firing rules as explained in the earlier chapters.

Main

This is the main program that coordinates the functioning of above modules and generates a status report of the system elements. The report is stored in an output file which is updated whenever a transition is fired in the PNM. Whenever an event occurs in the system, the output file is appended by the time at which the event occurred, marking of the PNM, F-function, Q-function, and A-function. As an example to demonstrate the format of input and output files for this software package, Tables 7.2 and 7.3 present the input and output files corresponding to the PN model shown in Fig. 7.3 (a) respectively.

Table 7.2 Input file corresponding to the PN model shown in Fig. 7.3 (a)

Number of places: 5
Number of transitions: 2
Petri net connectivity representation:
                     Input place ID (column #)
Transition ID: 1 1 1 0 0
(row #)   0 0 0 1 0
                     Output place ID (column #)
Transition ID: 0 0 0 1 0
(row #)   1 0 0 0 1
Transition duration vector (time units): 1 2
Initial Marking: 1 2 3

7.5 Other CASE Tools to Simulate and Analyze PNs

There are several CASE tools available to simulate and analyze various kinds of PNs. Note that there are several types of PNs - place/transition PNs, timed-PNs, colored PNs, predicate/transition PNs, extended PNs, etc. These CASE tools are developed either by universities or Industrial firms. A list of CASE tools available can be obtained from World Wide Web from the following address:

http://www.daimi.aau.dk/PetriNets/tools/crim/header.html

Another such address which gives PN CASE tool information is:

http://twilight.dsi.unimi.it/Users/Tesi/trompede/petri.html

7.5. Other Case Tools to Simulate and Analyze PNS

Table 7.3 Output file corresponding to the PN model shown in Fig. 7.3 (a)

Time: 0
Marking: 1, 2, 3, 0, 0
F-function: 1, 0
Q-function: 1, 0
A-function: 0, 0

Time: 1
Marking: 0, 1, 2, 1, 0
F-function: 0, 1
Q-function: 0, 2
A-function: 1, 0

Time: 2
Marking: 0, 1, 2, 1, 0
F-function: 0, 0
Q-function: 0, 1
A-function: 1, 1

Time: 3
Marking: 1, 1, 2, 0, 1
F-function: 1, 0
Q-function: 1, 0
A-function: 1, 2

Note that these lists are by no means the complete lists of available PN CASE tools world-wide. There may be other CASE tools that are not listed in the above URLs. In these URLs, the lists of CASE tools present the following information for each CASE tool:

1. Name of the CASE tool;
2. Types of PNs (Timed Petri nets, High Level PNs, Colored PNs, etc.) the CASE tool supports; the amount of fee to obtain the CASE tool;
3. Components of the CASE tool (support for graphical editing, token

game animation, simple performance analysis, invariant analysis, etc.);
4. Environments in which the CASE tool runs (Sun, MS Windows, Macintosh, etc.); Brief description of the CASE tool;
5. Contact information to get more information about the CASE tool; and
6. Documentation as well as other published references about the tool.

The second URL mentioned above also has a nice *tool search engine* that can be used to retrieve the CASE tools by supplying the such search criteria as:
- PN type;
- Extensions to PNs (time extensions, stochastic extensions, probabilities, priorities, queues, inhibitory arcs, object-orientation);
- Tool facilities (graphical editor, simulation, hierarchy/modularity features;
- Structural analysis, net transformation/reduction, performance analysis, code generation/prototyping); and
- Environment.

7.6 Summary

In this chapter, the procedure and steps involved for discrete event simulation of discrete event systems using Petri nets are discussed. The use of various modeling techniques such as queuing models, state-transition models, high-level Petri nets, object-oriented models for simulation are briefly explained. The event scheduling scheme and the corresponding Petri net algorithm used for the simulation process are described. The fundamentals on timed Petri nets and token game simulation of the Petri net models for performance the evaluation of FMSs are presented. For demonstrating these concepts, an example of the timed Petri net model of the simple assembly cell is developed and simulated. A software package that is used to simulate Petri net models is discussed. Also, the algorithm used for developing the software package that is used to implement the token game simulation of Petri net model is described along with the sample input and output files. Information that helps to explore other CASE tools to simulate and analyze Petri nets is also presented.

CHAPTER 8

PERFORMANCE EVALUATION OF PUSH AND PULL PARADIGMS IN FLEXIBLE AUTOMATION

8.1 Introduction

Even though Petri Nets (PNs) have been successfully applied to various problems related to FMSs [Murata, 1989; Silva and Valette, 1990; DiCesare and Desrochers, 1991; Cecil *et al.*, 1992; Zhou and DiCesare, 1993], there are still some areas where the power of PNs has not been exploited. For example, the application of PNs to study the performance of push and pull paradigms is not reported in the available literature. Such studies not only help to select between push and pull paradigms but also aid to widen the application of PNs for modeling and simulation of flexible manufacturing systems.

This chapter shows PNs as a powerful tool to investigate the problem often encountered in manufacturing systems management, namely comparing the performance of a factory automated system operating under *push* and *pull* paradigms [Venkatesh 1995; Venkatesh *et al.*, 1996]. The difficulty in solving this type of problem is compounded by many parameters such as processing times at work cells, number of automated guided vehicles and their routings, lot sizes, and setup times. The PN method to solve such a problem is illustrated by considering a manufacturing system. Its PN models are formulated and then analyzed to compare the performance of system with *push* and *pull* paradigms. The results show that for the particular system and operational parameters, the *push* paradigm outperforms the *pull* one. Manufacturing systems consist of machines, robots, and automated guided vehicles (AGVs) that are controlled by computers. Numerous asynchronous concurrent actions involved in these systems make their analysis difficult. This chapter focuses on the modeling and analysis issues related to manufacturing systems that are operated according to either *push* or *pull* paradigms.

The primary goals of a manufacturing system are to minimize the work in process inventory (WIP), and maximize the system utilization and production rate [Stecke and Solberg, 1981; Suri, 1984; Chang *et al.*, 1985; and Maione *et al.*, 1986].

In the manufacturing arena, it is well known that low WIP can be achieved by implementing just-in-time (JIT) which is based on a *pull* production paradigm. However, implementation of JIT may lead to lower system utilization [Monden, 1981]. In contrast to the *pull* paradigm, *push* production paradigm results in maximum system utilization and output rate, at the expense of higher WIP [Monden, 1989; Sarkar and Fitzsimmons, 1989]

Even though it is popular to divide production control systems into push and pull systems, there are no generally accepted definitions for these systems [Spearman *et al.*, 1990]. However, several authors distinguished push and pull paradigms [Kimura and Terada, 1981; Schonberger, 1983; Spearman *et al.*, 1990]. For example, Kimura and Terada (1981) discussed them as mechanisms of production orders. According to them, a push system's production and inventory control is based on the forecast value. In pull system, a certain amount of inventory is held at each stage and its replenishment is ordered by the succeeding process at the rate it has been consumed. Also, there are several ideas to implement push and pull systems [Schonberger, 1983; Spearman *et al.*, 1990; Lee, 1987]. Schonberger (1983) described the implementation of a pull paradigm using kanban techniques, and stated that push was simply a schedule based system which could be implemented using a master production schedule that was exploded by a computer into detailed schedules.

Spearman *et al.*, (1990) stated that push systems scheduled the start of jobs whereas pull system authorized the production. They reported that pull systems were not applicable to production environments which were controlled by job orders. Motivated by this, they described a pull-based production system called CONWIP and presented its advantages over push and some pull systems. Lee (1987) presented a parametric appraisal of a JIT system and concluded that, unlike the traditional push method, raising the pull demand in a JIT system did not ensure a high process utilization level. The details of differences between the *push* and *pull* ones were also discussed in detail by Sarkar and Fitzsimmons (1989).

To minimize WIP and maximize the system utilization and production rate of manufacturing systems simultaneously, it is often difficult to select between *push* and *pull* paradigms. This is because for certain system configurations and operational parameters, *push* may outperform *pull* and vice versa. To make the best decision, detailed design and performance analysis of a system has to be performed. This chapter shows PNs to serve for this purpose.

Another reason that adds to the difficulty of the present problem is the inadequacy of research on JIT in the area of flexible manufacturing systems (FMSs). In other words, even though there is much reported research in both the areas of JIT (in conventional manufacturing systems) and FMSs individually, there are only a few research studies on implementing JIT in FMSs [Venkatesh *et al.*, 1992]. Huang (1984) emphasized the uncertain consequences of the system output, the sensitivity in performance of a JIT production system to lot size, the impact of setup time reduction on the efficiency, and the effect of variability of processing

8.2. Push and Pull and Their PN Modeling

times on the JIT implementation and its subsequent performance. Sarkar and Fitzsimmons (1989) investigated the effects of variations in processing times on the performance of conventional *push* and *pull* systems. Yim and Linn (1993) analyzed an FMS with Petri nets and concluded that there was no significant difference in production rates between push and pull based AGV-dispatching rules for a busy FMS. However, none of the earlier papers have presented any methodology for studying the effects of variations of lot sizes and processing times on the performance of *push* and *pull* systems. This chapter presents PNs as a suitable tool to answer these questions.

The primary goal of this chapter is to show PNs as a powerful tool in investigating difficult problems in production management arena. It investigates a complex problem often faced in the management of manufacturing systems, i.e., comparing their performance functioning under *push* and *pull* paradigms. The specific objectives of this chapter are:

1. To present a PN approach to address a typical operations management problem stated earlier,
2. To formulate PN models (PNMs) considering important parameters in a manufacturing system example such as processing times at work cells, number of AGVs, routings of AGVs and their travel times among work cells, production and moving lot sizes, machine setup times, and machine loading/unloading times, and
3. To present the detailed analysis of PNMs to design and compare the performance of manufacturing systems operating under either *push* or *pull* paradigms.

8.2 Push and Pull and their PN Modeling

In this chapter, we regard push and pull paradigms as operational paradigms. In a push system, a preceding machine produces parts without waiting for a request from the succeeding machine. On the other hand, in a pull system a preceding machine produces parts only after it receives a request from the succeeding machine. In other words, in a pull system parts are produced only when there is a demand. The difference in modeling such systems is demonstrated through the following example.

Suppose that the machining sequence for a part is Machine 1 (M1) → Machine 2 (M2). Figure 8.1 presents PN models for push and pull paradigms. It is assumed that raw material is always available. This is modeled by depositing 'k' tokens in the place (p_2) modeling raw material. As illustrated in Fig. 8.1 (a), in a push system the buffer size (modeled by the marking p_5) may be increased when the processing time of M2 is greater than that of M1. However, in a pull system which

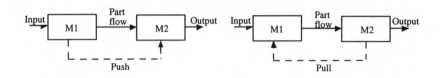

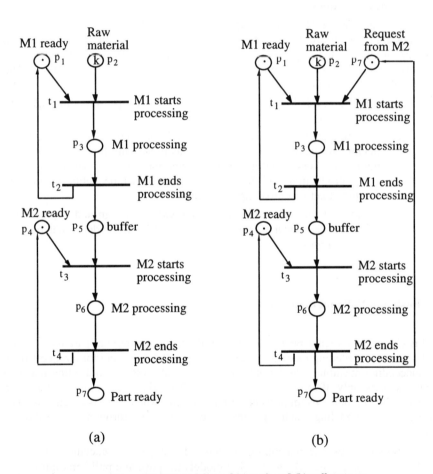

Figure 8.1 PN modeling of (a) push and (b) pull systems.

8.2. Push and Pull and Their PN Modeling

is modeled in Fig. 8.1 (b), the buffer size is always one. This is because M1 starts processing a part only when it obtains a request from M2 (modeled by firing of t_4). Since, at the beginning of production, M1 can not start processing without the request from M2, p_7 models the initial condition of request from M2.

While the above PN models illustrate the important difference between push and pull systems, there are several detailed issues such as lot sizes, processing times, setup times, and number of AGVs that have to be also included when a real-life production system is modeled. The detailed conventions of PN modeling useful for the design, performance evaluation, and monitoring of manufacturing systems are shown in Table 8.1. The subsequent sections present an application example which illustrates these modeling concepts.

Table 8.1 Conventions of Petri net modeling.

Implementation issue	PN modeling
Moving lot size or production lot size	Weight of the directed arc modeling the function of moving or production kanban
Routing of an AGV	Firing sequence of transitions
Number of AGVs for a transportation task, work stations, and pallets	Initial marking in the corresponding places modeling the availability of AGVs, workstations, and pallets
Values of in-process inventories	Number of tokens deposited in the places modeling input and output buffers of workstations
Utilization times of workstations, robots, AGVs	Active time durations of transitions modeling the activities of work stations, robots, AGVs
Production volume	Number of tokens deposited in the place modeling "counter for production volume"
Dynamic system state useful for design, performance analysis, and monitoring and control of the system	Marking of the PN model giving the token distributions, F, Q, and A functions giving the status of transitions with respect to time
Setup time	Firing duration of the transition modeling setup activity
Conveyance time	Firing duration of the transition modeling conveyance activity

8.3 Application Illustration

In this section a manufacturing system is modeled considering both push and pull paradigms.

8.3.1 System Configuration and Assumptions

A manufacturing system example as shown in Fig. 8.2 contains four work-cells (WCs) and an assembly shop (AS). Each WC consists of a machining cell (MC), an assembly cell (AC) and a robot (R). AGV(s) are present in the system to transport parts and subassemblies among work cells and the assembly shop (AS). The track layout of AGVs is also shown in Fig. 8.2.

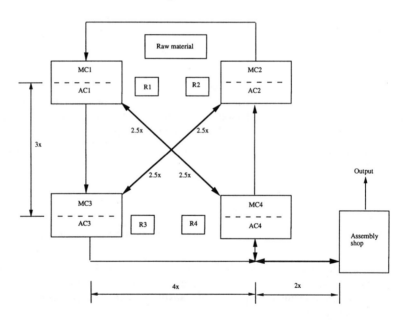

Figure 8.2 Layout of the manufacturing system investigated.

The conveyance times among work centers and assembly shop to produce products 1 and 2 (PR1 and PR2) are shown in Table 8.2. In the conveyance matrix show in Table 8.2, the rows and columns represent the sources and destinations of an AGV respectively. Each number shown in the cell of the table model the corresponding conveyance time units from the source to destination. For example, the conveyance time from WC1 to WC3 is 3 time units, from WC3 to AS is 6 time

8.3. Application Illustration

units. The entry, "--" shown in the table indicates that there is no need to convey material between the corresponding source and destination.

Table 8.2 Conveyance time matrix in the system for production of PR1 and PR2 (time units).

	WC1	WC2	WC3	WC4	AS
WC1	--	4	3	--	--
WC2	--	--	5	5	--
WC3	5	--	--	4	6
WC4	5	3	5	--	--
AS	7	--	2	--	--

There is a need for producing two product varieties, PR1 and PR2, both in the quantity of seventeen each (the demand of seventeen products is chosen arbitrarily). PR1 requires parts A, B, C, and D in the quantities of 2, 1, 3, and 2, respectively. These parts are machined by MC1, MC2, MC3, and MC4. PR2 requires subassemblies SA, SB, SC, and SD in the quantities of 2, 1, 3, and 2, assembled by AC1, AC2, AC3, and AC4 respectively. The machining/assembly sequence of above parts and subassemblies is same as shown in Table 8.3 along with the corresponding times. The values in parentheses in Table 8.3 show the assembly times for PR2.

Table 8.3 Processing times and the sequence of parts in the system (time units).

Part	Sequence	MC1 (AC1)	MC2 (AC2)	MC3 (AC3)	MC4 (AC4)
A (SA)	MC1--MC3 (AC1--AC3)	35 (8)	---	42 (6)	---
B (SB)	MC1--MC3 (AC1--AC3)	46 (10)	---	38 (9)	---
C (SC)	MC2--MC4 (AC2--AC4)	---	26 (7)	---	34 (11)
D (SD)	MC2--MC4 (AC2--AC4)	---	30 (9)	---	40 (12)

Since PR1 requires only machining operations, the manufacturing system becomes a flexible manufacturing system (FMS) when it is producing PR1. Similarly, the system is a flexible assembly system (FAS) when it is producing PR2. This schema allows for generally comparing the performance of FMS and

FAS when both of them function under a similar system configuration but with different processing times. The assumptions made about this system are as follows:
1. The outer path of the AGV track is unidirectional and the inner one is bi-directional.
2. The setup time for any MC or AC and loading/unloading a MC or AC is one time unit and sequence independent.
3. If there are more than one AGV in the system, the delays due to traffic are negligible.

8.3.2 Statement of the problem

The manufacturing system considered performs in four different system configurations, namely, FMS and FAS with the *pull* paradigm and FMS and FAS with the *push* one. In each configuration there are several parameters to be studied. The parameters whose impact on the system performance has to be investigated in this study are production lot sizes (PLSs), number of unique transportation tasks (N) which decides the moving lot size (MLS), and the number of AGVs assigned for each unique different transportation task (n). For example, PLS of Part A is defined as the number of A parts that are processed on MC1 before changing the setup of MC1 to produce another different part (here Part B). MLS is defined as the number of parts that an AGV carries corresponding to each unique transportation task. The routing of an AGV and the variation of MLS with respect to N are shown in Table 8.4. The goals are to minimize work-in-process-inventory and maximize system utilization and production rate. To achieve these goals, PNMs of the system for the four different configurations have to be formulated and the influence of the parameters on the goals has to be investigated. Also, the optimum values of N, n, and PLSs have to be determined for each system configuration. Finally, the performance of system under *push* and *pull* paradigms has to be compared.

8.3.3 PNM Formulation and Analysis

PNM for *pull paradigm*

To illustrate the concepts of PN modeling, Fig. 8.3 shows the PNM for MC1 and MC3 operating with the *pull* paradigm to produce parts A and B corresponding to PR1. In Fig. 8.3, dotted arcs model the *pull* paradigm. The modeling of production of parts C and D by MC2 and MC4 is similar to production of parts A and B by MC1 and MC3 except that the PLSs for C and D are different. However, the variation of PLSs can be easily modeled by changing certain arc weights in the

8.3. Application Illustration

Table 8.4 Variation of moving lot size with respect to number of AGVs and assigned tasks.

Number of unique different transportation tasks (N)	Task of AGV and its routing	Moving lot size (MLS)
1	AGV1: Starts from work cell 1 (WC1), delivers parts from WC1 and WC2 to WC3 and WC4, delivers parts to assembly shop (AS) and returns to WC1 Routing: WC1-WC2-WC3-WC4-WC3-AS-WC1	$2 + 1 + 3 + 2 = 8$
2	AGV1: Starts from WC1, delivers parts from WC1 and WC2 to WC3 and WC4 and returns to WC1 Routing: WC1-WC2-WC3-WC4-WC1	$2 + 1 + 3 + 2 = 8$
	AGV2: Starts from WC4, delivers parts from WC4 and WC3 to AS and returns to WC4 Routing: WC4-WC3-AS-WC4	$2 + 1 + 3 + 2 = 8$
3	AGV1: Starts from WC1 and delivers parts to WC3 and returns to WC1 Routing: WC1-WC3-WC1	$2 + 1 = 3$
	AGV2: Starts from WC2, delivers parts to WC4 and returns to WC2 Routing: WC2-WC4-WC2	$3 + 2 = 5$
	AGV3: Starts from WC4, delivers parts from WC4 and WC3 to AS shop and returns to WC4 Routing: WC4-WC3-AS-WC4	$3 + 2 + 2 + 1 = 8$

Chapter 8. Performance Evaluation

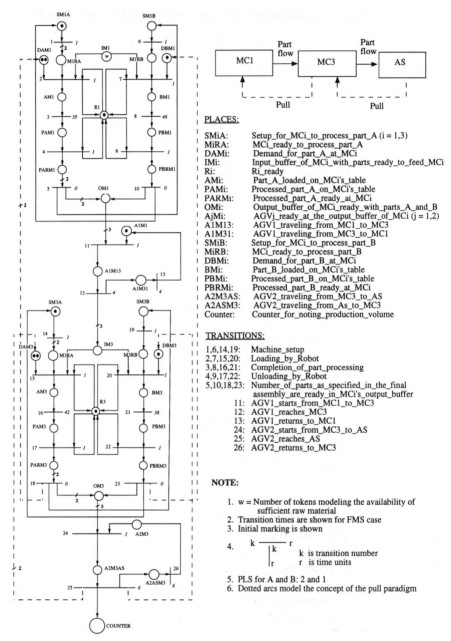

Figure 8.3 PNM for production of parts A and B under pull paradigm.

8.3. Application Illustration

PNM as illustrated in Fig. 8.3 and discussed in the next paragraphs. Hence, the discussion on modeling production of parts C and D is omitted here. Also, to simplify the modeling process, in Fig. 8.3, it is assumed that there is AGV1 which transfers parts from MC1 to MC3 and AGV2 which transfers parts from MC3 to assembly shop. Since time duration of operations in the system are deterministic, transitions are associated with deterministic firing durations (from Tables 8.3 and 8.4).

Processing of Part A

It is assumed that initially Part A has to be loaded on to MC1. This requires MC1 to be setup to process A modeled by depositing a token in place *setup_for_MC1_to_process_part_A*. Setup time is one unit of time and associated with t_1 modeling the activity *machine_setup*. After one time unit the completion of setup is modeled by firing t_1 and depositing the number of tokens equal to the production lot size (PLS) of part A in place *MC1_ready_to_process_part_A*. PLS is modeled as a weight on the output arc from t_1 to place *MC1_ready_to_process_part_A*. Since MC1 is functioning with the *pull* paradigm, MC1 cannot start processing until there is a demand for Part A from MC3. Hence, MC1 has to wait till MC3 requests MC1 to process Part A. The request for Part A at MC1 arrives when MC3 completes processing of Part A. The processing of Part A at MC3 is modeled by t_{18}, *number_of_parts_as_specified_in_final_assembly_are_ready_in_MC2's_output_buffer*.

Hence, the request for MC1 to process Part A from MC3 is modeled by an output arc from t_{18} to place *demand_for_part_A_on_MC1*. However, before the system begins production at MC1, i.e., before the processing of Part A by MC3 (firing t_{18}), demand for Part A should be present. This is modeled by depositing a certain number of tokens in place *demand_for_part_A_on_MC1*. The number of tokens in this place is controlled by production lot size (PLS) of Part A. Since, the PLS for Part A is 2, the initial marking of place *demand_for_part_A_on_MC1* is also two. Once the production starts, to continuously model the *pull* from MC3 to MC1, the output arc from t_{18} to place *demand_for_part_A_on_MC1* is drawn with weight two (PLS for Part A). Whenever MC3 completes processing it fires t_{18} and sends a request for MC1 to produce two parts of A. The raw material before MC1 is modeled by depositing sufficient number of tokens in place *input_buffer_of_MC1_with_parts_ready_to_feed_MC1* as its initial marking. Now processing of Part A on MC1 can start since there is demand for Part A, MC1 is ready to process Part A, and raw material is available. But, since robots are used to load the machining cells, MC1 starts processing as soon as Robot 1 (R1) loads the raw material. This is modeled by t_2, *loading_by_robot*.

After R1 completes loading, Part A is loaded onto the MC1's table. This is modeled by firing t_2 and depositing a token in place *part_A_loaded_on_ MC1's_table*. The completion of processing Part A on MC1 is modeled by firing t_3, *completion_of_part_processing* and putting a token in place *processed_part_A_ on_MC1's_table*. The unloading operation of the processed Part A by R1 is modeled by firing t_4, *unloading_by_robot* and depositing a token in place *processed_part_A_is_ready_at_MC1*. MC1 continues to produce another Part A since there is still one token left in place *demand_for_part_A_on_MC1*. When MC1 finishes processing of second Part A, there would be two tokens present in place *processed_part_A_is_ready_at_MC1*. Hence, two processed parts of A as specified in the final assembly (for the production of PR1, two parts of A are required according to the bill of materials of PR1) are available at the output buffer of MC1. This is modeled by t_5 and its input and output arcs with weight 2. Thus, firing t_5 takes two tokens from place *processed_part_A_ready_at_MC1* and deposits two tokens to place *output_buffer_of_MC1_ready_with_parts_ A_and_B*. The state changes in the system can be simulated and visualized by token movements in the PNM.

Processing of Part B

Upon the completion of processing two parts of A, MC1 starts to process Part B. This requires changing the setup of MC1 which is modeled by an output arc from t_5 to place *setup_for_MC1_to_process_part_A*. The modeling methodology for processing Part B is similar to that of Part A. Once Part B is processed, MC1 has to produce again Part A and the change-over for the setup is modeled by an output arc from t_{10} to place *setup_for_MC1_to_process_part_A*. After the completion of processing Part B, the output buffer of MC1 contains two parts of A and one part of B. This is modeled by firing t_{10} and depositing a token in place *output_buffer_of_MC1_ready_with_parts_A_and_B*. Now, these parts have to be transferred to MC3 for further processing. AGV1 is used to unload the parts from the output buffer of MC1 and transport to MC3. The presence of AGV1 at MC1, is modeled by depositing a token in place *AGV1_ready_at_MC1*.

Since AGV1 cannot move till the output buffer of MC1 contains two parts of A and one part of B, the input arc for t_{11} modeling *AGV1_unloads_and_ starts_at_MC1* is labeled with weight three (summation of the PLSs of parts A and B). As soon as place *output_buffer_of_MC1_ready_with_parts_A_and_B* contains three tokens, AGV1 starts conveying parts from MC1 to MC3. This is modeled by firing t_{11} and depositing a token in place *AGV1_travelling_from_ MC_to_MC3*. After its conveyance time between MC1 and MC3, AGV1 reaches MC3 and unloads the parts at the input buffer of MC3. This is modeled by firing t_{12} and

8.3. Application Illustration

putting three tokens in the place *input_buffer_of_MC3_with_parts_ready_to_feed_MC3*.

The modeling methodology for processing parts A and B by MC3 is similar to MC1 except that the demand for parts at MC3 is generated by the assembly shop and AGV3 is used to transfer parts between MC3 and the assembly shop (AS). Observe that PNMs model the functions of production and moving kanbans in JIT. For example, in the PNM shown in Fig. 8.4, only after MC3 finishes processing Part A (t_{18}), it gives a signal (moving kanban function modeled by the output arc from transition to the place DAM1) to MC1 to start processing. Then t_2 gives a signal (production kanban function) to MC1 to start production.

PNM for *push* paradigm

Fig. 8.4 shows the PNMs for MC1 and MC3 operating with push paradigm for parts A and B. In Fig. 8.4, dotted arcs model the *push* paradigm. The modeling methodology for *push* paradigm is similar to that of the *pull* one except the way tokens are deposited in places *demand_for_part_A_on_MC1* and *demand_for_part_B_on_MC1*. The same PNM used for the *pull* paradigm (shown in Fig. 8.3) is slightly modified to analyze the system operating with the *push* paradigm as shown in Fig. 8.4. Notice that in Fig. 8.4 the dotted arcs represent the only change compared to Fig. 8.3. That is, output arcs from t_{18} and t_{23} in Fig. 8.3 are removed and attached at t_{11} in Fig. 8.4. Also, output arcs from t_{25} in Fig. 8.3 are removed and attached at t_{24}. In the *push* paradigm MC1 produces parts without waiting for a request from a downstream machine. Hence, as soon as AGV1 leaves MC1, MC1 starts processing. This is modeled by each output arc from t_{11} to place *demand_for_part_A_on_MC1* and *demand_for_part_B_on_MC1*.

Similarly, MC3 produces parts without waiting for a request from AS. Hence, as soon as AGV2 leaves MC3, MC3 starts processing which is modeled by two output arcs from t_{24} as shown in Fig. 8.4.

The PNM for the whole manufacturing system is obtained by duplicating the similar modeling methodology to MC2 and MC4. The material/part transportation among work-cells is unique depending on the value of N and hence as N changes the routing of AGV changes as specified in Table 8.4, which in turn slightly changes the PN modeling of material transfer among work-cells in the system. The number of AGVs for each unique transportation task, n, is modeled by varying the initial marking of places, *AGV_ready_at_the_output_buffer_of_MC*. To avoid redundancy, only the PNM of the system functioning as FMS under the *pull* paradigm with the values of both N and n equal to one is shown in Fig. 8.5. The interpretations of places and transitions are shown in Table 8.5.

208   Chapter 8. Performance Evaluation

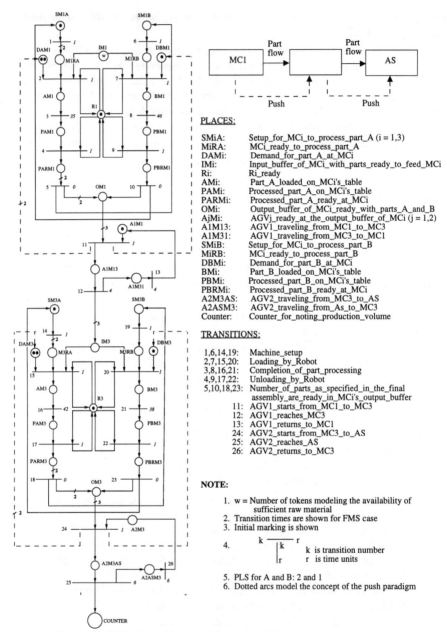

Figure 8.4 PNM for production of parts A and B under push paradigm.

8.3. Application Illustration

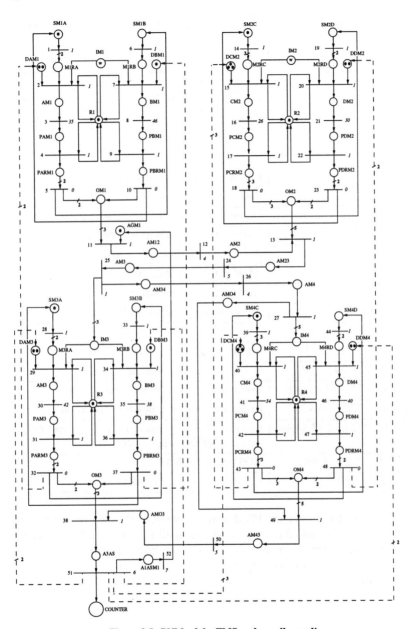

Figure 8.5. PNM of the FMS under pull paradigm.

NOTE:
1. Transition times are for FMS case
2. Initial marking is shown
3. $\begin{array}{c} k \overline{} r \\ \big| k \quad k \text{ is transition number} \\ \big| r \quad r \text{ is time units} \end{array}$
4. PLS for A,B,C&D : 2,1,3 & 2
5. $N = 1, n = 1$

Table 8.5 Explanation of typical places and transitions in PNMs shown in Figs. 8.5 and 8.6.

Place	Explanation
DAM1	Demand for Part A on MC1
IM1	Input buffer of MC1 with parts ready to feed MC1
SM1A	Setup for MC1 to process Part A
M1RA	MC1 ready to process Part A
AM1	Part A loaded on MC1's table
PAM1	Part A is being unloaded by R1
PARM1	Part A is ready at the MC1
OM1	Output buffer of MC1 ready with parts A and B
AGM1	AGV at the output buffer of MC1
AM12	AGV traveling from MC1 to MC2
AM1	AGV at the input buffer of MC1
AMO3	AGV at the output buffer of MC3
A1ASM1	AGV1 traveling from AS to MC1
Transition	
1,6,14,19,28, 33,39,44	Signal for machine setup
2,7,15,20,29,34,40,45	Robot finishes the loading operation
3,8,16,21,30,35,41,46	Completion of part processing
4,9,17,22,31,36,42,47	Robot finishes the unloading operation
5,10,18,23,32,37,43,48	Number of parts as specified in final assembly are ready in MC's output buffer
11,13,25,27,38,49	AGV starts from MC
12,24,26,50,51,52	AGV reaches its destination

Figure 8.6 shows the PNM corresponding to the system functioning with the *push* paradigm. Initial marking is shown in all PNMs. The PNMs used for FMS can also be used when the system functions as FAS. In such a PNM, the only difference is that the machining times associated with transitions are replaced by assembly times.

8.3. Application Illustration

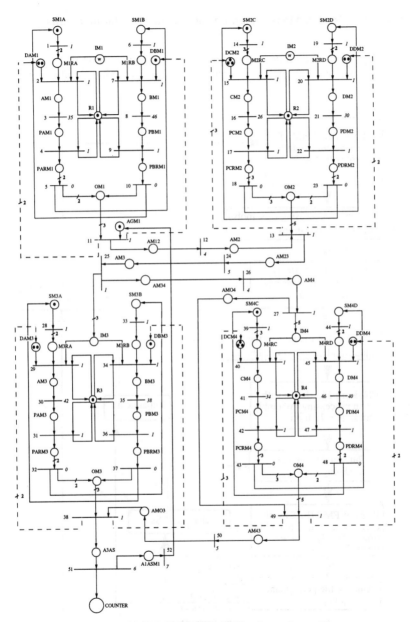

Figure 8.6. PNM of the FMS under push paradigm.

NOTE:
1. Transition times are for FMS case
2. Initial marking is shown
3. $\begin{array}{c} k \\ \mid k \\ \mid r \end{array}$ $\begin{array}{l} k \text{ is transition number} \\ r \text{ is time units} \end{array}$
4. PLS for A,B,C&D : 2,1,3 & 2
5. N = 1, n = 1

8.4 Procedure for PN Modeling and Analysis and Simulation results

Figure 8.7 presents the procedure for PN modeling and analysis that is adopted to evaluate the performance pull paradigm. The same procedure is also applied for push paradigm.

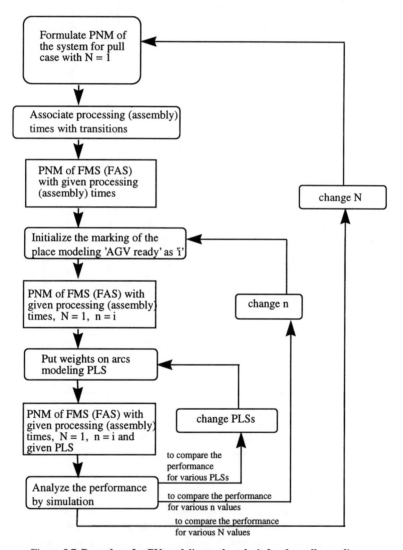

Figure 8.7 Procedure for PN modeling and analysis for the pull paradigm.

8.4. Procedure for PN Modeling and Analysis

From Fig. 8.7 it can be inferred that a PNM formulated for N = 1 can be slightly modified to study the performance of push and pull performance for both FMS and FAS cases corresponding to various values of N, n, and PLSs. This shows the reusability of the PNM obtained for a given value of N.

Simulation results

The software package developed in Venkatesh *et al.*, (1992) and presented in Chapter 7 is used for the quantitative analysis of PNMs. For each configuration of the system there exist twelve different combinations of the parameters, PLSs, N, and n. This is because there are three different values of N, two different values of n, and two different combinations of PLSs. The explanation for the various values of N is already discussed earlier (refer to Table 8.4). The results corresponding to two values of n only are presented because during simulation it is found that when n takes a value more than two, there is no improvement in system performance. The details of two combinations of PLSs are explained below.

There are two different combinations of PLSs for parts A, B, C, and D namely, 2, 1, 3, 2 and 1, 1, 1, and 1. In the first combination parts (subassemblies) A (SA), B (SB), C (SC), and D (SD) are produced in the lot sizes equal to their exact requirement as in the bill of materials of PR1 (PR2). In the second case, these are produced in unit lot sizes. In other words, the first combination corresponds to the loading sequences, part A → part A → part B on MC1 and part C → part C → part C → part D → part D on MC2. The second combination corresponds to part A → part B on MC1 and part C → part D on MC2. It is clear that in the former case, the setup time required to produce one finished product is less compared to the latter. However, the work in process inventory in the former may be more compared to the latter. Further, PLSs affect the utilization of machines, robots, and AGVs. The combination of the N, n, and PLSs, which results in the lowest minimum work-in-process inventory, the highest feasible utilization, and the highest production rate, is regarded as the optimal solution set for the system. In the following paragraphs, the results corresponding to four different configurations of the system are discussed.

8.4.1 FMS with the pull Paradigm

Table 8.6 shows the system performance when the system functions as an FMS with the *pull* paradigm. Consider PLSs of 2,1,3, and 2. In this case **N = 1** and **n = 1** is the solution set. This is because with an increase in either **N** or **n** or both, production rate, average machine cell utilization (AMU), and average robot utilization (ARU) are slightly increased, but average vehicle utilization (AVU) is significantly decreased. Now, consider the PLSs of 1,1,1, and 1. In this case, due to

Table 8.6 System functioning as FMS with the pull paradigm.

N	n	π	PR	AMU	AVU	ARU	BS1	BS2
PLS for A, B, C, and D: 2, 1, 3, and 2									
1	1	9898	0.00343	54.07	11.83	1.55	3	5
	2	9898	0.00343	54.07	5.92	1.55	3	5
2	1	9898	0.00343	54.07	5.95	1.55	3	5
	2	9898	0.00343	54.07	2.98	1.55	3	5
3	1	9708	0.00350	55.13	3.80	1.58	3	5
	2	9708	0.00350	55.13	1.90	1.58	3	5
PLS for A,B,C and D: 1, 1, 1, and 1									
1	1	5446	0.00624	51.15	21.50	1.43	3	5
	2	5446	0.00624	51.15	10.75	1.43	3	5
2	1	5446	0.00624	51.15	10.80	1.43	3	5
	2	5446	0.00624	51.15	5.40	1.43	3	5
3	1	5175	0.00657	53.89	11.12	1.49	3	5
	2	5175	0.00657	53.89	5.56	1.49	3	5

Legend:

- N   Number of unique different transportation tasks
- n   Number of AGVs assigned for each unique transportation task
- π   Time units required for production of 17 products of each PR1 and PR2
- PR   Production rate (Total number of products produced / Production time units = $34/\pi$)
- AMU   Average machine cell utilization in %
- AVU   Average AGV utilization in %
- ARU   Average robot utilization in %
- BS1,BS2 Maximum input buffer sizes at MC3 and MC4 respectively

the same reasons stated above, **N = 1** and **n = 1** is the solution set. After observing the system performance corresponding to the above two solution sets, it can be

8.4. Procedure for PN Modeling and Analysis

stated that the solution set corresponding to PLSs 1, 1, 1, and 1 results in the solution set. This is because with a decrease in the PLSs from 2, 1, 3, and 2 to 1, 1, 1, and 1, production rate is improved by 82%; AVU is increased by 82%; AMU and ARU are not much affected. Hence, when the system functions with the *pull* paradigm, PLSs of 1,1,1, and 1 with $N = 1$ and $n = 1$ is the optimal solution set.

8.4.2 FAS with the pull Paradigm

Table 8.7 shows the system performance when the system functions as an FAS with the *pull* paradigm.

Table 8.7 System functioning as FAS with the pull paradigm (the same legend as Table 8.6's).

N	n	π	PR	AAU	AVU	ARU	BS1	BS2
PLS for A, B, C, and D: 2,1,3,and 2								
1	1	3717	0.00915	36.95	31.61	4.15	3	5
	2	3495	0.00973	39.40	16.81	4.40	3	5
2	1	3495	0.00973	39.29	16.88	4.38	3	5
	2	3495	0.00973	39.29	8.44	4.38	3	5
3	1	3305	0.01030	41.52	11.18	4.66	3	5
	2	3305	0.01030	41.52	5.59	4.66	3	5
PLS for A,B,C and D: 1, 1, 1, and 1								
1	1	2409	0.01410	28.75	48.61	3.19	3	5
	2	1854	0.01830	37.35	31.74	4.22	3	5
2	1	1854	0.01830	37.35	31.88	4.15	3	5
	2	1854	0.01830	37.35	15.94	4.15	3	5
3	1	1657	0.02050	41.80	34.72	4.65	3	5
	2	1657	0.02050	41.80	17.36	4.65	3	5

Here again, when PLSs are 2, 1, 3 and 2, **N = 1** and **n = 1** is a solution set due to the same reasons listed above. But in the case of PLSs - 1, 1, 1, and 1, when **N = 1**, an increase in **n** from 1 to 2 resulted in the following: production rate is increased by 30%; AAU and ARU are increased by 30% and 32%, respectively; and AVU is decreased by 34.71%. With a further increase in either **N** or **n** or both, production rate, ARU, and average assembly cell utilization (AAU) are not much affected. Hence, **N = 1** and **n = 2** is a solution set for PLS values of 1,1,1, and 1. Now, it can be concluded that with a decrease in PLSs from 2, 1, 3, and 2 to 1, 1, 1, and 1, the production rate is improved by 100%; AAU, AVU, and ARU were slightly increased. Here PLSs of 1,1,1, and 1 with **N = 1** and **n = 2** is the optimal solution.

8.4.3 FMS with the push paradigm

Table 8.8 System functioning as FMS with the push paradigm (the same legend as Table 8.6's).

N	n	π	PR	AMU	AVU	ARU	BS1	BS2
PLS for A, B, C, and D: 2, 1, 3, and 2									
1	1	8566	0.00397	62.63	13.67	1.80	3	5
	2	7234	0.00470	74.56	16.37	2.18	3	5
2	1	7234	0.00470	85.11	9.65	2.49	36	60
	2	7234	0.00470	85.51	4.84	2.50	36	60
3	1	7229	0.00470	88.46	6.33	3.52	66	60
	2	7229	0.00470	88.46	3.16	2.73	66	60
PLS for A,B,C and D: 1, 1, 1, and 1									
1	1	4410	0.00771	63.46	26.55	1.75	3	5
	2	3485	0.00975	80.51	33.97	2.24	3	5
2	1	3337	0.01020	84.20	17.89	2.34	3	5
	2	3337	0.01020	84.20	8.95	2.34	3	5
3	1	3325	0.01020	91.73	12.46	2.60	3	32
	2	3325	0.01020	92.29	6.23	2.61	3	32

8.4. Procedure for PN Modeling and Analysis

Table 8.8 shows the system performance of the FMS functioning with the *push* paradigm. In this case, the solution set when PLSs are 2,1,3, and 2 is $N = 1$ and $n = 2$. This is because an increase in **n** from 1 to 2 (without changing **N**), improves the production rate by 19%; AMU by 19%; AVU by 20%; and ARU by 21%. Beyond this, with an increase in either **N** or **n** or both, production rate is not affected; AVU is reduced significantly; ARU, AMU increased slightly and buffer sizes increased significantly.

In the case of PLSs - 1,1,1, and 1, when $N = 1$, with an increase in **n** from 1 to 2, the production rate is increased by 26%; AMU by 27%; AVU by 28%; and ARU by 28%. Again due to the same reasons stated above, $N = 1$ and $n = 2$ is the solution set. Based on the system performance corresponding to the above two solution sets, it can be concluded that with a decrease in PLSs from 2, 1, 3, and 2 to 1, 1, 1, and 1 production rate and AVU are improved by 107% and 107% respectively. Also, there are slight increases in AMU and ARU but buffer sizes do not change. Hence, when the system functions as an FMS with the *push* paradigm, PLSs of 1,1,1, and 1 with $N = 1$ and $n = 2$ is the optimal solution set.

8.4.4 FAS with the push paradigm

Table 8.9 shows the system performance of FAS functioning with the *push* paradigm. In this case, the solution set when PLSs are 2,1,3, and 2 is $N = 1$ and $n = 2$. This is because an increase in **n** from 1 to 2 (without changing **N**), results in an increase in the production rate by 55%; an increase in AAU and ARU by 58% and 41% respectively; and a decrease in AVU by 21%. Beyond this, with an increase in either **N** or **n** or both, production rate is not affected. AVU shows decreasing trend; AAU and ARU are slightly increased; and buffer sizes are significantly increased. In the case of PLSs - 1,1,1, and 1, also when $N = 1$, with an increase in **n** from 1 to 2, production rate, AAU and ARU are increased by 91%, 94%, and 95% respectively; but AVU decreased by 48%. Again, due to the same reasons stated above, $N = 1$ and $n = 2$ is the solution set. Based on the system performance corresponding to the above two solution sets, it can be concluded that with a decrease in PLSs from 2, 1, 3, and 2 to 1, 1, 1, and 1, production rate and AVU are improved by 89% and 96% respectively; AAU and ARU are very slightly decreased and buffer sizes remains same. Hence, when the system functions as an FAS with the *push* paradigm, PLSs of 1,1,1, and 1, $N = 1$ and $n = 2$ is the optimal solution set.

8.4.5 Simulation Results

From Tables 8.6 and 8.7 when FMS and FAS function with the *pull* paradigm, the buffer sizes are minimum and constant irrespective of the values of PLS, N, and n.

Table 8.9 System functioning as FAS with the push paradigm (the same legend as Table 8.6's).

N	n	π	PR	AAU	AVU	ARU	BS1	BS2
PLS for A, B, C, and D: 2, 1, 3, and 2									
1	1	3717	0.00915	37.00	31.50	4.66	3	5
	2	2385	0.01420	58.49	24.82	6.59	3	5
2	1	2385	0.01420	69.18	31.11	7.93	51	82
	2	2385	0.01420	69.46	15.60	7.96	51	82
3	1	2380	0.01420	75.63	21.94	11.65	122	82
	2	2380	0.01420	75.63	21.94	11.65	122	82
PLS for A,B,C and D: 1,1,1, and 1									
1	1	2409	0.01410	28.75	93.82	3.19	3	5
	2	1262	0.02690	55.74	48.61	6.22	3	5
2	1	1077	0.03150	72.45	62.91	8.01	20	17
	2	1077	0.03150	72.45	31.45	8.01	20	17
3	1	1065	0.03190	74.70	40.53	8.41	19	27
	2	1065	0.03190	74.70	20.27	8.41	19	27

However, with the *push* paradigm from Tables 8.8 and 8.9, they increase with either PLS and/or N. For a given PLS combination with constant N, n does not affect the buffer sizes. This observation confirms to the general notion that higher lot sizes result in higher buffer sizes. When this system functions under the *pull* paradigm, the buffer sizes are same in both FMS and FAS. However, under the *push* paradigm when N is greater than one, the buffer sizes in case of FAS are larger compared to the FMS case because assembly times in FAS are less than processing times in FMS. Hence as the processing times vary, buffer sizes also vary in case of the *push* paradigm for higher values of N. The optimal solution sets for FMS and FAS functioning with *push* and *pull* paradigms are shown in Table 8.10.

8.4. Procedure for PN Modeling and Analysis

Table 8.10 Solution sets for FMS and FAS (the same legend as Table 8.6's).

N = 1 and PLS = 1,1,1, and 1, Demand: 17 products of each PR1 and PR2,
PR (Total production volume / Production time units) = (34 / π)

Operation Paradigm	π	PR	n	AMU or AAU	AVU	ARU	BS1	BS2
FMS case								
PUSH	3485	0.0097	2	80.51	33.8	2.24	3	5
PULL	5446	0.0062	1	51.15	21.5	1.42	3	5
FAS case								
PUSH	1262	0.0269	2	55.74	48.6	6.22	3	5
PULL	1854	0.0183	2	37.35	31.7	4.22	3	5

It is concluded that for both FMS and FAS, the optimal solution value of **N = 1** is same for both *push* and *pull* paradigms. In the case of FAS, **n = 2** is the best solution for both *push* and *pull* paradigms. But, in the case of FMS, the solution values of **n** are different for *push* and *pull* paradigms. In other words, the *push* paradigm in FMS requires two AGVs, but the *pull* paradigm requires only one AGV.

Figure 8.8 shows the performance of FMS and FAS for solution sets listed in Table 8.10. In both cases, the *push* paradigm yields better performance. This is because, in case of FMS (FAS), *push* paradigm results in an increase in production rate, AMU (AAU), AVU, and ARU by 56 (47)%; 57 (49)%; 58 (53)% and 58 (47)% respectively and buffer sizes remain same as shown in Table 8.10. Hence, for the particular system considered, the *push* paradigm helps achieve the goals of implementing JIT in flexible factory systems. Therefore, before adopting either *pull* or *push* as operational strategy for a particular system, it is important to evaluate thoroughly the system performance for both cases with an aim not only to reduce inventories but also to increase system production rate and resource utilization.

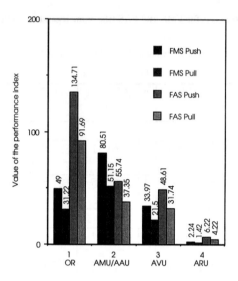

Figure 8.8 Performance of the system with different configurations.

Legend: OR   PR * 5,000 (finished products/production time units)
            AMU   Average machine cell utilization (%)
            AAU   Average assembly cell utilization (%)
            ARU   Average robot utilization

8.5 Summary

The performance of a particular FMS was evaluated using timed Petri nets by changing different operational parameters in the system operating under *push* and *pull* paradigms. The performance criteria are the buffer sizes, production rate, utilization of machines, AGVs, and robots. The configuration that results in the minimum buffer sizes, maximum system utilization and production rate is considered as the optimal solution. The manufacturing system considered is investigated as both FMS and FAS, by changing the production lot sizes, the number of unique transportation tasks (which decides moving lot sizes), and the number of AGVs for each transportation task. To achieve this, PNMs of the system are formulated and analyzed quantitatively through a PN driven simulation package.

8.5. Summary

The analysis shows that in both cases of FMS and FAS, the *push* paradigm performed better than the *pull* one for this system. This is because unlike the general notion that only *pull* paradigm results in minimum WIP, *push* may result not only in minimum WIP but also maximum system utilization and production rate for certain configurations and operation parameters.

The results may change if some of the system parameters such as processing and/or assembly times change. From the simulation results it can be observed that the utilization of system elements are generally low. This may be due to the parameter settings of machining and conveyance times. In other words, by changing these values, the system utilization can be increased. The same PNMs can be used by associating the new time values with the corresponding transitions modeling such activities. It is concluded that 1) before adopting either *push* or *pull* paradigm, the system should be evaluated with goals to reduce inventories and increase system utilization and production rate, and 2) PNs can be a very useful tool to perform such evaluation.

From the simulation results, many inferences are useful for the operations managers, system designers, manufacturing, industrial and software engineers. For example, the results show that average robot utilization is very low in all the cases studied in this chapter. This is because there is a dedicated robot for each work cell. To increase robot utilization, the possibility of sharing a common robot between two work cells should be studied further. However, scheduling of robot movements should be done to prevent impairing the performance of other system elements. This issue is another interesting topic for research in manufacturing systems [Venkatesh et al., 1992]. PNs can be used also to generate the supervisory control code of the systems as illustrated for manufacturing systems [Zhou et al., 1992; Srinivasan and Jafari, 1991]. The further enhancement of this work is to implement the discrete controller based on the control logic embedded in the PNMs as shown in the Chapter 10. Hybrid push-pull paradigms can be studied by changing the dotted output arc (in Fig. 8.5) from t_{32} to t_{29}. In this chapter, during the simulation of the system breakdowns are assumed not to occur. However, in the real-life production operations breakdowns do occur at shop-floor level. Hence, a class of PNs that are useful for detailed breakdown modeling and simulation are discussed in the next chapter.

CHAPTER 9

AUGMENTED-TIMED PETRI NETS FOR MODELING BREAKDOWN HANDLING

9.1 Introduction

Flexible manufacturing and assembly systems consist of machines, robots, and automated guided vehicles, aiming to meet the dynamically changing needs of the market. Numerous asynchronous concurrent actions involved in these systems make their analysis difficult. Breakdowns of the system components further complicate the investigation of the issues related to their design, performance optimization, and control. Furthermore, detailed breakdown handling helps implement the real-time shop-floor controller as reported by Zhou and DiCesare (1989). Even though there are several types of PNs available for discrete control [Crockett *et al.*, 1987, Murata *et al.*, 1986; Stefano and Mirabella, 1991; Valette *et al.*, 1983] there is a need for extending PNs to model the operations in factory floors effectively. More specifically, there is a need to enhance the power of timed PNs to model the breakdown situations in FMSs realistically.

This chapter presents a new class of modeling tools called Augmented-timed Petri nets (ATPNs) [Venkatesh *et al.*, 1994a; Venkatesh 1995] for modeling and analyzing manufacturing systems with breakdowns. These models aid designers in better understanding of concurrency, synchronization, and sequential relations involved in breakdown handling and in system simulation for performance analysis.

A flexible assembly system consisting of three robots with various breakdown rates is used to illustrate modeling, simulation and analysis with ATPNs. ATPN models for breakdown handling are presented and analyzed for estimating the system performance and designing the optimum number of assembly fixtures. The ATPN models can also be used for the real-time control of the system. In case of a breakdown many design and control issues have to be addressed. Considering the importance of breakdowns in production control, Gershwin and Berman (1981) presented the analysis of transfer lines consisting of two unreliable machines with random failures. Groenevelt *et al.*, (1992) investigated issues related to the estimation of economic lot size and safety stock levels for an unreliable manufacturing system with a constant failure rate. Glassey and Hong (1993) presented a model for the analysis of behavior of an unreliable n-stage transfer line with finite size buffers.

Most of the above researchers studied conventional manufacturing systems by estimating the economic safety stocks to handle breakdown situations. However, to implement Just-In-Time manufacturing and to increase the level of automation, there is a need to handle breakdowns by reducing the safety stock levels and implementing efficient breakdown handling procedures. Due to breakdowns, the optimal system operational parameters that are designed for the system without considering breakdowns have to be changed. Furthermore, these parameters may differ as the component breakdown rates vary. The issues to be specifically addressed when considering a system with breakdowns are detailed breakdown handling to ensure uninterrupted production, and estimation of the new optimum design parameters for different breakdown rates. Estimation of the new optimum design parameters for various breakdown rates aids system designers and production managers in achieving optimal system performance and is the focus of this chapter.

Although PNs are proved as a tool to solve a variety of problems relating to manufacturing systems, their full application to address design and analysis issues of FMS with breakdowns remains to be explored. PN modeling of breakdowns was considered in [Barad and Sipper, 1988; Sheng and Black, 1990]. The former presented a PN model considering a machine breakdown while illustrating the flexibility of modeling a flexible manufacturing system (FMS) with PNs. The latter modeled a PN for machine breakdown while demonstrating the application of PN in a cellular manufacturing system. Performance of a transfer line was analyzed by considering the breakdown of machines using stochastic PNs [Al-Jaar and Desrochers, 1990]. Stochastic PNs is also used for the performance analysis of a flexible assembly system considering various robot failure rates [Zhou and Leu, 1991].

None of the above presented the detailed breakdown handling procedures and performance optimization issues for various machine/robot breakdown rates. Furthermore, they considered only breakdowns that arrive before starting of an activity. However, in the real life situations breakdowns may arrive when an activity is in progress. In their models the transition time delays either follow either exponential distribution or are instantaneous. This restricts their accuracy for analysis of a realistic assembly system.

The goal of this chapter is to formulate graphical models that clearly capture the details of breakdown handling in FMSs to address issues related to their design, performance, and control. The objectives of this chapter are to:

1. Introduce a new class of PNs called Augmented Timed Petri nets (ATPNs) aimed to model breakdown handling in manufacturing systems,

9.2. Augmented Timed Petri Nets

2. Illustrate a methodology to formulate ATPN models for breakdown handling, and
3. Model, simulate, and analyze a flexible assembly system (FAS) using ATPNs for estimating the optimum number of assembly fixtures for various robot breakdown rates.

9.2 Augmented Timed Petri Nets

Before introducing Augmented-timed PNs, consider a timed PN modeling a machining operation as shown in Fig. 9.1.

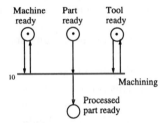

(a) Before firing of transition (before machining is started)

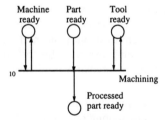

(b) During firing of transition (during machining is in progress)

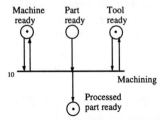

(c) After firing of transition (after machining is completed)

Figure 9.1 An example of a Timed Petri net model.

In Fig. 9.1 (a) before machining is started, machine, part, and tool availability are modeled by depositing tokens in places *machine_ready*, *part_ready*, and *tool_ready*. This triggers machining by absorbing the input tokens and firing the transition modeling the activity *machining*. During machining, both machine and tool are busy and there is no processed part available, as modeled in Fig. 9.1 (b). Once the machining is completed, the part is processed and the machine and tool are ready to process another part. This is modeled in Fig. 9.1 (c) by depositing each token in the places *processed_part_ready*, *machine_ready*, and *tool_ready*. Timed PN described above cannot easily model breakdowns in manufacturing systems. There exists an extension of PNs with inhibitor arcs which can model the breakdowns. However, they can only model breakdowns that arrive before the start of an activity. In real-life situations breakdowns may come at any time. Unlike the previous classes of PNs, ATPNs are proposed to model breakdowns which may occur before an activity starts and/or during an activity. Breakdowns may not only result due to the power or interface failure but also due to the sub-component failure in a component. For example, a machine breakdown may result due to a fault in cutting fluid lubrication or tool handling system and a robot may breakdown because of an error in its gripper. To model breakdowns, the following new constructs are added to TPNs, leading to ATPNs:

1. *Deactivation place*: This is used to model the message that is sent from a cell controller to stop the operation of the breakdown component and start the standby component. Deactivation places are pictured by two concentric circles.

2. *Deactivation transitions*: Two kinds of deactivation transitions are introduced. The former models the activity - change-over from the breakdown component to standby one and vice versa. The latter models an activity that is being executed by the component and immediately stopped at the time of breakdown. These transitions are pictured as shaded ones.

3. *Input and output deactivation arcs*: These are used to model the control and information flow among components and cell controllers. An input arc models control flow from a cell controller to the breakdowns component's controller (to stop its operation). An output arc models control flow from a cell controller to the standby component's controller (to start the operations of breakdown component).

4. *Secondary arc*: This is used to model the conditions that exist before and after change-over from one component to another. It is pictured by a dotted arc.

Formally, an ATPN, Z'' is an eleven tuple, $Z'' = (P, T, I, O, m, D, P^d, T^d, I^d, O^d, I^s)$ where the last five tuples are the new tuples proposed in this chapter as an addition to TPN. They are defined as follows:

9.2. Augmented Timed Petri Nets

1. $P^d \subset P$ is a finite set of deactivation places. $p^d \in P^d$ connects the two transitions defined in T^d which is described next.
2. T^d is a finite set of deactivation transition pairs. Each pair consists of two transitions: deactivating, t_I' that generates the deactivation command, and deactivated, t_I'' that gets deactivated. Denote

$$T^d = \{(t_1', t_1''), (t_2', t_2''), \ldots, (t_k', t_k'')\},$$

$$T^{d'} = \{t_1', t_2', \ldots, t_k'\}, \text{ and}$$

$$T^{d''}: \{t_1'', t_2'', \ldots, t_k''\}, \text{ where}$$

$T^{d'}, T^{d''} \subset T^d$, $t_i' \neq t_i''$, $1 \leq i \leq k$, and $k = |P^d|$,

3. $I^d: P^d \times T^{d''} \to \{0,1\}$, input function defining a set of deactivation arcs from P^d to $T^{d''}$,
4. $O^d: P^d \times T^{d'} \to \{0,1\}$, output function defining a set of deactivation arcs from $T^{d'}$ to p^d,
5. $I^s: P \times T \to \{0,1\}$, secondary input function defining a set of secondary arcs from P to T.

When a system is working normally without breakdowns, the initial marking of $p^d \in P^d$ is always zero. The firing rule of t_i' is same as a normal transition while that of t_i'' is different. The firing rule of t_i'' is same as a normal transition till the p_i^d contains a token. Once p_i^d contains a token, the firing of t_i'' is stopped (the activity modeled by t_i'' is stopped). When t_i'' is stopped because of p_i^d, it does not deposit the tokens in its output places in P. In other words, the tokens that t_i'' has taken from its input places for its normal firing are absorbed by t_i''. When a place is connected to a transition by I^s, the enabling rule of the transition is same. However the firing rule is different from the normal PN firing rule in updating the new marking. That is, if place p_i is connected to t_i by I^s, firing t_i does not take any token from p_i. In other words, during execution of t_i, p_i contains a token. Summarizing the above concepts, the firing rules of ATPN are given as follows:

1. $t \in T - T^{d''}$ is *enabled* if and only if $m(p) \geq I(p, t)$ and $m(p) \geq I^s(p, t), \forall p \in P$.

    $t \in T^{d''}$ is *enabled* if and only if $m(p) \geq I(p, t)$ and $m(p) \geq I^s(p, t), \forall p \in P$, and $\forall p^d \in P^d$.

2. $t \in T^{d''}$ immediately terminates its firing, if $\exists p^d \in P^d$, $I^d(p^d, t) = 1$ and $m(p^d) = 1$.

3. Enabled in a marking m, t *fires* and results in a new marking m' following the rule:

    $m'(p) = m(p) + O(p,t) - I(p,t) \; \forall \, p \in P$ and if during the firing process $\exists$

$p^d \in P^d \ni I^d(p^d,t)=1$ and p^d is not marked.

If $t \in T^{d"}$, $m'(p) = m(p) - I(p,t) + O(p^d,t) - I^d(p^d,t) \; \forall \; p \in P$, and if during the firing process there exists $p^d \in P^d \ni I^d(p^d,t)=1$ and p^d is marked.

The instantaneous description (ID) of ATPN is the same as that of TPN. Figs. 9.2 shows an example of ATPN. Fig. 9.2 (a) models machining before breakdown. The breakdown generation is modeled by the place *breakdown_generation*. Assume that the breakdown occurs four minutes after machining is started. This is modeled by associating four minutes to transition *breakdown_of_tool_occuring*. Fig. 9.2 (b) models the system when machining is in progress. Since the arc connecting place *signal_for_machine_to_do_its_task* and transition *machining* is secondary arc, there is a token present in this place during machining. Four minutes after the machining started, the tool breakdown arrives and hence machining should be stopped. This is modeled in Fig. 9.2 (c) by removing a token from place *breakdown_generation* and depositing in the place tool_breakdown. The actions involved to stop machining is represented transition *stop_machining* which removes tokens from places *signal_for_machine_ to_do_its_task* and *tool_breakdown* and deposits a token in *message_to_stop_ machining*. Once this place contains a token, it sends a message to the machine controller to stop immediately machining.

In the above example, breakdown time (four minutes) is assumed. However, breakdowns may occur at random in a shop-floor. To consider the random occurrence of breakdowns in the analysis, the breakdowns times should be generated by assuming proper probability distributions. Also, strategies to minimize the down time during the period of breakdowns should be modeled during the analysis. These issues and other related breakdown handling issues are elaborately addressed and modeled in the PNMs that are presented in the subsequent sections. For the analysis of an ATPN model, simply referred to as a Petri net model (PNM), the software package discussed in Chapter 7 was extended using the principles of ATPNs. The software package gives the following information with respect to real time:

- Marking of each place in the PNM
- Active time of each transition
- Remaining time of each transition
- Transitions which are enabled
- Conflicts between transitions (a conflict results when a single resource is required to serve more than one customer simultaneously).

9.2. Augmented Timed Petri Nets

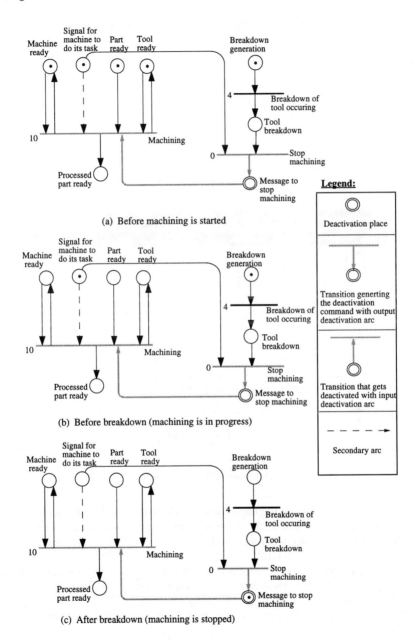

Figure 9.2 Example of an ATPN model.

9.3 Application Illustration: A Flexible Assembly System

9.3.1. System Description

An FAS is investigated to show the application of ATPNs. The system can be used for assembling a variety of products as shown in Fig. 9.3.

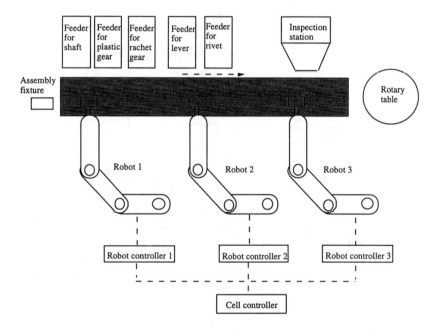

Figure 9.3 Layout of the flexible assembly system.

In order to focus on the objectives of this chapter, only one product, plastic ratchet assembly is considered. The system consists of three robots and an inspection station to do assembly and inspection. It is controlled by a cell controller. Each robot is controlled by its own controller. The functions of the cell controller are to give signals to robots to do their tasks and also signals to stop their functioning at the time of breakdown and change-over. The system described here is similar to industrial systems as Westinghouse robot assembly systems [Nof, 1985]. The assembly operation is split into tasks performed by three robots as follows:

Robot 1 (R1): Picks up and places a shaft in the assembly fixture; picks up and presses a plastic gear on the shaft; and picks up and presses a ratchet

9.3. Application Illustration

gear on the shaft.

<u>Robot 2 (R2)</u>: Picks up and transfers the subassembly from position 1 to position 2; picks up and keeps a lever in assembly fixture; and inserts a rivet in the lever and rivets the lever to the gear.

<u>Robot 3 (R3)</u>: Transfers the assembly to the rotary table and inspects position and operations of the lever.

After R3's operation, assembly fixture returns to R1 to start the assembly of a new product. Assembly time for R1, R2, and R3 are thirteen, eighteen, and ten minutes respectively. Various robot breakdown rates are considered to analyze the FAS. The system is evaluated for four values of mean time between failure (MTBF). The ranges of MTBFs are chosen randomly. The values of MTBFs are obtained by assuming that they follow the probability distribution, uniform distribution within the given range.

These ranges and the values of MTBFs for R1, R2, and R3 for different cases are shown in Table 9.1. Table 9.2 shows the exact time and breakdown sequence of robots for four different cases with different breakdown rates.

Table 9.1 MTBFs of Robots in the FAS (minutes).

Breakdown number	R1's range: 1000 to 10000	R2's range: 4000 to 8000	R3's range: 6000 to 14000
1	2,900	4,860	9,080
2	3,350	4,900	11,300
3	6,860	6,020	12,580
4	8,390	6,380	13,780

Table 9.2 Time and breakdown sequence of robots for various robot breakdown rates.

Case number	Time (minutes) and breakdown sequence of robots		
1	R1 at 2,900	R2 at 4,860	R3 at 9,080
2	R1 at 3,350	R2 at 4,900	R3 at 11,300
3	R2 at 6,020	R1 at 6,860	R3 at 12,580
4	R2 at 6,380	R1 at 8,390	R3 at 13,780

In each case, given the MTBF, the actual breakdown time is assumed to follow the probability distribution, truncated normal distribution to avoid negative values for time. Based on our experience, data with coefficient of variation below 20% are usually reliable and hence, we have assumed the standard deviation in normal distribution to be 10% of the mean.

The following assumptions are made in this model of the FAS to focus on the objectives of the chapter. All the parts required for assembling the product are always available in automatic part feeders and there always exists demand for finished products. There is a standby robot for each robot in the system that will come on-line when the corresponding robot fails. With this planning approach, unexpected manual intervention and subsequent productivity loss can be precluded. Moreover, it helps for the smooth control of the system during breakdown situations. Maintaining a standby robot for each robot may not be economically feasible. However, to demonstrate the application of ATPNs in the case of breakdown handling a standby robot is assumed for each robot. Change-over time is the time required to transfer the programs concerned with the breakdown robot to the standby robot (to carry subsequent assembly operations), and to remove the unfinished part from the assembly area of the breakdown robot. It is same for all robots and is assumed 20 minutes. The repair time for all robots is assumed to be the same and equal to 100 minutes. These time durations can be changed to random variables depending on the system under consideration.

9.3.2 ATPN Modeling of the System

Breakdown handling involves many concurrent actions like passing information to a standby robot or an operator, scheduling an unfinished part, repairing breakdown robot, etc. Figure 9.4 shows an ATPN model (PNM) for breakdown handling of R3. The modeling methodology of ATPN is explained and the activities modeled and controlled by the PNM are chronologically listed here. The initial state of the FAS is modeled by the initial marking shown in Fig. 9.4. The time durations for the activities in the system are modeled by associating times to the transitions modeling corresponding activities. These are shown on the left hand side of each transition. At the start of the system, R3 is ready to do its task (modeled by depositing a token in place *R3_ ready*) and the cell controller gives the signal to R3 to do its task (modeled by depositing a token in place *signal_for_R3_to_do_its_task*). R3 functions normally till a breakdown occurs. Breakdown occurs after a random time called breakdown time (BT, obtained from Table 9.2) which is associated to the transition *breakdown_of_R3_occuring*. When it occurs, a token is deposited in place *breakdown_of_R3*. Then, after a change-over time from R3 to R6 (standby for R3) the following concurrent operations are executed by the cell controller:

1. Sending a signal to the R3's controller to stop the operation and R6's to start. This is modeled by depositing a token in place *message_to_start_R6_and_stop_R3* and removing the token from the place *signal_for_R3_to_do_its_task*. Once place *message_to_ stop_R3* obtains a token, it stops the functioning of R3 by deactivating the transition *assembly_by_R3*.

9.3. Application Illustration

2. Activating the working of R6 by depositing a token in place *R6_ready*. The change-over between R3 and R6 is modeled by removing a token from place *signal_for_R3_to_do_its_task* and depositing it in place *signal_for_R6_to_do_its_task*.
3. Sending a message to the controller at the higher level to repair R3. This is modeled by depositing a token in place *R3_in_repair*.

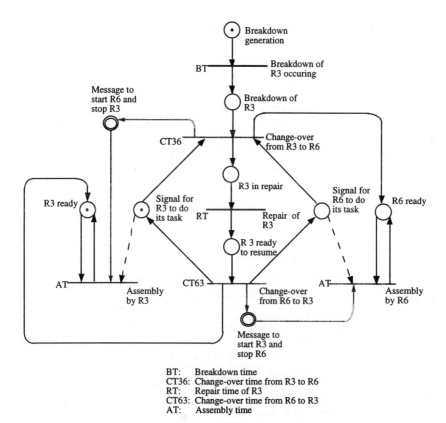

BT: Breakdown time
CT36: Change-over time from R3 to R6
RT: Repair time of R3
CT63: Change-over time from R6 to R3
AT: Assembly time

Figure 9.4 ATPN model for breakdown handling of Robot 3.

During the repair of R3, R6 performs assembly operations. After certain time, R3 is repaired, modeled by firing the transition *repair_of_R3*. This resumes R3's operation, modeled by place *R3_ ready_to_resume*. Now, the cell controller has to send signals for change-over from R6 to R3. Its functions for the change-over from R6 to R3 are similar to the previous change-over and hence the modeling is also similar. The various breakdown rates can be modeled by associating various

234   Chapter 9. Augmented-Timed Petri Nets

values of time duration to transition *breakdown_of_R3_occurimg*.

The above ATPN model for breakdown handling is exactly the same when a skilled operator is used to replace the functions of a standby robot. Furthermore, the same ATPN modeling approach can be extended for the breakdown handling of other robots, machines, AGVs, and cell controllers in a manufacturing system or other production interruption.

9.3.3 ATPN model for designing the optimum number of assembly fixtures

ATPN models can be used to address various design issues. As an example, this chapter determines the optimal number of assembly fixtures for various robot breakdown rates. To this end, FAS has to be modeled using ATPNs considering the breakdowns of all robots. Then, the obtained model has to be quantitatively analyzed to determine the effect of various MTBFs on the system performance. The ATPN model of FAS is shown in Fig. 9.5 and is obtained by duplicating the model shown in Fig. 9.4 for R1, and R2. The interpretation of places and transitions used in the PNM are listed in Table 9.3.

Table 9.3 Interpretation of places and transitions used in Fig. 9.5 ($i = 1$ to 3, $j = 4$ to 6, $k = 1$ to 2, and $u = 1$ to 6).

Place	Interpretation
AF	Assembly fixture(AF) before robot 1 ready, (n= # of AFs)
BGRi	Breakdown generation for robot i
RiB	Robot i breakdown
Ri	Robot i ready
MSRji	Message to start robot j and stop robot i
SRi	Signal for robot i to do its task
RiIR	Robot i in repair
RiR	Robot i ready to resume
AFk	Assembly fixture ready after robot k's operation
Counter	Counter for noting production output
Transition	
BRiA	Breakdown of robot i occurred
Crij	Change-over from robot i to robot j
Rri	Repair of robot i
ASRu	Assembly by robot u

9.3. Application Illustration

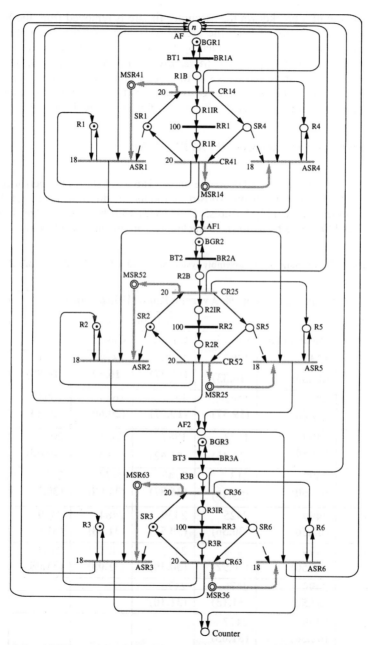

Figure 9.5 ATPN model of flexible assembly system with breakdown handling (n = number of assembly fixtures).

In case of a breakdown the assembly fixture is removed from the FAS, the unfinished part is removed from the assembly fixture, and the assembly process starts again from R1. This is modeled by output arcs from six transitions that are modeling change-overs to place *AF* (*assembly_fixture_ready_before_R1*). Once a finished product is transferred by R3 to the rotary table, the assembly fixture is to be send back to R1. This is modeled by output arcs from two transitions, one from R3 and another from R6 to place *AF* (*assembly_fixture_ready_before_R1*).

9.3.4 Simulation and Analysis of the ATPN Model

The system is simulated for 20,000 time units. Table 9.4 shows the performance of the system with and without breakdown of robots for different values of MTBFs and the number of assembly fixtures. Four cases with breakdowns are defined in Table 9.2 and the robot breakdown characteristics are given in Table 9.1.

Table 9.4 Performance of the assembly system with and without breakdown of robots.

Number of Assembly Fixtures	Production output (% **Average Robot Utilization**)				
	Without breakdown	With breakdown			
		Case 1	Case 2	Case 3	Case 4
1	1333 (33.33)	1302 (16.33)	1308 (16.37)	1318 (16.50)	1318 (16.50)
2	2499 (62.49)	1562 (19.57)	1571 (19.67)	1645 (20.60)	1664 (20.83)
3	2499 (62.50)	1873 (23.45)	1887 (23.62)	2353 (29.44)	2077 (26.83)
4	2499 (62.50)	213 (23.45)	2147 (26.87)	2478 (31.08)	2417 (30.25)
5	2499 (62.50)	2425 (30.36)	2439 (30.53)	2484 (31.08)	2479 (31.03)
6	2499 (62.50)	2469 (30.92)	2473 (30.10)	2484 (31.08)	2483 (31.08)
7	2499 (62.50)	2476 (31.13)	2476 (31.10)		
8	2499 (62.50)	2477 (31.03)			

9.3. Application Illustration

During simulation, the number of assembly fixtures is increased till the increase in production output is less than 1%. Figures 9.6 and 9.7 show the effect of MTBF on production volume and utilization.

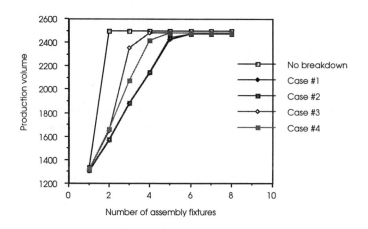

Figure 9.6 Effect of MTBF on production volume.

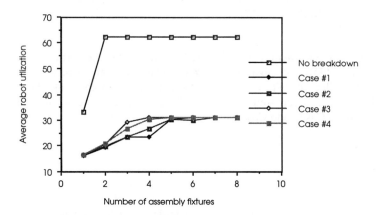

Figure 9.7 Effect of MTBF on average robot utilization.

From the simulation results the following conclusions are drawn. When there is no breakdown in the system, there is a significant increase in production output when the number of assembly fixtures is changed from one to two. However, after it increases to more than two, there is no increase either in production output or average robot utilization due to the deterministic nature of the FAS. Hence, the optimum number of assembly fixtures required without any breakdown in the FAS is two. Figures 9.6 and 9.7 also indicate the sensitivity of the production output and average robot utilization for different cases when there are robot breakdowns.

In the presence of breakdowns, to achieve the maximum production rate more assembly fixtures are required than that without breakdowns. The number of assembly fixtures does not increase linearly with the increasing value of MTBF. This is the conclusion drawn by observing the values of optimum assembly fixtures for various cases of MTBFs as shown in Table 9.5.

Table 9.5 Optimum number of assembly fixtures required for various robot breakdown rates.

Case #	Time and breakdown sequence of robots			n_{opt}
1	R1 at 2,900	R2 at 4,860	R3 at 9,080	6
2	R1 at 3,350	R2 at 4,900	R3 at 11,300	6
3	R2 at 6,020	R1 at 6,860	R3 at 12,580	4
4	R2 at 6,380	R1 at 8,390	R3 at 13,780	5

n_{opt} : Optimum number of assembly fixtures

From the results it can be inferred that the system performance depends not only on the MTBF but also on the exact breakdown sequence of robots. Even though the value of MTBF increases from Case #1 to case #4, the number of assembly fixtures decreased from Case #1 as well as Case #2 to Case #3 and increased from Case #3 to Case #4.

9.4 Summary

This chapter presented a class of modeling tools called Augmented Timed Petri nets (ATPNs) to model breakdown handling in manufacturing systems. ATPNs can model their operations in detail considering the breakdowns of various components. The methodology for formulating the ATPN models is illustrated by considering a flexible assembly system. Also, the application of ATPNs for optimization and design is shown by investigating the optimum number of assembly fixtures for the

9.3. Summary

system under various robot breakdown rates. The methodology proposed in this chapter can be extended to deal with breakdowns of several machines, AGVs, and cell controllers and other production interruptions. ATPNs provide rapid, flexible, and realistic modeling.

ATPN models can be extended for the real time control. In such cases, the transition modeling the occurrence of breakdown is not associated with random breakdown times. Instead, the sensors/limit switches that recognize the breakdown in the system are modeled as input places for this transition. When there is a breakdown in the system, these places obtain tokens and thus immediately fire the transition, modeling the occurrence of breakdown. Figure 9.8 shows an ATPN model that can be used for real-time control.

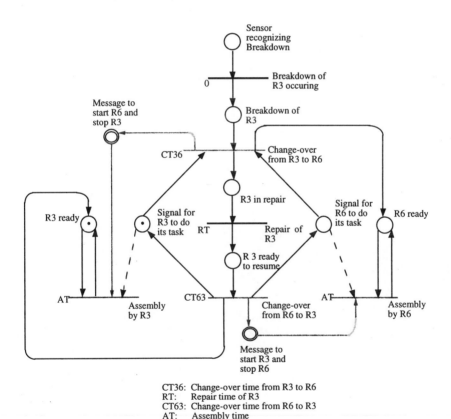

Figure 9.8 ATPN model for real-time control.

When the system contains several components, the size of ATPN models may grow. In such cases, *colored PNs* can be combined with the principles of ATPNs to formulate concise graphical models. This research can be extended to study some important issues such as robot scheduling during breakdowns when only a single robot exists as a standby to all three robots, system performance when several product varieties are produced simultaneously in the system with random routing of parts, and cost consideration for standby robots and breakdown handling. The effect of random distributions of repair and change-over times on the system performance can also be investigated by associating various time values to the transitions modeling repair and change-over activities.

CHAPTER 10

REAL-TIME PETRI NETS FOR DISCRETE EVENT CONTROL

10.1 Introduction

Several types of Petri nets (PNs) are available for supervisory control of FMSs. PNs have been augmented and implemented in a variety of ways to achieve real-time control [Chocron and Cerny, 1980; Valette *et al.*, 1983; Murata *et al.*, 1986; Crockett *et al.*, 1987; Boucher *et al.*, 1989; Stefano and Mirabella, 1991]. The classification of various PN-based control schemes is described later in this chapter. A new class of PNs for discrete-event control that closely resemble ordinary PNs is of the paramount importance for the development of control software because ordinary PNs are relatively simple and easier to understand.

Due to the close resemblance of ordinary PNs and ladder logic diagrams (LLDs), ordinary PNs are useful for the design recovery of ladder logic diagrams. In other words, LLDs can be converted into ordinary PNs. However, the design recovery of LLDs using earlier classes of PNs [Chocron and Cerny, 1980; Valette *et al.*, 1983; Murata *et al.*, 1986; Crockett *et al.*, 1987; Boucher *et al.*, 1989; Stefano and Mirabella, 1991] is relatively difficult as they do not closely resemble LLDs. The importance of design recovery of LLDs is emphasized in [Falcione and Krogh, 1993]. Motivated by these facts and based on the research in PN control literature, this chapter proposes and presents a new class of PNs called Real-time PNs (RTPNs) [Venkatesh *et al.*, 1994b; Venkatesh 1995] for sequence controller design.

10.2 Real-time Petri Nets

An RTPN can be obtained by associating timing, and I/O (input/output) sensory information to the untimed PN. It is an eight tuple and defined as: RTPN=(P, T, I, O, m, D, X, Y) where:

1. P is a finite set of places;
2. T is a finite set of transitions with P∪T_∅ and P∩T=∅;
3. I: P × T → N, is an input function that defines the set of directed arcs from P to T where N = {0, 1, 2, ...};
4. O: P × T → N, is an output function that defines the set of directed arcs from T to P;
5. m: P → N, is a marking whose i^{th} component represents the number of tokens in the ith place. An initial marking is denoted by m_0;
6. D: T → R+, is a firing time function where R+ is the set of non-negative real numbers;
7. X: P → {-, 0,1,2, ...K} and $X(p_i) \neq X(p_j)$, i≠j, is an input signal function, where K is the maximum number of input signal channels, and "-" is the dummy attribute indicating no assigned channel to the place.
8. Y: T → L, is an output signal function, where L is a set of integers.

In a RTPN, the first five tuples represent the untimed PN and the last three are extensions added to it and explained below:

1. Timing vector (D) is intended to associate time delays to transitions modeling the activities in the system,
2. Input signal vector (X) reads the state of the input signals from digital input interface. X associates attributes to every place. $X_i=X(p_i)$ and is an attribute associated with place p_i, representing the input channel number associated with p_i. For example, if p_i models a limit switch, the RTPN reads the status of that switch from the digital input interface through the channel number represented by X_i. The initial marking, $m_0(p_i)$ is considered as the first attribute of p_i and X_i is the second one. The contents of any input channel X_i are either 0 or 1.
3. Output signal vector (Y) is intended to send output signals through digital output interface. Y associates attributes to every transition. $Y_i=Y(t_i)$ is the attribute associated to transition t_i which represents the number that is to be sent to the digital output interface. For example, t_i may model the activity "send signal to actuate solenoid A," or "execute a procedure to control a robot." Each solenoid is activated by writing a specific number on to the digital output interface. During execution of the program, when a transition fires, RTPN writes the decimal number corresponding to the output channel to digital output interface. The contents of any output channel are either 0 or 1. The usage of this vector is later detailed in the example system.

10.2. Real-Time Petri Nets

There are two events for a transition firing, *start_firing* and *end_firing*. Between them the firing is in progress. The removal of tokens from a transition's input place(s) occurs at *start_firing*. The deposition of tokens to a transition's output places(s) occurs at the *end_firing*. While transition firing is in progress, the time to end firing, called the *remaining firing time*, decreases from firing duration to zero at which its firing is completed. The execution rules of a RTPN include enabling and firing rules:

1. A transition $t \in T$ is *enabled* iff $\forall\ p \in P$ and $I(p,t) \neq 0$, $m(p) \geq I(p, t)$ and $X(p)$ has content 1.
2. Enabled in a marking m, t *fires* and results in a new marking m' following the rule:

   $m'(p) = m(p) + O(p,t) - I(p,t), \forall\ p \in P.$

In the first rule, the first condition, $m(p) \geq I(p, t)$, ensures that the marking of each input place of a transition should be either equal to or greater than the input arc weight, and the second condition, "$X(p)$ has content 1" ensures that for each input place of a transition, the second attribute of that place is 1 indicating that the input signal modeled by that place is true. The second rule, after firing of a transition, updates the marking of its output places and is the same as in ordinary PNs. The design procedure for formulating a RTPN based controller is shown in Fig. 10.1 and is briefed in the following five steps:

1. Model the control sequence using PNs to obtain the PN model of the sequence controller.
2. Assign input channels to inputs of the system such as limit switches, sensors, etc. to formulate an input mapping table.
3. Assign output channels to outputs of the system such as solenoids, switches etc. Also, identify timing information for activities to obtain an output mapping table.
4. Based on the input mapping table, assign input channel number to each place (limit one channel per place) in PN based controller. The initial state of the system decides the initial marking of RTPN. In the PN model some places do not represent the inputs of the system as they represent the intermediate states of system or logical places to model counters in the sequence. Hence, no channel is assigned to these places which is represented by "-".
5. Using the output mapping table and the action(s) that are modeled by a transition, assign a number to each transition in a PN based controller. The operations and time delays given in the sequence to be controlled decide firing time function of RTPN. In the PN model some transitions may represent concurrent actions. Hence, care should be taken to assign the numbers for such transitions.

244   Chapter 10. Real-Time Petri Nets

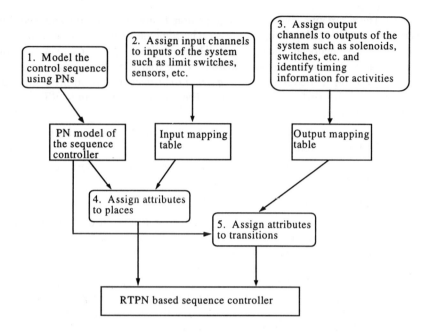

Figure 10.1 Procedure for formulating a RTPN based controller.

By following the above procedure a RTPN based controller can be formulated for a given sequence. The *instantaneous description (ID)* defined earlier in Chapter 8 in the case of timed PNs can also be extended for RTPNs. Observe that RTPNs can also be used for simulation and performance evaluation by associating times with transitions and dummy attributes for places modeling input signal information, and transitions modeling output signal information. There are several ways to use PNs to perform the sequence control. One is based on "token game" [Zhou *et al.*, 1992a] and the other converts the net into either Programmable Logic Controllers [Valette *et al.*, 1983] or control code directly [Zhou and DiCesare, 1993]. The first scheme, as illustrated in Fig. 10.2, is used here.

10.3 Real-time PNs and Other PN Extensions for Control

There are several types of PNs for control with various implementation schemes as shown in Table 10.1. The implementation schemes differ in terms of the extensions

10.3. Real-Time PNs and Other PN Extensions

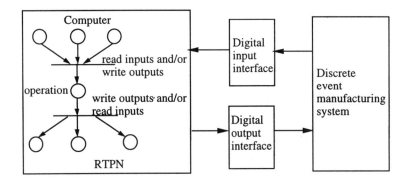

Figure 10.2 Controlling a system using a RTPN based controller.

Table 10.1 Various methods of Petri net based sequence control.

Implementation scheme of Petri net controller	Hardware used	Reference
Petri net → High-level language description → Compiler → Control sequence tables and decision programs → control	Intel 8085 Assembler	Chocron and Cerny (1980)
Petri net → High-level language description → Interpreter → Control tables → control	A PLC based on ZILOG 80A microprocessor	Valette, et al. (1983)
Control Petri net → Interpreter → control	16 bit microcomputer	Murata, et al. (1986)
Petri net → Execution algorithm → control	VAX 11/780 and IBM-PC/XT	Crockett, et al. (1987)
Petri net → Software → control (Note: Details of implementation are not given)	IBM-PC/XT	Boucher, et al. (1989)
Petri net → Petri net description language → Token player → control	IBM-PC/XT	Zhou, et al. (1992a)
Control Petri net → Petri net simulator → control	8086 CPU and DSP (TMS32020)	Stefano and Mirabella (1991)
Real-time Petri net → Software system based on token player → control	IBM-PC/AT	Venkatesh, et al. (1994)

added to ordinary PNs in order to model timers, counters, input/output signal information, process status functions, method of execution of PN model, the hardware on which the PN is executed, and the level of sophistication.

Even though RTPNs and earlier classes of PNs for control share similar principles, some of the differences between RTPNs and the others are listed below:

1. Earlier studies use a variety of places to model timers and counters [Murata *et al.*, 1986]. This might make the model difficult to understand. In RTPNs neither new places nor transitions are introduced for modeling them. In order to find the process I/O and process status, Murata *et al.*, (1986) uses new functions, in the definition of their PNs which are called as C-nets. They also define a new place called 'act box' to define process I/O functions. However, in RTPNs such new functions are not included in order to keep the definition of RTPN simple and easy to understand. In order to find the process I/O and process status, attributes are included in RTPN. C-nets are more suitable at the higher level of cell control. For example, C-nets have a new place called 'receive box' to detect the request signal to start a job and to generate a numbered token corresponding to each job variety. In other words, C-nets can distinguish the part-varieties that enter into the system. RTPNs do not have that level of sophistication. C-nets can resolve conflicts by using a new place called "conflict box" and the "process status function". C-nets introduce a new transition "count gate" to count the repetitive cycles. In RTPNs, ordinary transitions and places with multiple weights on arcs model such repetitive cycles. C-nets use a new place called "timer box" to model timing. In RTPNs, timing is modeled by assigning it as a second attribute to a transition.

2. In [Stefano and Mirabella, 1991] new sets of places are introduced for modeling I/O signals, which increases the number of places in PN model. In RTPNs, I/O signals are modeled as attributes for places and transitions respectively. Places are assigned with input signals because they model the limit switches. Transitions are assigned with output signals because they actuate solenoids or motors. Hence, due to the use of attributes RTPNs have fewer nodes and links compared to PNs in [Murata *et al.*, 1986; Stefano and Mirabella, 1991], thereby reducing the net size and hence the graphical complexity.

3. In earlier works [Chocron and Cerny, 1980; Boucher *et al.*, 1989; Stefano and Mirabella, 1991], the resetting of timers, counters, and emergency stop are not explicitly modeled. Furthermore, they often use additional functions to model and implement timers and counters [Stefano and Mirabella, 1991]. Using RTPNs all these features can be clearly modeled. The automatic resetting of timers and counters is also embedded in the execution of RTPNs.

10.3. Real-Time PNs and Other PN Extensions

4. In Crockett, *et al.*, (1987), actions (output signals) are associated with places and events (input signals) are with transitions. Macro places are introduced to model sub-PN, and switching places are introduced to resolve conflicts. In RTPNs, sub-PNs are modeled by ordinary places and transitions. Crockett *et al.*, (1987) have not mentioned how their PNs can be used for both control and simulation without changing the original PN that is aimed for control only.

5. Valette *et al.*, (1983) and Chocron and Cerny (1980) used PN description languages which are input to the translator or compiler. The translator or compiler generates the control tables that drive the actuators. In Valette *et al.*, (1983), the description language consists of instructions. These instructions specify 1) declarations of places and transitions and arcs among them, 2) VAR allowing variable declarations including input and output addresses corresponding to sensors and actuators and timer values, and 3) specification of boolean conditions attached to transitions and the operations that have to be executed. RTPNs can be implemented by a simple implementation scheme. For example, RTPNs eliminate the usage of description languages used in (Chocron and Cerny, 1980; Valette *et al.*, 1983] since the RTPN model can be directly used for control with the help of a token player. By adopting this implementation scheme, the need to translate the PN model to higher level net description language and the generation of control tables is avoided. The actual implementation of the token player in RTPNs is transparent to the users and hence their only task to control a system is simplified to model the control logic. Zhou *et al.*, (1992a) also used simple PN descriptive language as input to the PN controller. Descriptive languages are efficient in terms of the memory requirements. For a given PN model, RTPNs need more memory space than other classes of PN that use descriptive languages because in case of RTPNs, the PN model is stored in the form of input and output matrices and thereby consumes more memory space.

6. Even though Boucher *et al.*, (1989) reported PN based control by associating output signals to transitions, no explicit comments are made how the input signal information is mapped into their PN model. They also used ordinary PNs without the introduction of new places and transitions. The details on the method of PN execution are not given in Boucher *et al.*, (1989).

7. Zhou *et al.*, (1992a) used ordinary PNs for control. They also assign both input and output signal information to transitions. Even though, they use attributes for each place and transition, their attributes do not explicitly mention about how the input and output signal information is mapped on to places and transitions. Also, explicit modeling of timing information is not given in their PN implementation. In RTPNs such

mapping is made clear because of the attributes to places and transitions, and time can be modeled as the second attribute of a transition. Hence, like any timed PNs, the deterministic delays given in a control sequence can be clearly modeled in RTPNs.

8. Sheng and Black (1989) also used ordinary PNs and expert systems to control a manufacturing cell. They modeled the cell using PNs and generated the cell control rules from the firing sequence of the PN using the concept of forward chaining approach in expert systems. However, the details of the mapping of the input/output signal information and modeling of timers and counters are not reported.

To summarize, RTPNs are suitable at the lowest level of system control since they model the system more realistically by naturally mapping the limit switches, start, and stop buttons as places with attributes and actuators as transitions with attributes. RTPNs are easy to understand and require a simple procedure to implement. Since RTPNs and ordinary PNs have the same graphical representation, RTPNs are useful for the design recovery of ladder logic diagrams. This is because ladder logic diagrams can be converted into ordinary PNs which in turn can be converted into RTPNs by assigning proper attributes to places and transitions.

However, using earlier classes of PNs the design recovery is difficult. This is because they [Chocron and Cerny, 1980; Valette *et al.*, 1983; Murata *et al.*, 1986; Crockett *et al.*, 1987; Stefano and Mirabella, 1991] have special symbols for places and transitions to represent various functions in discrete-event control. The disadvantages of RTPNs include: 1) restriction of not allowing conflicts in a PN model, 2) lack of facility to distinguish product varieties, and 3) requirement of more memory space to store the structure of PN model because of the usage of input and output matrices.

10.4 Example: An Automatic Assembly System

In order to illustrate the control by RTPNs, a simple assembly system shown in Fig. 10.3 is considered. It consists of a pallet storage station, assembly station, an inspection station, an AGV, and a robot. Two workpieces are assembled by the robot at the assembly station. After the assembly is completed, the robot unloads a finished product on a work table. The AGV transports empty pallets from the pallet storage station to the assembly station.

10.4. Example: An Automatic Assembly System

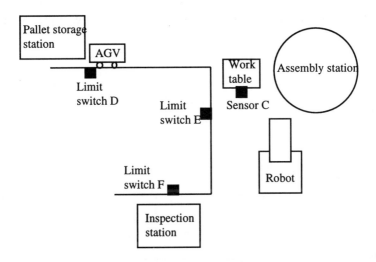

Figure 10.3 Schematic of an automatic assembly system.

Robot is also used to unload the finished product from work table and place it on the pallet which is on the AGV. Finally, the AGV transports the finished product loaded on pallet to inspection station for inspection. The availability of Workpieces 1 and 2 is sensed by two sensors A and B (not shown in figure), respectively. Limit switches D, E, and F sensing the presence of AGV are at pallet storage station, assembly station, and inspection station respectively. Sensor C recognizes the loading of finished product on work table by robot. Basically there are three operations to be executed in this system as follows:

1. Robot's assembly and transfer of a finished product on to the work table,
2. AGV's travel from the pallet storage station to the assembly station, and
3. Transfer of finished product by robot on to AGV and AGV's travel from the assembly station to the inspection station.

Assume that the time duration of operation i is τ_i time units and is executed when output relay i is high. There is a digital I/O interface to control this system with details as shown in Tables 10.2 and 10.3.

Table 10.2 The input mapping table where X_i is input channel number.

Limit switch/Sensor	A	B	C	D	E	F
X_i	0	1	2	3	4	5

Table 10.3 The output mapping table where Y_i is number sent to the digital output interface.

Output relay	Output Channel number
1	0
2	1
3	2

Table 10.4 Attributes of places and transitions in the RTPN.

Place	1st attri bute	2nd attri bute	Transition	Output relay	Output channel to be activated (x)	1st attri bute	2nd attribute (2^x)
A	1	0	t_1	1	0	τ_1	1
B	1	1	t_2	2	1	τ_2	2
C	0	2	t_3	3	2	τ_3	4
D	1	3					
E	0	4					
F	0	5					

Assume that this system is controlled by PN with places A, B, C, D, E, and F modeling the corresponding sensors, and transition t_i modeling operation i. The attributes of places and transitions for this RTPN are listed in Table 10.4. The first and second attributes of a place are initial marking and assigned input channel number, respectively. For a transition the first attribute is time duration and the second attribute is the number to be sent to the I/O interface to activate the corresponding output relay. Following the procedure illustrated in Fig. 10.1, the RTPN model obtained is shown in Fig. 10.4.

The initial marking $(1, 1, 1, 0, 0, 0)^T$ is shown in the RTPN model. When this RTPN model is started, it concurrently executes both t_1 and t_2. Before firing t_1, i.e., before starting assembly operation, RTPN checks for the status of sensors A and B for the presence of workpiece 1 and 2 respectively and D. When t_1 is fired, it sends an output signal to robot to perform assembly operation and transfer the finished product on to the work table.. Similarly, before firing t_2, i.e., before starting the transportation of AGV to assembly station, it checks for the status of the limit switch D for the availability of AGV at pallet storage station.

10.5. A Case Study: An Electro-Pneumatic System

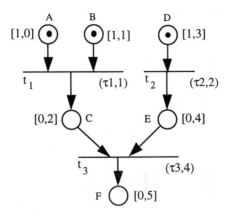

Figure 10.4 RTPN model for the system.

When t_2 is fired, it sends an output signal to AGV to move from pallet storage station to assembly station. Once firing of t_1 and t_2 is completed, places C and E obtain each token. Place C models sensor C that checks for the presence of the assembled part on the work table. Place E models the limit switch E that checks for the presence of AGV at assembly station. Hence, when both places C and E are marked and the status of these places obtained from digital input interface is true, t_3 is executed during which a signals are sent to robot to load the finished product on to AGV and AGV to move the finished product to the inspection station. Once firing of t_3 is completed, it deposits a token in place F modeling the limit switch F that checks for the presence of AGV at the inspection station. The final marking of the RTPN is $(0, 0, 0, 0, 0, 1)^T$ Observe that time durations for these transitions need not be known to execute RTPN (or to control the system). However, they are needed when the same RTPN model is used for performance evaluation.

10.5 A Case Study: An Electro-Pneumatic System

This section demonstrates the use of RTPNs for the control of concurrent tasks, through a practical electro-pneumatic system located in The Robotics Center, Florida Atlantic University. The system considered in this chapter is shown in Fig. 10.5. It consists of four pneumatic pistons (A, B, C, and D) which are operated by spring-loaded five ports and two-way solenoid valves. Note that these solenoids require sustained output signals to activate the forward motion of pistons because if the output signal is not sustained, the spring causes the solenoid to change its position resulting the backward motion of the piston.

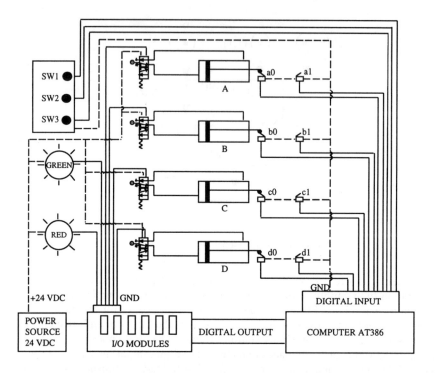

Figure 10.5 Schematic of an electro-pneumatic system considered.

More discussion on the differences between the solenoids that require sustained signals and those do not require sustained signals is presented later in Sections 11.4.3 and 11.6 of the next chapter. Each piston has two normally open limit switches. For example, when the end of piston A contacts limit switch a0 (a1), a0 (a1) is closed indicating that the piston A is at the end of its return stroke (forward stoke). The time that a piston takes to complete either forward or backward stroke is one second. In manufacturing, typical functions of these pistons can be to load/unload the part from the machine table, to extend/retract a cutting tool spindle, etc.

Three push buttons are provided to start the system (switch SW1), to stop the system normally (switch SW2) and to stop the system immediately in emergency (switch SW3, ES). The system has 11 inputs corresponding to 8 limit switches (two for each piston) and 3 push buttons. The system has 6 outputs corresponding to 4 solenoid valves and two lights that indicate the status of the system. In this work, PNs are implemented through IBM PC and digital I/O interface as shown in Fig. 10.5. The procedure given in Fig. 10.1 is followed to design the control logic by PNs. Tables 10.5 and 10.6 show the I/O mappings of the system to PN respectively. For more hardware details on this system and its

10.5. A Case Study: An Electro-Pneumatic System

implementation with PNs refer to [Venkatesh et al., 1993; Venkatesh et al., 1994b; Venkatesh and Ilyas 1995]. The following discussions focus on the sequence controller design.

Table 10.5 The input mapping table where X_i is input channel number.

Switch	a1	b1	c1	a0	b0	c0	SW1	ES
X_i	0	1	2	4	5	6	8	10

Table 10.6 The output mapping table.

Solenoid or Light	Output Channel number	Y_i	
		NA	ND
A	0	1	-1
B	1	2	-2
C	2	4	-4
D	3	8	-8
Light 1	4	32	-32
Light 2	5	64	-64

Note: 1. Y_i is the number sent to the digital output interface.
2. NA (ND) means the number that is to be written on digital output interface to activate (deactivate) the solenoid/light.

Sequence controller design

Consider that the system has to be controlled to execute the following sequence: START, A+, B+, {C+, A-}, {B-, C-}, where A+ represents that the piston has to do a forward stroke and A- return one. {C+, A-} represents two concurrent actions taking place simultaneously: Piston C to do a forward stoke and Piston A to do a return one. Fig. 10.6 shows the RTPN corresponding to this sequence. Note that in the RTPN, a place has attributes [n_1, n_2] where n_1 is the first attribute representing initial number of tokens and n_2 is the second one mapping input channel number. Similarly, a transition has attributes (n_1', n_2') where n_1' is the firing duration and, n_2' is the number to be written on the digital output interface. When concurrent actions such as {C+, A-} are to be modeled, care should be taken to associate the second attribute to transitions as shown in Table 10.7.

254   Chapter 10. Real-Time Petri Nets

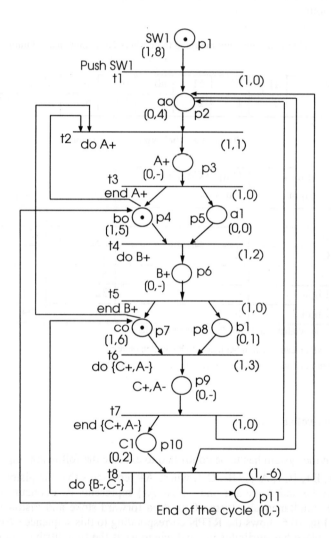

Figure 10.6 RTPN for sequence 1: ST, A+,B+,{C+,A-}, {B-,C-}.

10.6. Software Description to Execute RTPNs

Table 10.7 Attributes of transitions modeling actions.

Action	Output channel to be activated	Output channel to be deactivated	Transition modeling the action	2nd attribute of transition
do A+	0	-	t_2	$2^0 = 1$
do B+	1	-	t_4	$2^1 = 2$
do {C+, A-}	2	0	t_6	$2^2 - 2^0 = 3$
do {B-, C-}	-	1, 2	t_8	$-2^1 - 2^2 = -6$

10.6 Software Description to Execute Real-time Petri Nets

The algorithm used to develop the software package (Venkatesh and Ilyas 1995) to execute RTPNs is presented in Fig. 10.7. The software package runs on IBM compatible PCs by interacting with the digital input/output interface and the system under control. The biggest PN model that can be controlled using this package contains one hundred places and one hundred transitions. However, larger nets can be controlled when this software package is installed in a workstation with more memory. Notice that this software package is similar to the one described earlier in Chapter 7 except for the modules, "Read_Petri_Net" and "New_marking" which are describe below.

In addition to reading the input information - Petri net structure, initial marking, and transitions time durations as described in Chapter 7, the module "Read_Petri_Net" also reads the information corresponding to input signal vector and output signal vector. The module "New_marking" contains two sub-modules: 1) "Read_marking" checks for the second attributes of all places that are inputs for enabled transitions. 2) "Update_marking" fires the transitions and changes the current marking. Read_marking uses F-function as input. If a transition t_i is enabled it checks for the second attribute of all input places of t_i. If the second attribute of an input place is high it sets the variable "flag" corresponding to that place as TRUE. After flags corresponding to all the input places of t_i are set to TRUE, it removes the tokens from these places and sends a signal to "update_marking" to fire t_i. After receiving the signal corresponding to the transition to be fired from "read_marking", "update_marking" sends the second attribute of the transition to be fired to the digital output interface and deposits tokens in all the output places of the transition fired.

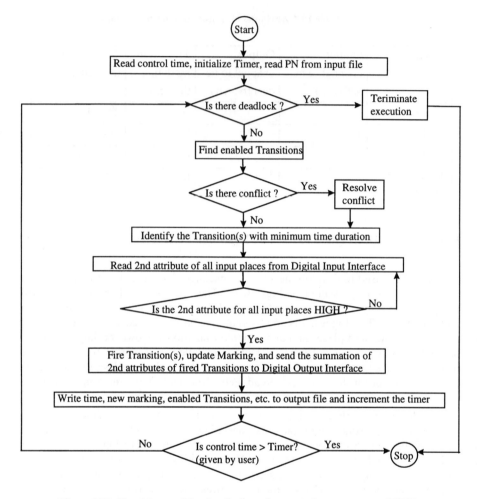

Figure 10.7 Algorithm used for developing software package to execute RTPNs.

Note that by including a dummy value for the second attribute of each place and associating timing values to each transition, this package can also be used for simulation and performance evaluation [Venkatesh and Ilyas, 1995]. In the next chapter, Real time PNs and ladder logic diagrams are compared for discrete-event control of industrial automated systems.

10.7 Summary

In this chapter, Real-time PNs (RTPNs) are proposed for the discrete-event control of FMSs. The detailed theory of RTPNs, the procedure of formulating a RTPN based controller, and the method of controlling a system using an RTPN based controller are described. Several types of Petri nets and various implementation schemes of the Petri net based controllers are briefly presented. The differences between RTPNs and earlier class of Petri nets for control are explained. In order to demonstrate the use of RTPNs for discrete-event control, an RTPN to control a simple assembly system is developed and its execution procedure is discussed. Also, the use of RTPNs for the control of concurrent tasks in an FMS is illustrated by considering an electro-pneumatic system and developing an RTPN to control the same. Finally, a software package that can be not only used for the simulation of a Petri net but also for the real-time control of any discrete-event system using RTPNs is described.

CHAPTER 11

COMPARISON OF REAL-TIME PETRI NETS AND LADDER LOGIC DIAGRAMS

11.1 Introduction

According to a given logic specification, a sequence controller synchronizes and coordinates the operations of various units in a manufacturing system. The sequence controller in this chapter means a class of discrete event controllers without choices in executing the operations/activities. Design methods for sequence controllers play a prominent role in advancing industrial automation. A comparison of Real-time Petri nets (RTPNs) and ladder logic diagrams (LLDs) is presented in this chapter.

Traditionally, LLDs are used to capture the sequence of operations executed by the system's control software. They specify the I/O procedures of the Programmable Logic Controller (PLC) that drive and perform the repetitive operations of the system. These diagrams grow so complex that locating the cause when a problem is detected becomes extremely difficult [Chaar et al., 1991]. Furthermore, their usage is limited to control the system; they can not analyze and evaluate the qualitative and performance characteristics. They often need significant changes as the specification changes resulting in difficult maintenance problems. The increasing complexity and varying needs of modern discrete manufacturing systems have challenged the use of LLDs for programmable logic controllers. The methodologies based on research results in computer science have recently received growing attention by academic researchers and industrial engineers in order to design flexible, reusable, and maintainable control software. Particularly, Petri nets (PNs) are emerging as a very important tool to provide an integrated solution for modeling, analysis, simulation, and control of industrial automated systems. However, in order to establish PNs as alternative to LLDs [Venkatesh et al., 1994e, 1995] there is a need for benchmark studies to compare them formally.

It is observed that none of the earlier studies on PN-based control has compared LLDs and PNs for the design of sequence controllers in detail and formally [Venkatesh et al., 1994b; Venkatesh 1995]. Boucher et al., (1989) used LLDs and PNs to control a same manufacturing system and reported the graphical representation by PNs makes the controller more tractable than that of LLDs. However, they have not formally quantified the comparison between PN and LLDs to design sequence controllers. Some of the problems associated to compare PNs and LLDs are: (1) unlike in case of LLDs, there exist several classes of PNs with various implementation schemes for discrete control as shown in Table 10.1, and (2) the criteria with respect to which the comparison should be performed need to be identified.

First, certain criteria are identified to compare LLDs and PNs in designing sequence controllers subject to the changing control requirements. The comparison is performed through a practical industrial automated system. Secondly, some analytical formulas and a methodology are developed to estimate the number of basic elements used in the PN and LLD designs prior to their constructions. The results will be useful for researchers and engineers to design control systems for complex industrial automated systems [Venkatesh et al., 1994d; Venkatesh 1995].

The goal of this chapter is to compare LLDs and PNs when they are used to design discrete event controllers for manufacturing systems. The objectives of this chapter are:

1. To identify the criteria to compare LLDs and RTPNs for design of sequence control,
2. To compare LLDs and RTPNs in designing sequence controllers that respond to specification changes,
3. To formulate mathematical formulas to calculate the number of basic elements to model certain building blocks of logic models using PNs and LLDs, and
4. To present a methodology that synthesizes these analytical formulas for estimating the total number of basic elements required to design sequence controllers using PNs and LLDs.

The comparison criteria between LLDs and RTPNs are identified in Sec. 11.2. The comparison between LLDs and RTPNs is performed through a practical system as a case study in Sec. 11.3. Section 11.4 presents analytical formulas to estimate the basic elements required to implement certain building blocks of control logic. Section 11.5 presents a methodology that uses these formulas to estimate the number of basic elements required to implement a given control logic specification using PNs and LLDs. In Sec. 11.6, this methodology is illustrated through two examples and the case study considered in Sec. 11.3. Section 11.7 concludes this chapter.

11.2 Comparison Criteria for Control Logic Design by PNs and LLDs

The application of LLDs for sequence control is widely known as they are used by several industries [Pessen, 1989; Michel, 1990]. Graphically, RTPNs are same as PNs except that in the former, places are given attributes to model input sensory information and transitions are associated with attributes to model output and timing information. Hence, the comparison presented in this chapter corresponds to RTPNs which are referred to PNs hereon for short. In order to start an operation in a system some conditions have to be fulfilled which are called as pre-conditions. Upon its completion, an operation results in some conditions which are called as post-conditions. A transition in a PN models an operation. The input and output places of a transition model the pre- and post- conditions respectively.

In LLDs, an operation is modeled by activating an output coil corresponding to a relay or a solenoid. In addition to an output coil a relay consists of number of contacts, some normally open (N.O) and some normally closed (N.C). In order to energize the output coil of a relay, the corresponding contacts of that relay should be switched, i.e., an N.O. contact closes, while an N.C. one opens. The LLD also uses various input and output elements. A typical output element is a solenoid which is usually used to actuate pneumatic or hydraulic solenoid valves. Push-button and limit switches constitute input elements. The contacts of relay along with the input elements constitute pre-conditions. The solenoids and output coils are referred to as post-conditions. The logic and other basic building blocks used in sequence control are modeled by PNs and LLDs as shown in Table 11.1. The explanations for these symbols is given below.

The first four rows show the basic PN elements to model conditions, status, activity, information and material flow, and resources. Note that LLDs do not have the corresponding explicit representations. Logical AND and logical OR can be easily modeled by both PN and LLD with similar complexity. Other important concepts, e.g., concurrency, time delay, and synchronization are also illustrated in Table 11.1. The methods of developing LLDs can be seen in [Pessen, 1989; Michel, 1990]. Two important factors for comparison of PN and LLD for discrete event control are identified as design complexity and response time.

Design Complexity

Design complexity is defined as the complexity associated in designing the control logic for a given specification. Since design complexity is influenced by many factors, e.g., the experience of designers, size of a control program, and number of dynamic steps necessary for coding or changing control program, it is very hard to

Table 11.1 Representations by Petri nets and ladder logic diagrams.

Logic constructs	Petri nets	Ladder logic diagrams
Condition or the status of a system element	○ Place	No explicit representation
An activity	—— Transition	No explicit representation
Flow of information or material.	——▶ Directed arc	No explicit representation
Objects such as machines, robots, pallets, etc.	○ Token(s) in place(s)	No explicit representation
Logical AND IF A = 1 and B = 1 and C = 1 THEN D = 1		
Logical OR IF A = 1 or B = 1 or C = 1 THEN D = 1		
Concurrency IF A = 1 and B = 1 THEN C = 1 and D = 1 and E = 1		
Time delay IF A = 1 THEN delay "τ time units" B = 1		

Synchronization
IF A = 1
THEN
  delay "τ1 time units"; D = 1
IF B = 1 and C = 1
THEN
  delay "τ2 time units"; E = 1
IF D = 1 and E = 1
THEN
  delay "τ3 time units"; F = 1		

quantify formally. However, it can be characterized by two factors namely graphical complexity and adaptability for change in specification.

Graphical complexity: It is mainly determined by the number of nodes and links for a given graphical control logic design. Graphical complexity influences the understandability of control logic by people who do not have knowledge of either PN or LLD.

Hence, graphical complexity is an important factor in designing the logic at the initial stages and subsequently debugging the errors during its implementation.

11.2. Comparison Criteria

The graphical complexity in terms of the net size is a major issue in manufacturing systems [Zhou et al., 1992a] and it was reported that the simpler the graphical representation of control logic, the easier to track the controller [Boucher et al., 1989]. Graphical complexity may also influence response time as described later. For example, in the case of LLD, the response time depends on the size of the LLD because as the number of rungs increases scan time also increases. Hence, a shorter LLD results in a faster controller [Pessen, 1989].

Adaptability to change in specifications: This factor is gaining much importance in the context of agile manufacturing in which control sequences need to be changed often to meet the dynamically changing requirements of the market. The control software should be easily adaptable to changes in specifications in order to improve the software productivity and thus keep minimal development time. One of two designs is said to be more adaptable if it needs fewer changes compared to another in order to fulfill a specification change.

Response Time

Response time is termed as scan time in LLD literature and execution time in PN. Response time can be defined as the minimum time that a control model takes to respond to an external event. Its importance to control real-time systems is clear since it decides how fast the control system responds to an event in the system/process under control.

An important factor that influences graphical complexity and adaptability is the physical appearance (size) of the model, whereas the response time is influenced by not only physical appearance but also the method of implementation. Methods of implementation constitute the software and hardware used to control the system using either PNs or LLDs. Graphical complexity and adaptability cannot be quantified whereas response time can be measured accurately, given a logic design and implementation. However, since there are several ways to implement PNs as shown in Table 10.1 and LLDs [Michel, 1990; Pessen, 1989] both in terms of hardware and software, it is very difficult to make a fair comparison of LLDs and PNs solely on the basis of response time criterion.

Hence, we propose some common measures that give an idea about the graphical complexity, adaptability, and response time. One such measure is the number of nodes and links used in a control logic model. For PNs nodes are places and transitions and links are arcs; whereas in LLDs, nodes are normally opened/closed switches, timers, counters, relays, and push buttons, and links are connections. If more nodes and links are used in a design, it is graphically more complex and thus may need more response time. In the similar manner, a control logic is more adaptable if it needs fewer changes in the number of nodes and links compared to another logic to meet a change in specification. Hence, this study uses the number of nodes and links in an LLD and a comparable PN as a measure to compare their design complexity and response time. For the sake of convenience, nodes and links are called basic elements.

11.3 Comparison Through An Electro-Pneumatic System

One effective way to perform the comparison between LLDs and PNs is through an actual industrial automated system. The system shown in Fig. 10.5 is used for this comparative study along with the same input/output mapping tables.

11.3.1 Sequence Controller Design

Sequence 1: START, A+, B+, {C+, A-}, {B-, C-}

Consider that the system has to be controlled to execute the above sequence where A+ represents that the piston has to do a forward stroke and A- a return stroke. {C+, A-} represents two concurrent actions taking place simultaneously: Piston C a forward stoke and Piston A a return one. Fig. 11.1(a) shows the LLD and Fig. 11.1(b) shows the PN corresponding to this sequence. In the LLD, CR-1, CR-2, CR-3, and CR-4 represent controlled relays that activate relays R1, R2, R3, and R4 respectively. For more detailed description of the symbols used in this LLD and the procedure for developing the LLDs, refer to [Michel, 1990; Pessen, 1980]. In both LLD and PN, SW1 represents the push button which is used to start the sequence. The description of symbols, a0, b0, c0, a1, b1, and c1 that appear both in LLD and PN model is already presented in Section 5 of Chapter 10.

In LLD, the left hand side vertical line represents the input power line and the right hand side vertical line represents the output power line. Current flows from input line to the output line depending upon the logic modeled using the combination of various push buttons and limit switches. The logic of LLD executing the sequence is briefly described below:

- 1st and 2nd rows:

  When SW1 is pushed, and the normally opened switches a0, b0, and c0 are closed, CR-1 is activated which in turn will activate the relay coil R1 because current flows back from CR-1 to normally-closed R3 to R1.

11.3. Comparison Through an Electro-Pneumatic System

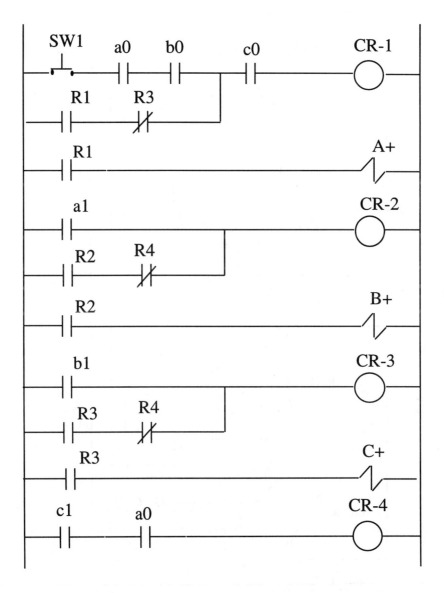

Figure 11.1(a) LLD for Sequence 1.

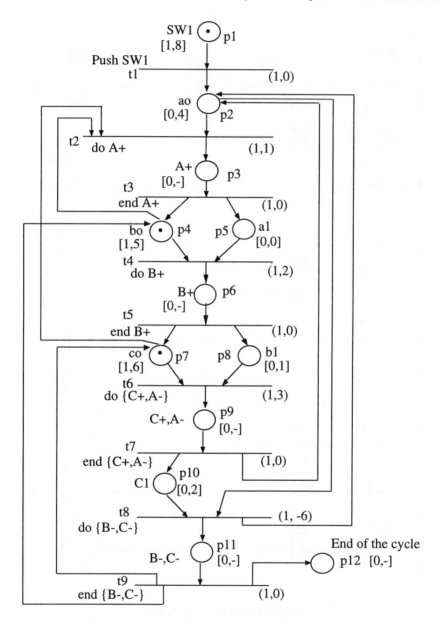

Figure 11.1(b) PN for Sequence 1.

11.3. Comparison Through an Electro-Pneumatic System

R1 will be in active state and thus sustain the output signal required to move piston A forward as long as the normally closed R3 is in the closed state. Hence the logic modeled in rows 1 and 2 triggers the action, A+.

- 3rd row:

  Since, R1 is in active state, the output signal A+ is sent to the digital output interface to move piston A forward. Piston A will be in this state as long as R1 is not deactivated. Note that since the solenoid switches used in the system require sustained signals to activate the forward motion of the pistons, we need to sustain R1's signal until piston A is ready to start backward motion, i.e., A-. Hence the logic modeled in row 3 executes the action, A+. More discussion on the differences between the solenoids that require sustained signals and those do not require sustained signals is presented later in Sections 11.4.3 and 11.6.

- 4th and 5th rows:

  After piston A completes its forward motion, it hits the limits switch a1 thereby changing the state of a1 from opened to closed. As soon as a1 is closed, CR-2 is activated which in turn will activate the relay coil R2 because current flows back from CR-2 to normally-closed R4 to R2. R2 will be in active state and thus sustain the output signal required to move piston B forward as long as the normally closed R4 is in the closed state. Hence the logic modeled in rows 4 and 5 triggers the action, B+.

- 6th row:

  Since, R2 is in active state, the output signal B+ is sent to the digital output interface to move piston B forward. Piston B will be in this state as long as R2 is not deactivated. Again, note that since the solenoid switches used in the system require sustained signals to activate the forward motion of the pistons, we need to sustain R2's signal until piston B is ready to start backward motion, i.e., B-. Hence the logic modeled in row 6 executes the action, B+.

- 7th and 8th rows:

  After piston B completes its forward motion, it hits the limits switch b1 thereby changing the state of b1 from opened to closed. As soon as b1 is closed, CR-3 is activated which in turn will activate the relay coil R3 because current flows back from CR-3 to normally-closed R4 to R3. R3 will be in active state and thus sustain the output signal required to move piston C forward as long as the normally closed R4 is in the closed state.

  Also, note that as soon as R3 is activated, the state of normally closed R3 in row 2 changes from closed to open, thereby breaking the current flow from CR-1 to R1. Hence, R1 will be deactivated which means now there is no signal to sustain piston A in its forward position. In other words, the spring of the solenoid activating piston A changes the position of the solenoid which in turn results piston A's to retreat backward, i.e., A-. Hence the logic modeled in rows 7, 8, 1 and 2 simultaneously triggers two simultaneous actions, namely {C+, A-}.

- 9th row:

Since, R3 is in active state, the output signal C+ is sent to the digital output interface to move piston C forward. Piston c will be in this state as long as R3 is not deactivated. Again, note that since the solenoid switches used in the system require sustained signals to activate the forward motion of the pistons, we need to sustain R3's signal until piston C is ready to start backward motion, i.e., C-. Hence the logic modeled in row 9 executes the action, C+.

- 10th row:

After piston C completes its forward motion, it hits the limits switch c1 thereby changing the state of c1 from opened to closed. Also, completion of piston A's backward motion hits the limit switch a0 thereby changing the state of a0 from opened to closed. When both c1 and a0 are closed, CR-4 is activated. As soon as R4 is activated, the state of normally closed R4 in rows 5 and 8 changes from closed to open, thereby breaking the current flow from CR-2 to R2 and CR3 to R3 respectively. Hence, both R2 and R3 will be deactivated which means now there is no signals to sustain piston B and piston C in their forward position. In other words, the springs of the solenoids activating piston B and piston C changes the position of the solenoid which in turn results piston B and piston C to retreat backward, i.e., B-, and C-. Hence the logic modeled in rows 10, 4, 5, 7, and 8 simultaneously triggers two simultaneous actions, namely {B-, C-}.

The PN shown in Figure 11.1(b) is modeled by using the concepts of RTPNs. The detailed procedure of modeling a discrete-even controller using RTPNs is already discussed in the previous chapter. The input mapping table used to model attributes for places is the same as shown in Table 10.5. Also, the attributes of transitions are the same as given in Table 10.7. The description of PN modeling the control logic to execute the sequence is self explanatory. The PN model executes the control sequence using the "token game" concept. In other words, as tokens move in the PN, they enable certain transitions to execute appropriate actions, firing of the transition causes to execute the corresponding action modeled by that transition. Note that in the PN, a place has attributes $[n_1, n_2]$ where n_1 is the first attribute representing initial number of tokens and n_2 is the second one mapping input channel number. Similarly, a transition has attributes

11.3. Comparison Through an Electro-Pneumatic System

(n_1', n_2') where n_1' is the firing duration and, n_2' is the number to be written on the digital output interface.

As discussed earlier, basic elements in a LLD or PN are nodes and links. In LLD, nodes are push buttons, normally opened switches, normally closed switches, output relay coils, timers, and counters and links are connections which connect these nodes. In PN, nodes are places and transitions and links are arcs connecting them. The LLD shown in Fig. 11.1(a) has 58 basic elements (24 nodes and 34 links), whereas the PN shown in Fig. 11.1(b) has 50 basic elements (21 nodes and 29 links). Even though PN uses fewer basic elements, it looks more complex due to the fact that all loops have to be closed to represent repetitive processes. This complicates the graphical appearance of PN compared to LLD. However, when the specification is changed the complexity of LLD grows faster than PN and this is illustrated below.

11.3.2 Control for other sequences

In order to compare the LLDs and PNs various sequences with increasing complexity are considered. These sequences involve emergency stop, counters for counting the number of repetitive operations, and timers for providing delays between certain operations. In order to highlight the changes in a model (PN or LLD) from one sequence to next sequence, given a model for a sequence, the additional basic elements needed to model the next sequence are shown by bold lines.

Sequence 2: START, 5 [A+, B+, {C+, A-}, {B+, C-}] (with emergency stop and counter)

Now, consider that the specification is changed such that the new control sequence is indicated as above. In this sequence, there is a need to provide emergency stop and a counter. The number 5 at the beginning of the sequence indicates that the sequence enclosed in the square brackets is to be repeated first times. Both LLD and PN are implemented such that when emergency stop switch, ES is pressed, the whole system including the active elements are immediately stopped. In other words, when switch ES is pressed all the solenoids are immediately deactivated. However, there exist several ways of modeling an emergency stop. Fig. 11.2(a) shows the LLD and Fig. 11.2(b) the PN corresponding to this sequence. In order to incorporate an emergency stop, PN uses a place p13 with an inhibitory arc as an input place for t2-t9 whereas LLD uses an output coil CR-5. When ES is pressed, CR-5 would become active and deactivates the coils CR-1, CR-2, and CR-3 which drive pistons A, B, and C respectively.

270   Chapter 11. Comparison of RTPNs and LLDs

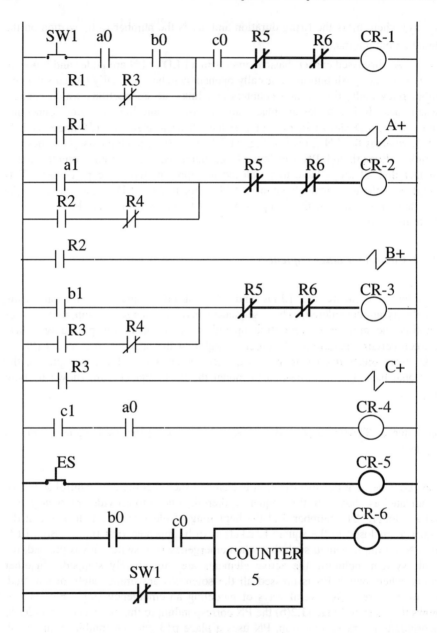

Figure 11.2(a) LLD for Sequence 2.

11.3. Comparison Through an Electro-Pneumatic System

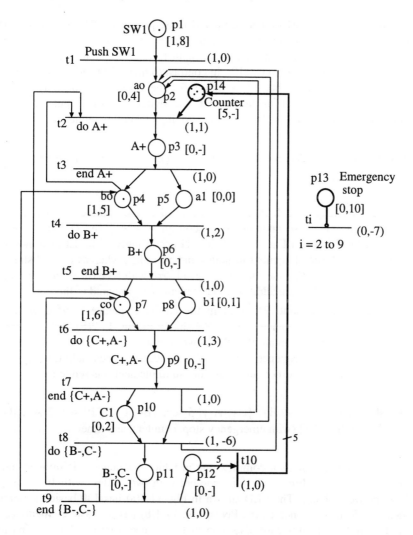

Figure 11.2(b) PN for Sequence 2.

In the PN, when the ES is pressed, the output transition corresponding to the place modeling ES writes an integer on the digital output interface which deactivates all the solenoids. The LLD shown in Fig. 11.2(a) has 87 basic elements (37 nodes and 50 links), whereas the PN shown in Fig. 11.2(b) has 64 basic elements (24 nodes and 40 links). Notice that there is no significant change in the physical appearance of LLD or PN compared to Sequence 1. Observe that LLD needs more additional basic elements compared to the PN. This is because the PN needs only one place p14 with an arc as an input to t2 to implement the counter. The counter resetting is modeled by t10. In contrast to this, LLD additionally needs 2 normally opened switches b0, c0 as inputs for the counter, 4 normally closed switches (SW1 as reset signal for the counter, and switch R6 as an input for CR-1, CR-2, and CR-3), a counter, and 16 more links to implement this sequence.

Sequence 3: START, 5 [A+, B+, {C+, A-}, 5 sec, {B+, C-}] (with emergency stop, counter, and timer)

The sequence is changed such that there is a need to incorporate a timer in the control logic to provide 5 seconds delay in between {C+, A-} and {B+, C-}. Fig. 11.3(a) shows the LLD and Fig. 11.3(b) the PN accordingly. The LLD shown in Fig. 11.3(a) has 94 basic elements (40 nodes and 54 links), whereas the PN shown in Fig. 11.3(b) is same as the one shown in Fig. 11.2(b) with 64 basic elements. It is observed that LLD needs more additional basic elements compared to the PN.

This is because the same PN used in the earlier sequence is used without changing the physical appearance. In the PN shown in Fig. 11.2(b), only the first attribute of t8 is changed to obtain the PN shown in Fig. 11.3(b) to incorporate time delay of 5 seconds in the sequence. On the other hand, LLD needs additionally two normally closed switches, a timer, and four links to implement this sequence.

Sequence 4: START, 3[A+, B+, {C+, A-}, 5 sec, {B+, C-}], 10 sec, 2[A+, B+, {C+, A-}, 5 sec, {B+, C-}] (with emergency stop, counters, and timers)

This new sequence represents a complex one in which Sequence 3 is divided into two segments (one with three cycles and another with two cycles) with 10 seconds time delay between them. The LLD shown in Fig. 11.4(a) has 130 basic elements (54 nodes and 76 links), whereas the PN in Fig. 11.4(b) has 69 basic elements (26 nodes and 43 links). In this case also note that LLD needs more additional basic elements compared to PN. This is because the PN needs only one additional transition t11 with an input arc from a new place p15 and an output arc to p14 to reset the counter.

11.3.3 Discussions

As mentioned in Section 11. 2 the number of basic elements is a common measure that gives an idea about graphical complexity, adaptability, and response time. It is

11.3. Comparison Through an Electro-Pneumatic System

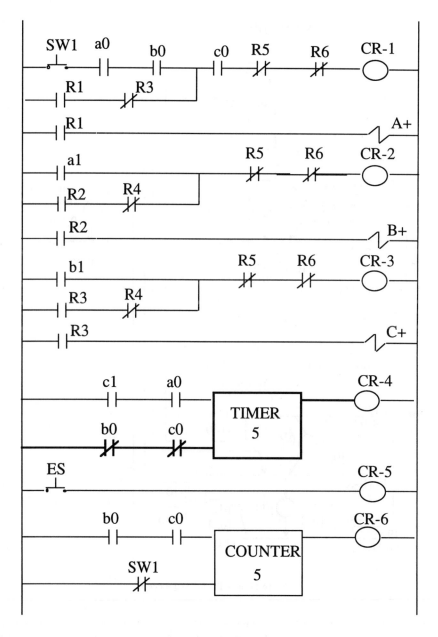

Figure 11.3(a) LLD for Sequence 3.

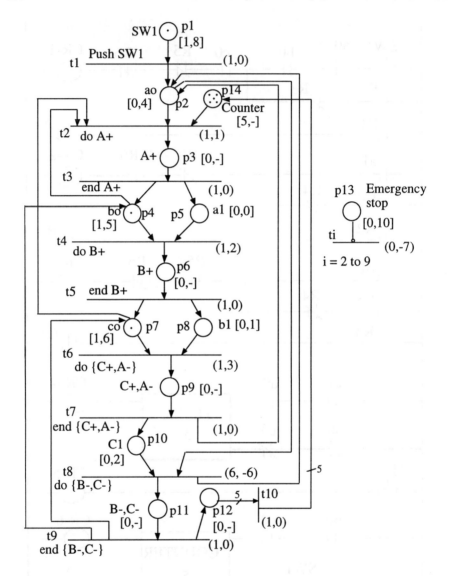

Figure 11.3(b) PN for Sequence 3.

observed that as the specification changes, PN requires fewer changes compared to LLD. Table 11.2 summarizes how the number of basic elements increases in LLD and PN as the complexity of sequence control specifications increases.

11.3. Comparison Through an Electro-Pneumatic System

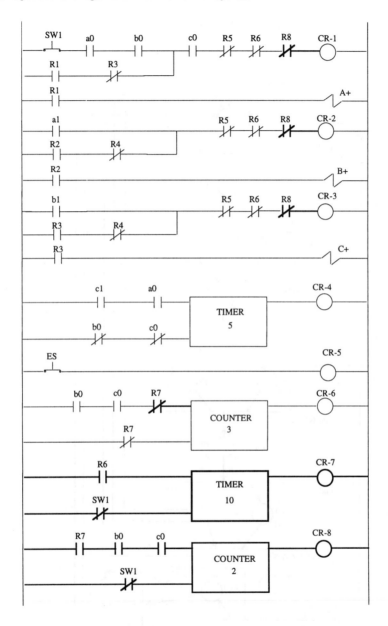

Figure 11.4(a) LLD for Sequence 4.

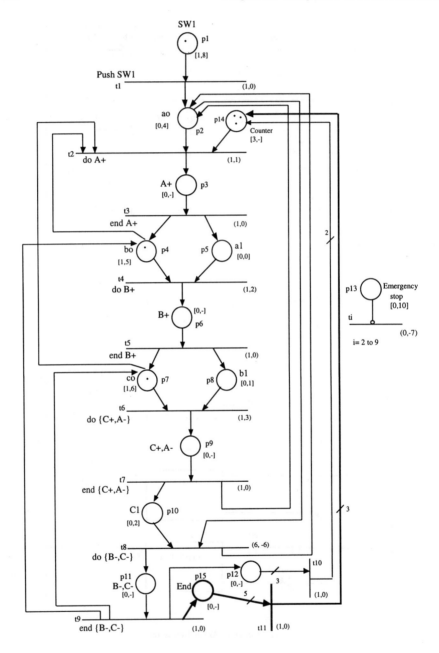

Figure 11.4(b) PN for Sequence 4.

11.3. Comparison Through an Electro-Pneumatic System

Table 11.2 Comparison of the basic elements in LLD and PNs.

Sequence #	Basic elements in LLD			Basic elements in PN		
	Nodes (NLLD)	Links (LLLD)	**Total (TLLD)**	Nodes (NPN)	Links (LPN)	**Total (TPN)**
1	24	34	**58**	21	29	**50**
2	37	50	**87**	24	40	**64**
3	40	54	**94**	24	40	**64**
4	54	76	**130**	26	43	**69**

It can be inferred that the PN shown in Fig. 11.1(b) used to execute the first sequence is slightly modified to get the PN shown in Fig. 11.4(b) corresponding to the last sequence. However, the LLD shown in Fig. 11.1(a) is significantly modified to get the LLD shown in Fig. 11.4(a). The modifications in terms of basic elements are quantified in Table 11.2 and shown graphically in Figure 11.5.

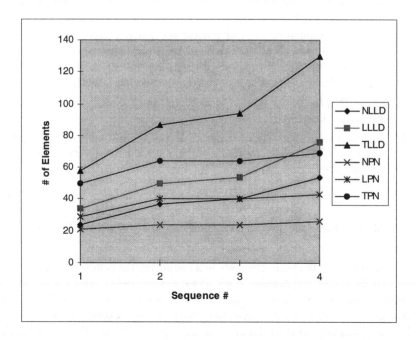

Figure 11.5 Comparison of basic elements in LLD and PN with respect to sequence number.

As shown in the graph, number of total basic elements for LLD increases significantly with respect to the sequence number. Also, observe that the physical appearance of PN is preserved (with slight modifications) starting from the first sequence to the last one. This is not true in the case of LLDs as in the previous figures. Furthermore, this case study reveals that PNs and LLDs do not differ much when the control sequence is relatively simpler as seen in the first sequence. In fact, PN model may appear more complex than LLD at the first sight as shown for the first sequence. However, when this sequence is gradually modified to result in a complex one, PNs are more easily modifiable and hence maintainable than LLDs.

Ease in modifiability and maintainability yields several advantages such as improvement in readability, understandability, and reliability as concluded in (Murata *et al.,* 1986). In LLDs, a same node may appear multiple times which may lead to difficulty in understanding the logic and cause errors in developing the logic. LLDs needs more basic elements to model timers and counters compared to PNs. In addition to these findings, the following points are experienced during the design and implementation of sequence controllers using LLDs and PNs:

1. Using PNs the control logic can be qualitatively analyzed to check properties such as absence of deadlocks and presence of reinitilizability in the system. Using LLDs qualitative analysis is not possible until it is simulated or implemented,

2. During the implementation of control sequences 3 and 4, it is found that debugging of the control logic with LLDs is difficult compared to PN. This is because PNs help to track the system dynamically with the help of the states of places and transitions [Venkatesh and Ilyas, 1995],

3. Using PNs, the initial state of the system can be directly represented by its initial marking, and

4. The reason for the complexity of LLD and the difficulty in debugging is that, in an LLD, a node appears more than once in the diagram. Due to this, tracking of control from one node to other becomes very difficult.

11.4 Analytical Formulas to Evaluate the Complexity of PNs and LLDs

In general, the fewer the basic elements in a controller, the better the model used to implement the controller [Boucher *et al.,* 1989; Pessen, 1989]. The number of basic elements decides the length of the control model. A shorter model uses less number of basic elements and is usually easier to understand, check, diagnose, and maintain. It also takes less time to enter the controller/computer [Pessen, 1989; Venkatesh *et al.,* 1994d]. Moreover, when low cost controllers with short memory are used it is

11.4. Analytical Formulas to Evaluate the Complexity

possible that they may run out of memory if the control model is too large [Pessen, 1989]. Counting of basic elements in a PN or LLD becomes cumbersome when their control specification becomes complex. Also, before physically modeling the given control logic with either LLD or PN, there is a need for a method that can help control engineers select between PN and LLD by estimating the basic elements used in these two models.

Motivated by these reasons, this section presents some analytical formulas to estimate the basic elements used in a PN or LLD [Venkatesh et al., 1994d; Venkatesh 1995]. Most of the control specifications can be modeled by using the logic constructs such as logical AND (NAND), logical OR (NOR), sequential model, and timed sequential model. It is obvious that AND/OR basically requires the same number of basic elements as NAND/NOR. Hence, the analytical formulas presented for AND/OR can also be applied to NAND/NOR cases. The formulation of these analytical formulas consists of their derivation and verification. This is done by physically modeling each logic construct using PN and LLD as described below. For the sake of convenience, the number of basic elements used in PN and LLD are represented by α and β, respectively. α is the summation of α_n and α_l, where α_n and α_l represent the number of nodes and links used in PN respectively. Similarly, β is the summation of β_n and β_l. We also define a function Δ which is defined as $\Delta = \alpha - \beta$ which helps to decide whether PN is better or LLD is better in terms of design complexity. For example, if Δ is negative, PN is preferred as it uses a smaller number of basic elements. On the other hand LLD is preferred when Δ is positive. In all the formulas presented in this chapter, m stands for the number of pre-conditions, and n for the number of post-conditions.

11.4.1 Logical AND, Logical OR, and Sequential Modeling

Logical AND

Figure 11.6 shows the PNs and LLDs for deriving the formulas to obtain α and β for logical AND. They are given as follows:

For PN: $\alpha = 2(m+n) + 1$, where

$\alpha_n = m + n + 1$,

$\alpha_l = m + n$.

For LLD: $\beta = 2m + 3n$, where

$\beta_n = m + n$,

$\beta_l = m + 2n$.

280   Chapter 11. Comparison of RTPNs and LLDs

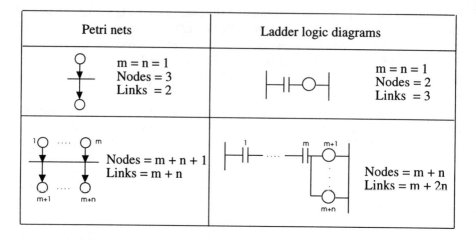

Figure 11.6 PNs and LLDs modeling logical AND.

Here, $\Delta = 1 - n$. This indicates when $n = 1$, both PN and LLD yield the same number of basic elements and if $n > 1$, PN is preferred.

Logical OR

Figure 11.7 shows the PNs and LLDs for deriving the formulas to obtain α and β for logical OR.

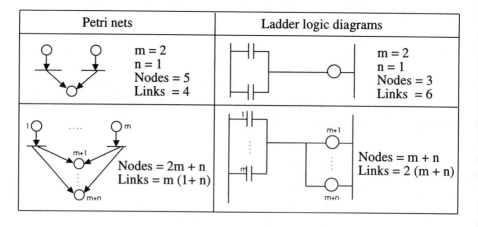

Figure 11.7 PNs and LLDs modeling logical OR.

11.4. Analytical Formulas to Evaluate the Complexity

They are given as follows:

For PN: $\alpha = 3m + n + mn$, where

$$\alpha_n = 2m + n,$$
$$\alpha_l = m(n + 1).$$

For LLD: $\beta = 3(m + n)$, where

$$\beta_n = m + n,$$
$$\beta_l = 2(m + n).$$

$\Delta = n(m - 2)$. LLD is better than PN for $m > 2$.

Sequential modeling

A condition is called a sequential condition when it acts as a pre-condition for its output operation (say operation $i+1$ and post conditions for its input operation (say operation i). The assumption here is that all operations except the first and last operations in the sequence have only one pre-condition and post-condition.

For sequences that violate this assumption, the number of basic elements can be found by decomposing the sequence with logical AND where some portions of the sequences have $n = 1$ and others have $n > 1$. Figure 11.8 shows the PNs and LLDs to derive the formulas to obtain α and β for sequential models.

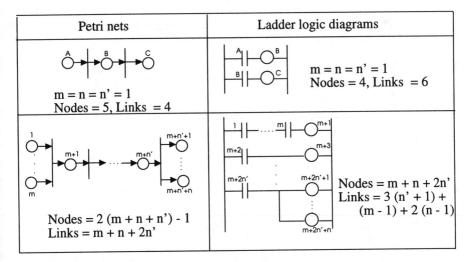

Figure 11.8 PNs and LLDs for sequential modeling.

For PN: $\alpha = 3(m + n) + 4n' - 1$, where

$\alpha_n = 2(m + n + n') - 1$,

$\alpha_l = m + n + 2n'$.

n' represents the number of sequential conditions.

For LLD: $\beta = 2m + 3n + 5n'$, where

$\beta_n = m + n + 2n'$,

$\beta_l = 3(n' + 1) + (m- 1) + 2(n - 1)$.

$\Delta = f(m,n,n') = m - n' - 1$. This indicates that PN is preferred if $m < (n + 1)$, i.e., the number of pre-conditions is less than the number of post-conditions plus one.

11.4.2 Timed logical AND, Timed logical OR, and Timed sequential models

Timed logical AND

In this model delays are associated with operations of the logical AND considered before. Figure 11.9 shows the PNs and LLDs to derive the analytical formulas to obtain α and β for timed logical AND. They are given as follows:

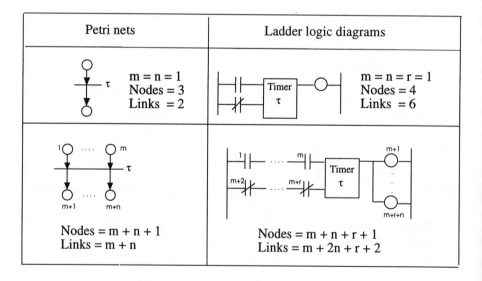

Figure 11.9 PNs and LLDs for timed logical AND.

11.4. Analytical Formulas to Evaluate the Complexity

For PN: $\alpha = 2(m + n) + 1$, where

$$\alpha_n = m + n + 1,$$
$$\alpha_l = m + n.$$

For LLD: $\beta = 2m + 3n + 2r + 3$, where

$$\beta_n = m + n + r + 1,$$
$$\beta_l = m + 2n + r + 2.$$

$\Delta = -n - 2r - 2$. This indicates that PN is always preferred to model timed logical AND.

Timed Logical OR

In this model it is assumed that for each timer there is only one reset signal. Figure 11.10 shows the PNs and LLDs to derive the analytical formulas for α and β.

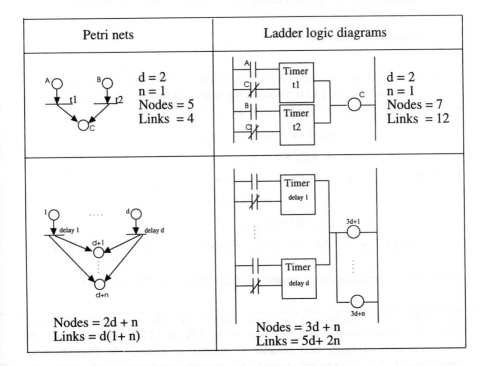

Figure 11.10 PNs and LLDs for timed logical OR.

α and β are given as follows:

For PN: $\alpha = 3d + n + dn$, where

$$\alpha_n = 2d + n,$$
$$\alpha_l = d(1 + n),$$

d represents the number of delays in the sequence.

For LLD: $\beta = 8d + 3n$, where

$$\beta_n = 3d + n,$$
$$\beta_l = 5d + 2n.$$

$\Delta = f(d,n) = -5d - 2n + dn$.

This indicates that for d and n values which satisfy $5d + 2n > dn$, PN is preferred. For d and n values that do not satisfy the above condition, LLD is preferred. For example, for the pairs, $d = 3$, $n = 4$; and $d = 4$, $n = 3$, PN is preferred. On the other hand for $d = 7$, $n = 8$, LLD is preferred. Similar analysis can be performed for any given d and n values using the above Δ function.

Timed Sequential Modeling

In this model it is assumed that for each timer there is only one reset signal. Figure 11.11 shows the PNs and LLDs to derive the analytical formulas for α and β.

For PN: $\alpha = 4d + 2(m + n) - 3$, where

$$\alpha_n = 2d + m + n - 1, \text{ and}$$
$$\alpha_l = 2d + m + n - 2.$$

For LLD: $\beta = 10d + 2m + 3n - 5$, where

$$\beta_n = 4d + m + n - 2,$$
$$\beta_l = 6d + m + 2n - 3.$$

$\Delta = f(d,m,n) = -6d - n + 2$ implies that PN is always better than LLD.

11.4.3 *Other Formulas for Estimating Basic Elements in PN and LLD*

The models presented till now are common for both PNs and LLDs. However, there are certain models that are specific to PN or LLD. For example, the implementation of timer and counter is similar in LLDs. In PNs, timer is implemented by associating delays to certain transitions and hence the implementation of timer and counter are not the same. Hence, there is a need for

11.4. Analytical Formulas to Evaluate the Complexity

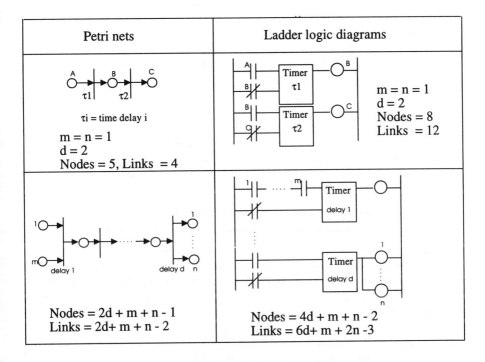

Figure 11.11 PNs and LLDs for timed sequential modeling.

separate formulas to estimate the number of basic elements to model a counter in case of a PN. Also, to control certain systems, relays have to be used in LLDs. This is explained here. A discrete event system can be controlled by controlling two types of elements: those that require sustained actuating signals, and those that need only a momentary or pulsed actuating signal [Pessen 1989b]. The first type of elements can be exemplified by an on-off solenoid valve with return spring and shown in Fig. 11.12(a). As long as the solenoid is actuated, the valve remains open (or closed, as the case may be). As soon as the solenoid is released, the return spring returns the valve to its original position. Hence, the solenoid needs a sustained actuating signal to keep the valve open. This requires the use of relays in LLDs. The second type of elements can be exemplified by a solenoid valve with two opposing solenoids but without a return spring and is shown in Fig. 11.12(b).

A momentary solenoid signal is sufficient to shift the valve into its other position, and the valve remains there until the opposing solenoid is actuated. It is strictly not allowed to actuate both solenoids at the same time. Actuating both solenoids simultaneously results in an undefined state and termed as "interlock" as

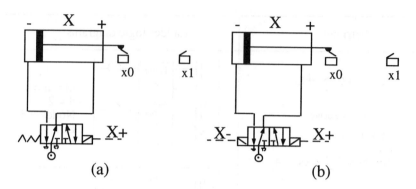

Figure 11.12 Typical cylinder-actuating circuit [Pessen, 1989b].
(a) requiring and (b) not requiring sustained solenoid signals.

the two solenoids fight each other causing heat up and burn out. Using a single PN model, both types of systems in Fig. 11.12 can be controlled. This can be accomplished by changing the corresponding attributes for transitions as reported in PN-based control using Real-time PNs. However, when using ladder logic diagrams, two separate design procedures have to be followed to design two separate diagrams as shown in [Pessen 1989b]. The formulas presented earlier are relatively simple compared to the ones described below.

Emergency stop modeling

Emergency stop can be implemented in two different ways. In the first, when the push button corresponding to emergency stop is pushed, all the operations that are in progress and those that are ready to start are immediately stopped. In other words, the pistons executing the operations are stopped at the position where they are. In the second implementation, when the push button is pushed, the system is brought to initial condition. That is, all the pistons that are executing operations are retracted. The second method of implementing emergency stop is very common and hence it is implemented and studied in this work. Denote the number of basic elements needed to model emergency stop (ES) modeling by PN and LLD as α_{ES} and β_{ES} respectively. In a PN, emergency stop is modeled by a place and inhibitory arcs from this place to transitions that model execution of actions. Figure 13 shows the implementation of ES using PN and LLD.

Given any sequence, the additional basic elements needed are shown with bold lines. α_{ES} is given as $\alpha_{ES} = 1 + k\Gamma$ where $k = 1$ if each action is modeled by one transition, and $k = 2$ if each action is modeled by two transitions (one to

11.4. Analytical Formulas to Evaluate the Complexity

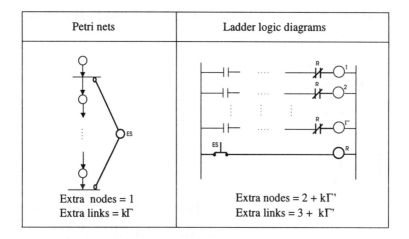

Figure 11.13 PNs and LLDs modeling an emergency stop.

represent start and second to represent end) and Γ stands for the number of actions in a given specification. β_{ES} is given as $\beta_{ES} = 5 + 2k\Gamma'$ where $k' = 1$ if each action is executed by a sustained actuating signal, and $k' = 2$ if each action is executed by a pulsed actuating signal and Γ' stands for the number of pistons in a given specification.

Basic elements needed to model a counter

Irrespective of the given sequence, the implementation of a counter is unique as shown in Fig. 11.14. In Fig. 11.14, given any sequence, the additional basic elements to implement a counter are shown with bold lines. As it can be observed from this PN model three new arcs (one output arc from place modeling the counter, one output arc from the place modeling the end of the cycle, and one input arc to the place modeling start of the cycle) and two new nodes (one place and one transition) are needed to incorporate a counter in a sequence. Hence, the number of basic elements needed to model a counter using PNs is five.

In LLDs, even though the implementation of timer and counter are similar, the output of the counter may have to be used as a precondition to execute certain actions. This is usually done by adding the output of the counter as a precondition for certain actions with normally closed contact. This leads to an extra node and link for each action. Hence, The number of basic elements to implement counter in LLD is given as $\beta = (2m + 3n + 2r + 2k\Gamma' + 3)$, where r = *number of reset signals*, $k' = 1$, if each action is executed by sustained actuating signal, $k' = 2$, if each action is executed by a pulsed actuating signal, and Γ' stands for the number of pistons in a given specification.

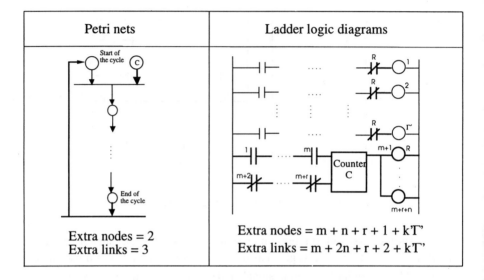

Figure 11.14 PNs and LLDs modeling a counter.

Basic elements needed to model a relay

Figure 11.15 shows the implementation of a relay. When LLDs are used to control

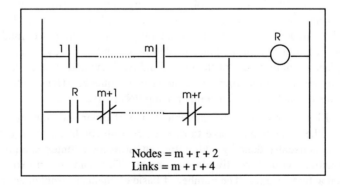

Figure 11.15 LLD modeling a relay.

11.5. Methodology to use the Analytical Formulas

a system with sustained actuating signals, relays are needed to implement the sustained actuating signals. It can be observed that the number of basic elements needed to model a relay is $2m + 2r + 6$, where r is the number of reset signals.

11.5 Methodology to use the Analytical Formulas

Figure 11.16 illustrates the method to obtain the total number of basic elements to model a given control logic using the analytical formulas.

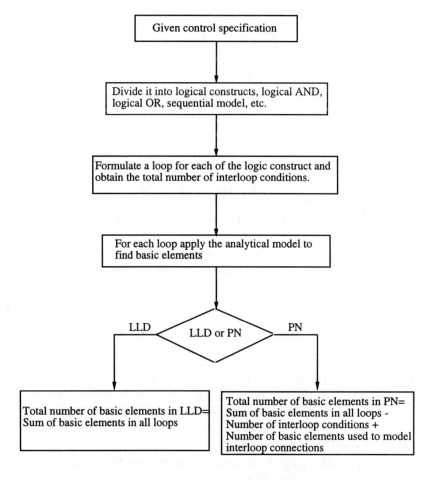

Figure 11.16 Method to estimate basic elements.

Most of the control logics can be represented using the basic building blocks of logic constructs described in the earlier section. Hence, in order to find the basic elements to model a given control logic using the models developed, first it has to be decomposed into the logic constructs or segments. Then for each segment, the corresponding analytical formula is applied to find the number of basic elements.

The basic elements corresponding to all such segments are then added to obtain the total number of basic elements (β) needed to model the given control logic using a LLD.

$$\beta = \sum_{i=1}^{k} \beta_i$$

where β = Total number of basic elements to model the control logic using PN,

β_i = Number of basic elements in segment i, and

k = Total number of segments.

The final step (as shown in Fig. 11.16) is slightly different in the case of PN due to its physical model. This is because when modeling a control logic, intersegment conditions and connections exist in a PN model. They are explained below. For some cases in the given control logic, the output condition for segment i may be one of the input conditions for segment j. These conditions are called as intersegment conditions. In LLD, such conditions are separately modeled in both segments i and j. Even though such conditions appear in more than one segment, the place modeling such condition physically appears only once in the PN. Therefore, in case of PN to account for the repetitive count of these intersegment conditions, the total number of such conditions has to be subtracted from the summation of the number of basic elements in all segments.

For some control specifications it may possible that an output condition of segment i causes an action which produces an input condition(s) for segment j. That means intersegment connections exist between segments i and j. Basically, intersegment connections model the power/control flow from one segment to another. In LLD these connections need not be explicitly modeled since the vertical line at the left hand side of LLD always models the existence of electric power. Depending upon the logic of each rung, the current flows from this power line and energizes the output of the rung connected to the right hand side line in LLD. However, in case of PN the power flow (called as control flow in PN terminology) across segments is modeled by the movement of tokens in PN. Since, transition firings cause the flow of tokens through arcs, the intersegment connections have to be explicitly modeled. The total number of basic elements to model intersegment connections is estimated as shown below.

Let θ_{ij} = The number of basic elements used to model intersegment connections between segments i and j,

11.5. Methodology to use the Analytical Formulas

$\theta_{ij} = \theta_{ijn} + \theta_{ijl}$,

θ_{ijn} = Number of Nodes (typically transitions) used to model intersegment connections between segments i and j,

θ_{ijl} = Number of Links used to model intersegment connections between segments i and j, then

$\theta_{ijn} = q$, where q represents the total number of output conditions in segment i producing inputs for segment j.

Sometimes, it may happen that an action corresponding to the output condition in segment i produces several inputs for segment j. Considering this, q_{ijl} is given as follows:

$$\theta_{ijl} = q + \sum_{r=1}^{q} w_r$$

where w_r represents the number of inputs for segment j produced by output condition r in the segment i. The first term in θ_{ijl} corresponds to the output arcs from the places modeling output conditions in segment i and the second term corresponds to the input arcs to the places modeling input conditions in segment j. Hence, θ_{ij} is given as:

$$\theta_{ij} = 2q + \sum_{r=1}^{q} w_r$$

The total number of basic elements modeling intersegment connections in the whole control logic is then given as follows:

$$\theta = \sum_{i=j}^{k} \theta_{ij}$$

Considering intersegment connections, in case of PN, the total number of basic elements used to model the total number of intersegment conditions has to be added for the summation of basic elements in all segments. Hence, considering both intersegment conditions and connections, the total number of basic elements needed to model the control logic using PN is termed as α and given as follows:

$$\alpha = \sum_{i=1}^{k} \alpha_i - \phi + \theta$$

where, α_i = Number of basic elements in segment i,

k = Total number of segments,

ϕ = Total number of intersegment conditions,

292   Chapter 11. Comparison of RTPNs and LLDs

θ = Total number of basic elements needed to model intersegment connections.

Once, α and β are known, we can select PN or LLD based on $\Delta = \alpha - \beta$. If Δ is positive, LLD is preferred, otherwise, PN is preferred from the design complexity view point.

11.6 Illustration of the Methodology Through Examples

In order to make the application of the methodology more clear, this section presents three examples. In the first example, the PN has intersegment conditions without intersegment connections. In the second example, the PN has intersegment connections without intersegment conditions. Also, to control this system using LLDs, sustained signals are not required. Finally, the electro-pneumatic system earlier considered in Section 11.3 is adopted in Example 3. The PN models of this system have both intersegment connections and intersegment conditions. To control this system using LLDs, sustained signals are also needed. The systems considered in these examples are very common in flexible automation.

11.6.1 An automatic assembly system

Consider the automatic assembly system shown in Fig. 10.3. The functions of the elements present in the system have to be synchronized by a control logic. With an objective to use a smaller control program, we need to find which one between PN and LLD is better to model this control logic. The methodology described in the above section is used to estimate the number of basic elements needed for PN and LLD. Table 11.3 shows the results obtained.

Since C and E are intersegment conditions, $\phi = 2$, and since there are no intersegment connections, $\theta = 0$. Hence, $\alpha = 7 + 5 + 7 - 2 = 17$; $\beta = 12 + 10 + 12 = 34$, and $\Delta = \alpha - \beta = -17$. This indicates for this system, PN gives 50 % shorter model than LLD.

This can also be validated by manually counting the basic elements after the physical models of PN and LLD are formulated. The PN and LLD to control the system is shown in Figs. 10.4 and Fig. 11.17 respectively. It can be easily observed that PN and LLD uses 17 and 34 total basic elements to model the control logic.

11.6. Illustration of the Methodology

Table 11.3 Required basic elements to control the system in Fig 10.3.

Segment	Description of logic construct	Models applied	α_i	β_i
1	A.B, τ_1 -> C	Timed logical AND $m = 2, n = 1, r = 1$	$2m+2n+1 = 7$	$2m+3n+2r+3 = 12$
2	D, τ_2 --> E	Timed logical AND $m = 1, n = 1, r = 1$	$2m+2n+1 = 5$	$2m+3n+2r+3 = 10$
3	C.E, τ_3 --> F	Timed logical AND $m = 2, n = 1, r = 1$	$2m+2n+1 = 7$	$2m+3n+2r+3 = 12$

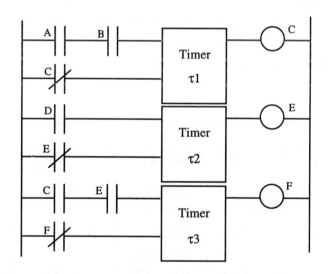

Figure 11.17 LLD model for the system shown in Fig. 10.3.

11.6.2 An electro-pneumatic system without sustained signals

Now, consider another system where intersegment connections exist in the control specification. Low cost automated systems with hydraulic/pneumatic pistons actuated by solenoids exhibit such type of intersegment connections. This is illustrated in this example. The system consists of four pneumatic pistons (A, B, C, and D) which are to be sequenced by double activated five ports and two-way solenoid valves. Each piston has two normally open limit switches. Also, consider that all pistons are of the type shown in Fig. 11.12 (b), which require momentary actuating signals. When the end of piston X contacts limit switch x1 (x0), x1 (x0) is closed indicating that the piston X is at the end of its forward stroke (return stoke). A push button, START is provided to start the system. It is given that pistons A, B, C, and D have to be sequenced according to the following sequence: START, {A+, B+}, {A-, D+}, {B-, D-, C+}, C.

The methodology earlier described is followed to estimate the number of basic elements needed for PN and LLD. Table 11.4 shows the results obtained. Since there are no intersegment conditions, $\phi = 0$,. θ is calculated as follows: Upon observation it can be seen that intersegment connections exist between Segments 1 and 2; 2 and 3; 3 and 4; and 4 and 1. For example, A+ and B+ in Segment 1 results in a+ and b+ in Segment 2. Therefore, for Segments 1 and 2, q = 2.

Assuming A+ is action 1 and B+ as 2, $\theta_{12} = 2q + w_1 + w_2 = 2.2 + 1 + 1 = 6$. Similarly, for Segments 2 and 3, $\theta_{23} = 6$. For Segments 3 and 4 assuming B-, D-, C+ as first, second, and third actions, $\theta_{34} = 2q + w_1 + w_2 + w_3 = 2.3 + 1 + 1 + 1 = 9$. For Segments 4 and 1, $\theta_{41} = 2q + w_1 = 2.1 + 1 = 3$. Hence, $\theta = \theta_{12} + \theta_{23} + \theta_{34} + \theta_{41} = 6 + 6 + 9 + 3 = 24$.

Table 11.4 Required basic elements to control the system in Example 2.

Seg-ment	Description of logic construct	Models applied	α_i	β_i
1	START.c- --> {A+,B+}	Logical AND m = 2, n = 2	2m+2n+1 = 9	2m+3n = 10
2	a+.b+ --> {A-,D+}	Logical AND m = 2, n = 2	2m+2n+1 = 9	2m+3n = 10
3	a-.d+ --> (B-,D-,C+}	Logical AND m = 2, n = 3	2m+2n+1 = 11	2m+3n = 13
4	b-.d-.c+ --> {C-}	Logical AND m = 3, n = 1	2m+2n+1 = 9	2m+3n = 9

Hence, α, β, and Δ are given as follows:
$\alpha = (9 + 9 + 11 + 9) + 24 = 62$, $\beta = 10 + 10 + 13 + 9 = 42$, and $\Delta = \alpha - \beta = 20$. This indicates for the example system considered, LLD gives 32.25 % shorter program than PN. This can also be validated by manually counting the basic elements after

11.6. Illustration of the Methodology

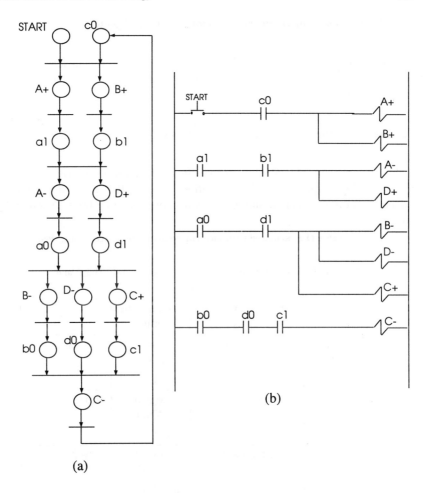

Figure 11.18 (a) PN and (b) LLD for Sequence: {A+, B+}, {A-, D+}, {B-, D-, C+}, C-.

the physical models of PN and LLD are formulated. The PN and LLD to control the system are given Fig. 11.18(a) and (b) respectively. After counting the basic elements, it can be easily observed that PN and LLD uses 62 and 42 total basic elements to model the control logic. This indicates that when the system contains only pneumatic pistons that do not require sustained signals for solenoids, LLD would yield a shorter program than PN. However, if time delays are added in the above sequence, PN structure will not be changed as time can be modeled as an attribute to a transition. On the other hand the length of LLD will be increased to incorporate the timers.

11.6.3 An electro-pneumatic system with sustained signals

The electro-pneumatic system considered in Sec. 11.3 is considered here. The number of basic elements in PNs and LLDs is estimated using analytical formulas instead of manual counting as we did in Sec. 11.3.

Estimation of the number of basic elements in PNs

Table 11.5 presents the calculations to estimate the number of basic elements in the PN modeling Sequence 1. In calculation, observe that x0 is a precondition to execute action X+ where x and X stands for corresponding limit switch and piston identity.

Table 11.5 Required basic elements to model Sequence 1 using PN.

(a) Basic elements in each segment without considering θ and ϕ.

Segment	Description of logic construct	Model to be applied	α_i
1	ST- --> a0, b0, c0	Logical AND m = 1, n = 3	2m+2n+1 = 9
2	a0 --> A+	Logical AND m = 1, n = 1	2m+2n+1 = 5
3	a1.b0 --> B+	Logical AND m = 2, n = 1	2m+2n+1 = 7
4	b1.c0 --> {C+, A-}	Logical AND m = 2, n = 1	2m+2n+1 = 7
5	c1.a0 --> {B-, C-}	Logical AND m = 2, n = 1	2m+2n+1 = 7
6	{B-, C}- --> End	Logical AND m = 1, n = 1	2m+2n+1 = 5

11.6. Illustration of the Methodology

Table 11.5 (b) Basic elements representing intersegment conditions.

Segments j and k	Intersegment condition	ϕ_{jk}
1 and 2	a0	1
1 and 3	b0	1
1 and 4	c0	1
1 and 5	a0	1
5 and 6	{B-, C-}	1

Table 11.5 (c) Basic elements representing intersegment connections.

Segments j and k	Intersegment connection	Values of variables	θ_{jk}
2 and 3	A+ --> a1	$q = 1, w_1 = 1$	$2q + w_1 = 3$
3 and 4	B+ --> b1	$q = 1, w_1 = 1$	$2q + w_1 = 3$
4 and 5	C+, A- --> c1, a0	$q = 1, w_1 = 2$	$2q + w_1 + w_2 = 4$
5 and 3	B- --> b0	$q = 1, w_1 = 1$	$2q + w_1 = 3$
5 and 4	C- --> c0	$q = 1, w_1 = 1$	$2q + w_1 = 3$

For simplicity, denote the number of basic elements in PN corresponding to Sequence i as α_{si}. Now, following the formulas presented earlier, α_{si} can be estimated as follows:

$$\alpha_{s1} = (9+5+7+7+7+5) - (1+1+1+1+1) + (3+3+4+3+3) = 40 - 5 + 16 = 51.$$

Observe that a0 is presented as a precondition in both Segments 2 and 5. However, in PN each limit switch appears physically once (with several input and output arcs). Hence, to compensate this the previous value of α_{s1} has to be subtracted by one. Then, the final value is 50 which exactly matches with the value obtained in Sec. 11.3. Sequence 2 is obtained by adding a counter and emergency stop to Sequence 1. Hence, α_{s2} can be obtained by adding the basic elements needed to model a counter (α_C) and emergency stop (α_{ES}) to α_{s1}. Then, $\alpha_{s2} = \alpha_{s1} + \alpha_C + \alpha_{ES}$. By following the models presented earlier, $\alpha_C = 5$. The number of basic elements to model ES can be calculated using the earlier developed formula.

Since there are four actions (recall that {C+,A-} and {B-,C-} are treated as two concurrent actions) present in the sequence and each action is modeled by two transitions (one to start and second to end), $\alpha_{ES} = 1 + k\Gamma = 1 + 2.4 = 9$. Therefore, $\alpha_{s2} = 58+5+9 = 64$ as obtained earlier in Sec. 11.3.

Sequence 3 is obtained by adding delay before {B-,C-} of Sequence 2. However, in PNs this delay is implemented by changing the second attribute of transition modeling "do {C+,A-}". Hence, there is no need for extra basic elements. In other words, $\alpha_{s3} = \alpha_{s2} = 64$ as given in Sec. 11.4. Sequence 4 is obtained by adding a counter and timer to Sequence 3. Again, we know that $\alpha_C = 5$, and there is no need for extra basic elements to model timer. Hence, $\alpha_{s4} = \alpha_{s3} + \alpha_C = 64 + 5 = 69$ as obtained in Sec. 11.3.

Estimation of the number of basic elements in LLDs

When sustained actuating signals are used, the following simple three step procedure is developed to estimate β.

1. Assign a relay for each segment in the given sequence except the segment containing 'Start' command,
2. Identify set and reset signals for each sustained signal implemented by a relay. Note that, setting (resetting) of a relay executes forward (return) stroke of a piston,
3: Treat set signals for a relay as preconditions, and apply the formulas to calculate the number of basic elements.

For the case study considered in Section 11.3, the results obtained after each step by following the above procedure are as follows:

After Step 1: Relay 1 (R1), R2, R3, and R4 are assigned to A+,B+,{C+,A-}, and {B-,C-}, respectively.

After Step 2: Set signals for the sustained signal by R1 are ST, a0, b0, R1. Since A- is in Segment 3, R3 is a reset signal for R1. Since R2 has to wait till R1 completes all of its actions, the set signals corresponding to R2 are a1 and R2 and reset signal is R4 because B- is in Segment 4. Similarly, the set signals for R3 are b1 and R3 and reset signal is R4 because C- is in Segment 4. Set signals for R4 are c1 and a0.

After Step 3: Table 11.6 shows the calculations to estimate β.

From the table, $\beta_{s1} = 16+5+10+5+10+5+7 = 58$. In Sequence 2 counter and emergency stop are added to Sequence 1. Denote β_C and β_{ES} the number of basic

11.6. Illustration of the Methodology

elements needed to model a counter and emergency stop, respectively. $\beta_{s2} = \beta_{s1} + \beta_C + \beta_{ES}$. β_C and β_{ES} are calculated by using the derived formulas as follows:

m is 2 because b0,c0 resulting from executing the last action in the given sequence are to be taken as the set signals for the counter. n is 1 because there is only one output from the counter. r is 1 because start switch ST can be taken as a reset signal for the counter. k' is 1 because in the system under study each action that drives the piston forward is executed by sustained actuating signal and Γ' is 3 because three pistons are present.

Table 11.6 Required basic elements to model the Sequence 1 using LLD.

Segment	Description of logic construct	Model to be applied	β
1(a)	ST.a0.b0.c0.R3 --> R1	Relay m = 4, r = 1	2m+2r+6 = 16
1(b)	R1 --> A+	Logical AND m = 1, n = 1	2m+3n = 5
2(a)	a1.R4 --> R2	Relay m = 1, r = 1	2m+2r+6 = 10
2(b)	R2 --> B+	Logical AND m = 1, n = 1	2m+3n = 5
3(a)	b1.R4 --> R3	Relay m = 1, r = 1	2m+2r+6 = 10
3(b)	R3 --> C+	Logical AND m = 1, n = 1	2m+3n = 5
4	c1.a0 --> R4	Logical AND m = 2, n = 1	2m+3n = 7

Hence, $\beta_C = 2m + 3n + 2r + 2k\Gamma' + 3 = (2.2 + 3.1 + 2.1 + 2.1.3 + 3) = 18$. $\beta_{ES} = 5 + 2k\Gamma' = 5 + (2)(1)(3) = 11$. Hence, $\beta_{s2} = 58 + 18 + 11 = 87$ which is exactly the same as obtained in Section 11.3. Also, since Sequence 3 is obtained by including a delay after {C+, A-} in Sequence 2. Hence, the timer is to be set with c1 and a0 as input signals and b0 and c0 as reset signals. There is only one output from timer. Hence, m = 2, n = 1, and r = 2. The number of basic elements needed to implement the timer, $\beta_T = 2m + 3n + 2r + 3 = 2.2 + 3.1. + 2.2. + 3 = 14$. Segment

4 in Table 11.6 already contains the elements and links corresponding to c1, a0, and R4. Hence, to compensate these, $\beta_{s3} = \beta_{s2} + \beta_T - \beta_4$ where $\beta_4 = 7$ denotes the number of basic elements in Segment 4. Therefore, $\beta_{s3} = 87 + 14 - 7 = 94$ as obtained in Section 11.3.

Since Sequence 4 is obtained by adding a counter and delay to Sequence 3, it can be implemented by adding a counter and timer to the LLD of Sequence 3. Even though the number and identity of set and reset signals for the new counter and timer are different from those used in the earlier sequences, as an approximation we can still use the number of basic elements that were calculated earlier. In other words, $\beta_{s4} = \beta_{s3} + \beta_C + \beta_T$. Since, β_C and β_T are 18 and 14 respectively, $\beta_{s4} = 94 + 18 + 14 = 126$ which is close to 130 as in Section 11.3.

Even though we could accurately estimate the number of basic elements in the last case by accurately identifying the set and reset signals for timer and counter, it may take significant effort to analyze the logic. However, the motivation behind proposing the formulas presented here is to not to exactly calculate the number of basic elements to match that result from manually counting an LLD or PN, but to give an idea of how many basic elements are needed to model the control logic. Using the approach shown here these formulas can be incrementally applied even for complex sequences consisting of timers and counters. In other words, they are first applied to a simple sequence to estimate the number of basic elements. Then the number of basic elements needed for counters and timers is estimated and added to the previous value. The approach allows designers to select between systems that use sustained actuating signals and momentary/pulse actuating signals as well as PNs and LLDs. For example, for the system considered in Section 11.6.2, consider that the pistons are controlled by solenoids that require sustained actuating signals. Then by applying the formulas as shown earlier, the number of basic elements in LLD can be found as 76. Recall that when momentary/pulse actuating signals are used, the basic elements in LLD are earlier calculated as 42. Also, recall that PN required 62 basic elements. Hence, by using the formulas and following the proposed methodology to estimate the basic elements, control engineers can select between PN and LLD even before they start to build the control model in detail.

11.7 Summary

Development of flexible, reusable, and maintainable control software is important in implementing advanced industrial automated systems. Traditional methods of using ladder logic diagrams (LLDs) to design sequence controllers are being challenged by the needs in flexible and agile manufacturing systems. On the other

11.7. Summary

hand, Petri nets (PNs) are an emerging tool that needs to be established for the control of discrete manufacturing systems. A class of PNs called real-time PNs that resemble ordinary PNs were introduced to design sequence controllers. This chapter identified design complexity and response time as the criteria to compare LLDs and PNs. Design complexity is defined and characterized by two factors namely graphical complexity and adaptability to meet changes in control specifications. By designing and implementing the control of an industrial automated system subject to changing control requirements, this chapter compares LLDs and PNs in terms of a common measure namely, the numbers of basic elements, which signify both design complexity and response time. Motivated by the fact that a sequence controller can be designed by synthesizing the building blocks of logic models [Venkatesh *et al.*, 1994e, 1995], this chapter proposed analytical formulas to estimate the number of basic elements to model the most commonly used building blocks of logic modeling by both PN and LLD. Furthermore, this chapter presented a methodology that uses the developed analytical formulas to estimate the total number of basic elements to model a control logic even before physically modeling it using PNs and LLDs. The concepts developed in this chapter are demonstrated by considering several examples of sequence controllers. The examples considered here demonstrate the potential for practical applications of the research results.

The methodology presented provides an accurate quantitative comparison of PN and LLD in terms of their basic elements by precluding the need for physically building the controllers by either PN or LLD and serves as an effective aid for a control engineer to select between PN and LLD even before starting to write the control program. The methodology developed is simple and straightforward to apply. However, in case of complex control specifications, decomposing the control specification in terms of the logic constructs may be difficult. This problem can be solved using the traditional methods such as Karnaugh maps, Huffman method, etc. Other factors which have impact on the selection of LLDs and PNs should also be explored in the future work. For example, similar to graphical complexity, irrespective of the implementation scheme an effort to quantify the response time complexity should be made.

Recently there have been many studies similar to this chapter that compare Petri Nets with LLDs and Sequential Function Charts. A comparison between LLDs and RTPNs is recently reported in [Zhou and Twiss, 1995]. Zhou and Twiss (1996) have also presented a survey comparing six techniques for discrete event control. They considered techniques such as direct implementation of relay ladder logic diagrams, Instrumentation Society of America (ISA) logic diagrams, timing/sequence diagrams, state diagrams, Petri nets, and sequential function charts. Realizing the power of PNs in discrete-event systems, Stanton *et al.*, (1996) focused on the implementation of PNs on PLCs using LLDs.

CHAPTER 12

AN OBJECT-ORIENTED DESIGN METHODOLOGY FOR DEVELOPMENT OF FMS CONTROL SOFTWARE

12.1 Introduction

In order to develop integrated control software, there is a need for integrated tools and a systematic design methodology to utilize those tools. The integrated tools should aid in not only real-time control but also simulation. The systematic design methodology should aim to develop reusable, modifiable, and extendible control software in order to meet the changing control specifications. Earlier chapters have shown the application of Petri nets (PNs) for both real-time control and simulation. This chapter makes an attempt to introduce PNs into object-oriented design (OOD) of integrated control software development. A flexible manufacturing system (FMS) consists of machines, robots, automated guided vehicles, programmable logic controllers, and computers, all of which can be viewed as objects in OOD. The system control software has to be designed to meet the real time constraints of a production line and the dynamically changing needs of the market. Because of the complexity, systematic software development is very important in realizing the full benefits of agile and flexible manufacturing.

This chapter highlights the difficulties in developing such software and proposes a systematic OOD methodology for its development using object modeling technique (OMT) diagrams and Petri nets (PNs). OOD is used to design reusable and easily maintainable software. OMT diagrams are used to represent explicitly different kinds of static relations such as generalization, aggregation, and association among the objects in FMS. PNs are used as a tool to model the dynamic behavior of the objects and FMS. By adopting a bottom-up approach of PN modeling, this chapter shows that PNs can support important characteristics of OOD, namely, reusability, extendibility, and modifiability. They can also be used for the FMS performance analysis. Furthermore, with PN models, a method to identify the data structures and operations of objects is presented. The proposed methodology is illustrated through an example of FMS. The issues related to the reusability, extendibility, and modifiability of the resulting control software are also discussed by changing the initial specifications of the FMS and thus its OMT diagram and PN model.

12.1.1 Background

Flexibility in the functioning of FMSs is mainly imparted by the use of computers and robots which are programmable according to the shop floor needs. Systematic control software development for FMSs is very important in realizing agile and flexible production [May, 1986; Marinov and Todorov, 1988]. Control software that not only supports simulation but also helps in FMS implementation is of paramount importance to exploit the full benefits of FMSs [May, 1986]. The advantages and the importance of integrated software packages that can be used for both simulation and control are discussed in [Bruno and Marchetto, 1986; May, 1986; Glassey and Adiga, 1990; Venkatesh *et al.*, 1991, Johnson, *et al.*, 1992].

Traditionally the function of FMS control software was to coordinate and control different elements in a manufacturing system. However, recently there are several efforts to integrate the control software and simulation software in order to expedite the development of control software and system design [Naylor and Voltz, 1987; Chaar *et al.*, 1991, 1993a,b; Venkatesh *et al.*, 1991; Johnson *et al.*, 1992; Venkatesh and Fernandez, 1993]. Changes in the production facilities may also require redesign and rewriting of large amount of software. The difficulty in developing the FMS control software is further compounded by the inherent complexity of FMS execution since it is a complex asynchronous concurrent system aiming to produce several product types with different production processes. Hence, the traditional definition of control software has been extended to handle simulation, planning, monitoring, rapid prototyping, and scheduling [Jain, 1986; Naylor and Voltz, 1987; Chaar *et al.*, 1991, 93a,b; Venkatesh *et al.*, 1994c].

A design methodology for FMS control software should at least deal with the issues related to 1) modeling, simulation, and analysis, and 2) real-time control implementation of FMS. The first one typically involves building models of an FMS and estimating measures of system performance such as throughput rate, utilization of robots, machines, and queue lengths at each machine. The aim of this activity is to suggest the optimum configuration of FMS for the given specifications. The optimum configuration may include the layout of the FMS, the routings of automated guided vehicles (AGVs) among machines, the machine scheduling polices, and other tasks. The second one is to implement effective real-time control of FMS which typically involves coordination, on-line scheduling and

12.1. Introduction

monitoring of system resources. The functional objective of control software is to maintain high system utilization and throughput as well as to satisfy the real-time production deadlines. In addition to this in FMSs, it is very common to find similar machines and robots grouped in cells which function in a similar way. The only difference between the functioning of these cells from discrete control point of view is the routing of parts among machines in cells. Hence, the control software should be written for a cell such that it can be used to control another cell with minimum or no change.

Based on the above discussion, it can be concluded that the control software should be reusable, modifiable, and extendible to: 1) adapt to changes in the system configuration and specifications, and 2) to deal with a complex shop-floor system which often consists of numerous similar components. A systematic design methodology is obviously needed to develop such FMS control software [Naylor and Voltz, 1987; Glassey and Adiga, 1990; Chaar et al., 1993b; Venkatesh and Fernandez, 1993].

12.1.2 Literature Review and Motivation

Realizing the importance and complexity of FMS software many researchers are investigating different issues related to it. Naylor and Voltz (1987) proposed an approach for designing integrated manufacturing system control software. They use three software concepts, the use of software components extended to include hardware, a common distributed language environment, and generic software. Smith and Joshi (1992) described reusable software concepts for developing scaleable FMS control architecture, automatic generation of control code, and object-oriented design of equipment controllers. Hsu (1992) described how object-oriented programming is used for FMS control software development along with its advantages.

Stuznebecker (1991) proposed extensions to C++ language with an aim to formulate it as a base language for developing an object-oriented environment for distributed manufacturing software. Char et al., (1991, 1993a,b) presented excellent reviews on recent methods for developing manufacturing control software and concluded that current methodologies are not sufficient to support planning, scheduling, and monitoring activities involved in manufacturing. Another research area related to FMS control software is the application of Ada as an implementation language. An approach is proposed for rapid prototyping of control software using Ada and PNs [Sahraoui and Ould-Kaddour, 1992]. Ada is used for building reusable software components and PNs for specifying task communication and synchronization. Ould-Kaddour and Courvoisier (1989) presented a method for real-time software prototyping using PNs for specification, Ada for specification description, and Modula-2 for implementation. Ada was selected as a suitable language for software development of fully automated manufacturing systems [Voltz *et al.*, 1984; Venkatesh *et al.*, 1991]. The research on FMS control software has been concentrated on combining OOD concepts, PNs, and Ada as shown in Table 12.1.

Table 12.1 Research related to OOD, PNs, and Ada.

Research related to:	Selected references
FMS and OOD	Garg (1985); Glassey and Adiga (1990); Graham *et al.*, (1991); Sturzenebecker (1991); Hsu (1992); Smith and Joshi (1992)
FMS and PN	Bruno and Marchetto (1986); Cecil *et al.*, (1992); Jafari (1992); Proth (1992); Venkatesh and Ilyas (1995); Zhou and DiCesare (1989); Zhou *et al.*, (1991); Zhou *et al.*, (1992)
FMS and Ada	Voltz *et al.*, (1984), Venkatesh *et al.*, (1991)
Ada and PNs	Murata *et al.*, (1989)
FMS, OOD and PN	Bruno and Marchetto (1986); Ould-Kaddour and Courvoisier (1989); Booch (1991); Sahraoui and Ould-Kaddour (1992); Wang (1996)
FMS, OOD and Ada	Chaar *et al.*, (1993b); Naylor and Voltz (1987)
FMS, OOD, PN, and Ada	Venkatesh and Fernandez (1993)

The benefits of OOD such as reusability, extendibility, and modifiability of a software system were not sufficiently demonstrated with a particular example of FMS in previous works [Glassey and Adiga, 1990; Sturzenbecker, 1991; Graham *et al.*, 1991; Hsu, 1992]. They did not discuss in detail the relations among several objects present in an FMS, which are essential for OOD. Furthermore, the earlier research [Glassey and Adiga, 1990; Sturzenbecker 1991; Graham *et al.*, 1991; Hsu 1992; Char *et al.*, 1993b] that used OOD for design of FMS software did not pay much attention to describing the dynamic behavior of the objects present in a software system and to analyze quantitatively the system performance. Realizing the importance of these issues, Venkatesh and Fernandez (1993) first proposed an approach for object oriented simulation and control of FMSs using timed PNs and Ada. Subsequently, Fernandez and Han (1993) illustrated the concepts of OOD design by developing Object Modeling Technique (OMT) diagram of a simple assembly cell and showed how OOD can adapt to changing conditions in the manufacturing cells.

More recently, Wang (1996) combined PNs and object-oriented concepts to develop object-oriented PN cell control models. Stulle and Schmidt (1996) have presented object-oriented PNs for modeling of FMSs by constructing a reusable OOPN that modeled a typical production cycle. The modeled OOPNs are then interconnected in a predefined way to model FMS similar to the way shown in [Venkatesh et al., 1994c]. The goal of this chapter is to formulate an OOD methodology for developing reusable, extendible, and modifiable control software for FMSs using OMT diagrams and PNs. The objectives of this chapter are:

1. To present a OOD methodology combining the concepts of OMT diagrams and PNs by discussing the rationale and advantages of selecting OOD and PNs respectively,
2. To emphasize the use of PNs as a dynamic tool in OOD by discussing the advantages of PNs over earlier used dynamic models,
3. To demonstrate how PNs can support the concepts of reusability and extendibility by adopting a bottom-up approach of FMS modeling,
4. To illustrate the methodology by developing the object modeling technique (OMT) diagrams and PN models of an FMS example, and
5. To demonstrate the benefits of the methodology to support reusability, modifiability, and extendibility of a software system when the configuration and specifications of the FMS are subject to change.

12.2 Methodology for FMS Control Software Development

12.2.1 Methodology

It is assumed that the reader has basic knowledge on OOD [Booch, 1994; Rumbaugh et al., 1991]. While the advantages of OOD and PNs are shown in Fig. 12.1 and will be discussed later, the rationale why they are combined is discussed here. The motivation for combining OOD and PNs is that in a broader sense they are complementary to each other to achieve the goal of system development at incremental stages.

OOD is an appropriate design methodology to support system design [Booch, 1994; Rumbaugh et al., 1991]. PNs are a versatile hierarchical modeling tools that support all the stages of system development including its specification, planning, design, evaluation, monitoring, control, and implementation [Zhou and DiCesare, 1989; Zhou et al., 1991; Proth 1992].

Earlier techniques of OOD that do not use PNs [Glassey and Adiga, 1990; Booch, 1994; Rumbaugh et al., 1991; Smith and Joshi, 1992] have some disadvantages since they cannot explicitly represent detailed dynamic interactions that involve concurrency and synchronization among objects present in a system. Also, they are difficult to consider timing information in the design for qualitative analysis and performance evaluation of an FMS. PNs can be used to augment the applicability of OOD by eliminating these disadvantages because they are primarily a tool that models and analyzes complex asynchronous concurrent interactions in distributed systems. They can also help select data structures and operations for objects. On the other hand, PNs have some disadvantages since they cannot clearly represent the static relationships among objects, which is essential for reusable software development [Garg, 1985; Boucher et al., 1989; Rumbaugh et al., 1991; Venkatesh and Fernandez, 1993]. However, these relationships can be modeled by using object modeling technique (OMT) diagrams in OOD [Rumbaugh et al., 1991].

Hence, by using OMT to represent the static relations and PNs to represent dynamic relations, a powerful OOD methodology [Venkatesh 1995; Venkatesh and Zhou, 1998] for the development of FMS control software can be developed. The methodology proposed is illustrated in Fig. 12.1. In this approach, OOD is used as a software design tool and PNs are selected mainly as a dynamic modeling and performance evaluation tools. The methodology is explained in the following steps:

1. Apply OOD concepts to find objects in an FMS and to show the explicit relations among them by developing an object modeling technique (OMT) diagram [Rumbaugh et al., 1991]. The objective of an OMT diagram is to show the static relations among objects in the FMS.

2. Use PNs for modeling the FMS in order to show the dynamic relations among objects.

3. Use the formulated PNM in the above step to analyze, simulate, and evaluate the FMS performance and ensure the absence of deadlocks and capacity overflows. The performance criteria can be production rate, machine utilization, work-in-process inventory, etc. At this stage, if its performance is found unsatisfactory, it should be improved by changing some parameters, the system configuration, and/or operational policies. In order to accommodate these changes PN and/or OMT diagrams need to be modified. For example, if a change is just for parameters such as timing of events, only PN needs to be modified; but, if the change is for system configuration (adding new resources with new responsibilities), then both OMT and PN needs to be modified. After the analysis of the final PNM is completed, the design parameters of specific interest in controlling FMS can be decided. Furthermore, with the aid of PNM,

12.2. Methodology for FMS Control Software

data structures and operations for objects present in FMS can be systematically identified.

4. Combine OMT and PNM to design the complete structure of objects with their static and dynamic relations and elect an appropriate language such as Ada, C++, or Smalltalk to implement objects.

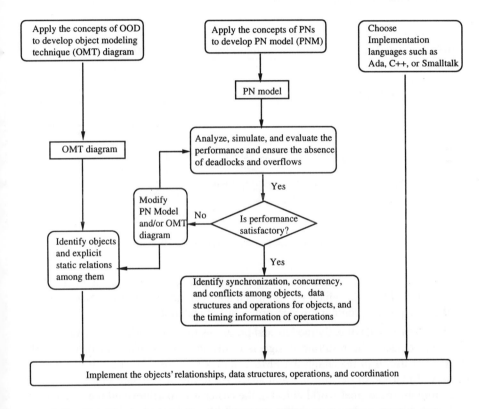

Figure 12.1 Proposed systematic methodology for FMS control software development.

12.2.2 Fundamentals of OOD

Each unit in an FMS interacts with several other objects to complete a set of production tasks. Further, objects can be added or removed from the FMS thereby demanding the software to be extensible, modifiable, and reusable. This makes OOD particularly suitable for designing FMS control software. Fig. 12.2 shows how the concepts of OOD can be used for developing FMS control software. All real-world objects have some properties and behavior. By using OOD concepts the properties and behavior of the real-world object are modeled by the *data/attributes* and *methods/operations* of the corresponding software object. For example, in Fig. 12.2, Machine has properties such as type, length, and cost, and behavior such as machining a work-piece, raising alarm, and sending signal to robot to unload the work-piece, which are modeled in the corresponding software object Machine as data/attributes and methods/operations.

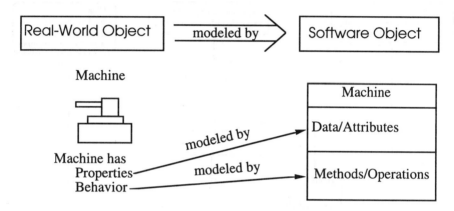

Figure 12.2 Principle of object-oriented design of FMS control software.

Note that real-world objects include not only physical objects such as machine, robot, AGV, PLC, etc. which we can touch and feel but also logical objects such as scheduling process, routing process, order, bill of materials, etc.

Since OOD directly models and programs the real-world entities in terms of software objects, the resulting design is easily understandable and, therefore, the method is useful in managing the FMS complexity. From Fig. 12.2, it can also be observed that the closer the solution space (in terms of software objects) maps to the problem space (real-world objects), the easier it is to achieve all the goals of the software system such as reusability, extendibility, and modifiability. An interested reader is referred to (Booch 1994, Rumbaugh, *et al.* 1991) for further discussion on OOD and its advantages. Its brief discussion is given below. Note that OOD is one of the many phases of object-oriented software development in addition to phases such as object-oriented analysis, testing, etc. An interested reader is referred to [Booch, 1994; Rumbaugh *et al.*, 1991, Jacobson *et al.*, 1995] for more discussion on the development of OOD and its advantages. The following paragraphs present a brief discussion on the OOD concepts used in this chapter.

The fundamental building block of OOD is an *object* that contains both a data structure and a collection of related procedures. Procedures are also called as

12.2. Methodology for FMS Control Software

operations or methods. Objects interact with each other by sending messages or by calls to their interfaces. Objects with the identical data structure (attributes) and behavior (operations) can be grouped into a class. Each object is an instance of its class. An important feature of OOD is *inheritance*. Inheritance is a mechanism whereby one class of objects can be defined as a special case of another class, automatically including the data structure and behavior of that class. The special cases of a class are known as subclasses. For example, machine_controller (subclass) is a special class of controller (superclass). *Polymorphism*, another important feature of OOD is a property in which the same operation may behave differently on different classes. For example, an operation called process_part behaves differently on two different classes, milling_machine, and drilling_machine. There are several approaches to develop OOD systems [Booch, 1994; Rumbaugh *et al.*, 1991]. The complete OOD methodology consists in developing three orthogonal models [Rumbaugh *et al.*, 1991]:

1. Object model: It divides the application into object classes and shows the static data. The relationships (interfaces) among the classes are described by the class structure. The behavior of the objects are defined by operations associated with the object class.
2. Dynamic model: It shows the way the system behaves with internal and external events by capturing the time-dependent behavior of the system.
3. Functional model: It shows how to process the data flow in the system during each event or action.

Since the functional model deals with the lower level implementation of a software system, it is not considered here. This chapter uses object modeling technique (OMT) diagram [Rumbaugh *et al.*, 1991] as an object model and PNs as a dynamic model. There are few other approaches for object modeling [Monarchi and Puhr, 1992; Booch, 1994]. Even though the notations used in these methods are different, they are conceptually similar in their way to represent the object structure.

12.2.3 Object Modeling Technique Diagram as a Static Modeling Tool

Figure 12.3 shows the definition of a class with three separate areas: *class name*, *class attributes*, and *operations*. The OMT diagram [Rumbagh, 1991] allows three basic types of relations among classes, i.e., *generalization*, *aggregation*, and *association* which are explained below:

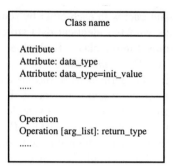

Figure 12.3 A class definition in OMT.

1. Generalization implies the definition of a superclass that collects the common characteristics of several subclasses. It describes an "is-a" association between a subclass and its superclass. The symbol for generalization is a triangle. For example, in Fig. 12.4, the superclass *conveyance equipment* is a generalization of the classes *robot*, *conveyor*, and *AGV*; *robot* in turn is a generalization of *mobile robot*.

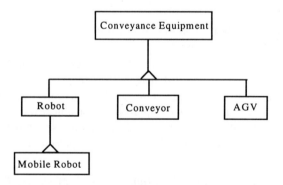

Figure 12.4 OMT diagram for Generalization.

2. Aggregation implies the description of a class in terms of its constituent parts. The concept of aggregation defines an "is-a-part-of" relationship between a subclass and its superclass. For example, Fig. 12.5 shows that an *assembly system* consists of assembly cells, which in turn consists of *robots* and part feeders. Aggregation is denoted by a diamond. The black dot indicates multiplicity, e.g., an *assembly system*

12.2. Methodology for FMS Control Software

is composed of several assembly cells, an *assembly cell* is composed of several robots, and part feeders.

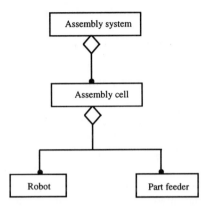

Figure 12.5 OMT diagram for Aggregation.

3. Association describes how objects belonging to different classes are related to each other. The OMT symbol for an association is a line connecting two classes labeled with the association name. Associations can be binary, ternary, or even higher order, and can have any number of attributes. For example, in Fig. 12.6, *sensor* monitors *machine*, and *machine* machines *part*.

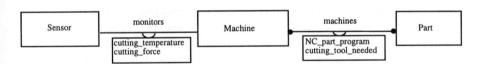

Figure 12.6 OMT diagram for Association.

Associations can have attributes, e.g., cutting temperature, cutting force in Fig. 12.6. Since a machine can process several parts, and a part can be processed by several machines, block dots are present at the both ends of the link connecting *machine* and *part* in Fig. 12.6. For more details on developing OMT refer to [Rumbaugh *et al.*, 1991]. After the OMT diagram is developed, the next step is to formulate the dynamic model of FMS. State diagrams were used to develop a dynamic model as shown in [Rumbaugh *et al.*, 1991]. However, they are not convenient to deal with timing aspects and the dynamic behavior of objects in FMS which must be considered to [Rumbaugh et al., 1991; Fernandez and Han, 1993] evaluate FMS performance. Furthermore, they cannot explicitly represent important features such as concurrency and synchronization in FMS which makes it difficult to visualize the functioning of FMS. The state diagrams become complex in case of FMS [Crockett et al., 1987; Zhou and DiCesare, 1993]. Hence, PNs are chosen to develop a dynamic model of FMS as discussed next.

12.2.4 Petri Nets as a Dynamic Modeling Tool

In this chapter PNs are used as a dynamic model in object-oriented design (OOD). In contrast, other researchers use two different kinds of diagrams for representing the dynamic behavior of objects. Rumbaugh et al., (1991) uses state diagrams and event trace diagrams; Booch (1994) uses state transition diagrams and interaction diagrams. State diagrams or state transition diagrams are used to represent how objects respond to the internal and external events in the system. Event trace or interaction diagrams are used to study the synchronization aspects and to trace the execution of events in the system.

However, in order to develop systematic control software for FMSs, a more expressive dynamic modeling tool is needed. This is because the coordination of the individual units in FMSs is important. Hence, a dynamic modeling tool should model in detail the concurreny and synchronization in the system with respect to time. Furthermore, such a tool should help to analyze the system behavior to check for aspects such as deadlocks. Since it is very common in FMSs to share certain resources (e.g. a robot is shared by more than one machine to load/unload), a dynamic modeling tool should represent these aspects to analyze the conflicts during the system execution. In addition to all these requirements, a dynamic modeling tool should support the system designer for system performance evaluation and assist control engineers in controlling and monitoring the FMS. PNs have all these capabilities and hence are suitable as dynamic modeling tool irrespective of the various methods used for object model [Rumbaugh et al., 1991; Monarchi and Puhr, 1992; Booch, 1994). Also, unlike previous works which use two different kinds of diagrams for representing system states and tracing events [Rumbaugh et al., 1991 and Booch 1994], PNs can be used as a single tool to represent both the system states and to trace the events in the system when time durations of activities are associated with transitions.

Compared with the previous techniques for a dynamic model in OOD, PNs have the following advantages:
1. PNs can explicitly and realistically represent concurrent operations, synchronization activities, and conflicts.
2. They can be easily associated with timing information for performance

analysis of the system. Both analytical and simulation methods are available depending on the system complexity and accuracy needed.
3. They allow to check the system behavioral properties such as deadlock and capacity overflow.
4. By using PNs, a more compact model can be obtained and thus avoiding painstaking enumeration of all the states at the design stage.
5. The developed PN models can also be extended for real-time control and monitoring.
6. The attributes and operations of potential objects can be selected from Petri net models (PNMs) of the FMS. This advantage of PNs reaches far significance when the quote from [Rogers, 1991] is recalled: "the potential of OOD is impeded due to the lack of an established methodology for object identification."

After the OMT diagram and PN model (PNM) of FMS are developed, the software system can be implemented by selecting a proper computer language. Some researchers use Ada [Bruno and Marchetto, 1986; Naylor and Voltz, 1987; Venkatesh et al., 1991; Sahraoui and Old-Kaddour, 1992] and others prefer C++ for implementing OOD [Sturzenbecker, 1991; Smith and Joshi, 1992]. Detailed comparison between Ada and C++ for OOD implementation of FMS control software falls beyond the scope of this book.

12.3 Illustration of the Methodology with an FMS

To illustrate the proposed methodology, the FMS discussed in the Chapter 8 is considered. The FMS is assumed to function under push paradigm with the same production parameters and requirements (i.e., conveyance time matrix as shown in Table 8.2 and process sequence and timings as shown in Table 8.3). In this section first, the control system of the FMS is described. Next, the OMT diagram of the FMS is developed. Then, the PNM is formulated and finalized after the quantitative analysis of the system. OMT diagram and the final PNM are combined to identify objects and the static and dynamic relations.

System description

The FMS considered is shown in Fig. 12.7 and consists of four machines. In this system each machine is served by a robot and processes either raw material or intermediate parts.

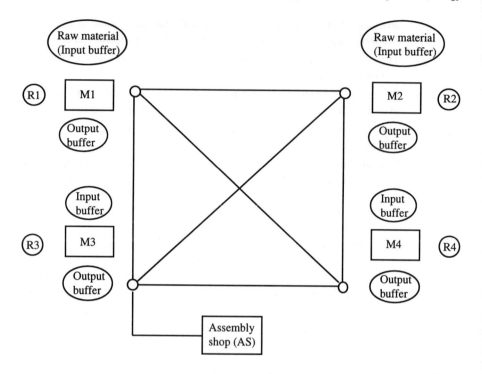

Figure 12.7 The configuration of FMS.

Once a part is processed, the machine setup is needed to process another part type. The activities of the robots are to load tools and parts to and unload used tools and processed parts from the machines. Input and output buffers are provided at machines to store raw material, the work-in-process inventory, and processed parts. Sensors monitor the functioning of machines and robots. For example, if the temperature during machining exceeds its maximum value, the sensor monitoring the corresponding machine will send a signal to the cell controller to stop machining. AGVs are used to convey parts and subassemblies among the machines and to the assembly shop (AS). The AGV track layout is shown in Fig. 12.7. The system is used to produce finished product, PR1 that is considered in Chapter 8. A machine and a robot constitute one flexible manufacturing cell called "cell" for short. Thus there are in total four such cells.

There is a cell controller responsible for controlling the operations in each cell. There may be one or more PLCs to coordinate the sequencing of different elements present in each cell. Each PLC may control more than one machine, robot, and sensor. It receives signals from sensors and accordingly controls the functions of robots and machines. There is a main controller for the FMS to control

12.3. Illustration of the Methodology with an FMS

the cell controllers and schedule production tasks among cells. During the production if any malfunction/exception occurs at the PLC level (due to the factors such as breakdown of a tool and excessively high machining temperature) an exception is raised by the PLC and passed to the cell controller. Then the cell controller handles that exception and passes the resume signal to PLC to continue the cell operations. Similarly, if malfunction/exception occurs at the cell controller level (due to the factors, e.g., introducing of a new product variety), it raises an exception and sends to the main controller. Then the main controller handles that exception and passes the resume signal to cell controller. There is also a business host which is on the top of the main controller to deal with higher level issues, e.g., production control and corporate policies.

12.3.1 OMT diagram and PNM of the FMS

OMT diagram

Fig. 12.8 shows the OMT diagram corresponding to the FMS under investigation.

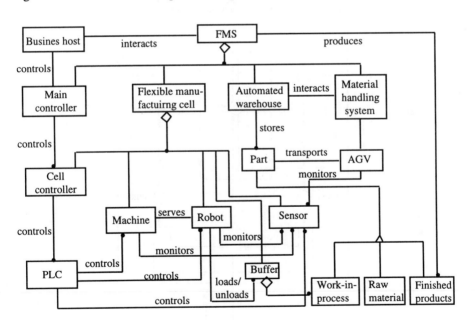

Figure 12.8 OMT diagram of the FMS.

318   Chapter 12. An Object-Oriented Design Methodology

Even though the OMT diagram appears complex, the design shown is believed to be realistic and captures the relevant real world properties of the objects and their functions in this FMS. For example, this diagram models that the FMS *produces* many finished products, *interacts* with a business host, and *consists of* a main controller, several flexible manufacturing cells, an automated warehouse, and a material handling system. This OMT diagram can be easily extended or modified when the configuration of FMS changes, as shown later.

PNM

To formulate the PNM of FMS, first the PNM corresponding to machine 1 (MC1) is formulated as shown in Fig. 12.9.

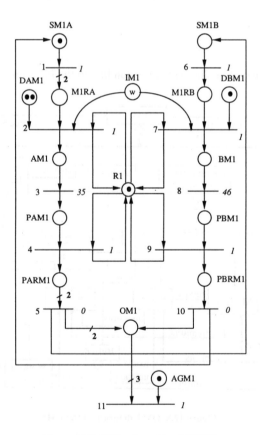

Figure 12.9 PNM of machine 1 (MC1).

12.3. Illustration of the Methodology with an FMS

The interpretation of its places and transitions is listed in Table 12.2. Note that in this table only interpretation for places corresponding to Part A is listed. Interpretation of the remaining places of the PN corresponding to Part B is similar to that given for corresponding places of Part A. For simplicity, objects like business host, automated warehouse, and sensors are not considered in our models. Here, a bottom-up approach is adopted to construct the PNM of FMS.

Table 12.2 Interpretation of typical places and transitions shown in Figs. 12.9 and 12.10

Place	Interpretation
DAMi (i = 1 and 3)	Demand for Part A on Mi
IM1	Input buffer where "w" is the initial marking representing the number raw-material pieces
SM1A	MC1 being setup to process Part A (B)
M1RA	MC1 ready to process Part A
R1	R1 ready to load/unload
AM1	Part A loaded on MC1's table and MC1 processing
PAM1	MC1 finished processing Part A and Part A is being unloaded by R1
PARM1	Part A is ready at MC1
OM1	Output buffer of MC1 ready with parts A and B
AGM1	AGV at the output buffer of MC1
AMij (i,j = 1, 2, 3, and 4)	AGV traveling from MCi to MCj
AMi (i=1)	AGV at the input buffer of MCi
AMOi (i = 1, 2, 3, and 4)	AGV at the output buffer of MCi
A1ASM1	AGV traveling from AS to MC1
Transition	
1,6,14,19,28,33,39,44	Signal for machine setup
2,7,15,20,29,34,40,45	Robot finishing the loading operation
3,8,16,21,30,35,41,46	Completion of part processing
4,9,17,22,31,36,42,47	Robot finishing the unloading operation
5,10,18,23,32,36,43,48	Number of parts as specified in final assembly ready in machine's output buffer
11,13,25,27,38,49	AGV starting from MC
12,24,26,50,51,52	AGV reaching its destination

As the functioning of all machines is almost similar, the same PN modeling methodology can be duplicated to formulate the PNMs of other machine cells. After formulating the PNMs of all cells, they are connected to each other according to the process sequence of parts (given in Chapter 8) as shown in the Fig. 12.10. Observe that transferring parts in between cells is also modeled in Fig. 12.10. Since, the functioning of all cells is similar, the same PN modeling methodology can be duplicated to formulate the PNMs of other cells in FMS.

Figure 12.11 shows the PNM of the FMS control system. In the PNMs of Figs. 12.10 and 12.11, some places are pictured as concentric circles. The reasons and motivation for this are discussed later. In PNMs shown in Figs. 12.9 and 12.10, places and transitions represent different states and operations related to each object respectively. It can be observed that these PNMs clearly model the dynamic interactions among the objects in the FMS. While modeling MC3, the PN model of MC1 shown in Fig. 12.9 is reused. In case of modeling MC2 and MC4, the PN model of MC1 is not only reused but also extended to model the machining of a different part variety.

The PNM corresponding to M1 in Fig. 12.9 can be used as a basis to design/code a software module that controls M1. In other words, the PNM of M1 can be extended to develop control software for M1 that can be implemented using high level languages and/or PLC ladder logic. For examples of PNMs converted to controllers using high level languages, refer to Desrochers and Al-Jaar (1995). For examples of PNMs converted to controllers using PLC ladder logic, refer to Taholakian and Hales (1997), Boucher (1996), and Jafari and Boucher (1994). After developing the software module for M1 using the PNM of M1, it can be reused and extended to produce software modules corresponding to M2, M3, and M4 because the PNM of M1 is similar to that of M2, M3, and M4. These modules are then combined to generate the software module corresponding to the FMS. Similarly, the control software corresponding to this FMS can be reused and extended to produce software modules of other systems. From this example, it can be said that the PN modeling of similar components in FMS supports the concepts of reusability and extendibility which are two essential characteristics of the software generated by OOD methodology [Meyer, 1988; Rumbaugh *et al.*, 1991]. Reusability and extendibility are defined and illustrated in the next section.

Figure 12.11 shows the PNM for the FMS control system for normal production cases (without breakdowns). It models the hierarchical control of the FMS. For simplicity sake, during modeling it is assumed that there is a main controller (M), a cell controller (CC), and a PLC to control the FMS considered. In this PNM, the places and transitions represent different states and operations related to each object respectively. It can be observed that this PNM clearly models the dynamic interactions among the objects in the control system. It can be easily extended to consider more cell controllers and PLCs.

12.3. Illustration of the Methodology with an FMS

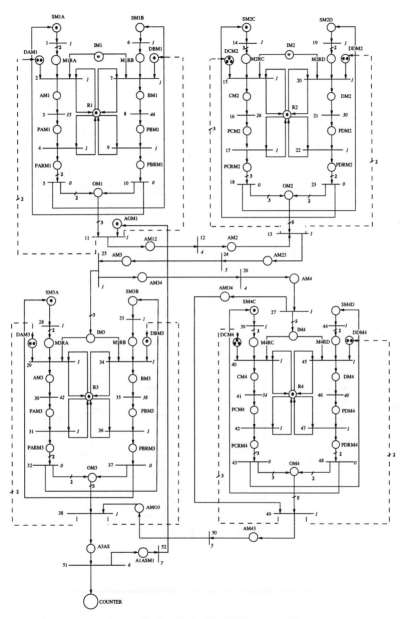

Figure 12.10. PNM of the FMS under push paradigm.

NOTE:
1. Transition times are for FMS case
2. Initial marking is shown
3. $\overset{k\ \overline{}\ r}{\underset{r}{\big| k}}$ k is transition number r is time units
4. PLS for A,B,C&D : 2,1,3&2
5. $N = 1, n = 1$

322   Chapter 12. An Object-Oriented Design Methodology

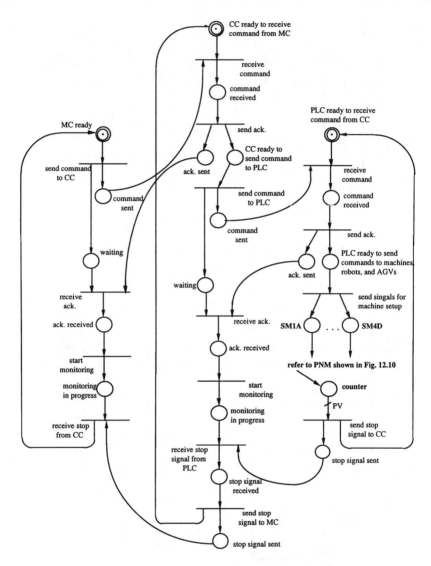

Note: MC --> Main controller; CC --> Cell controller; PLC --> Programmable logic controller; PV --> Production volume

Figure 12.11 PNM for the FMS control system.

The execution principles of the PNM fulfill the requirements of the system operations. For example, in Fig. 12.10, consider "transition 2 (t_2)" that models the

12.3. Illustration of the Methodology with an FMS

activity "Robot finishes the loading operation". In other words, t_2 models an activity: Robot 1 (R1) loads Part A on MC1. The constraints that have to be fulfilled to fire t_2 (to enable the activity modeled by t_2) are modeled by the input places corresponding to t_2. For example, marked places: "M1RA" models the condition that "MC1 should be ready to process Part A", "DAM1" there should be demand for Part A, "IM1" there should be raw material needed to produce Part A, finally "R1" should be ready to load the tool on MC1. The time that R1 takes to finish this operation is assumed as 1 time unit and associated with t_2 (shown at right side of t_2 in Fig. 12.10). From the preceding discussion, objects involved in firing t_2 can be easily recognized as R1, M1, and Part A. A similar discussion can be given for other places and transitions in this PNM, i.e., there is an one-to-one correspondence between the actions in the FMS and transitions in its PNM, and thus the execution of the PNM precisely specifies the operations involved in FMS. In Figs. 12.10 and 12.11 the modeling of exception handling is not shown. However, exceptions do occur sometimes during production. Hence, Fig. 12.12 shows the PNM for exception handling by the main controller (M).

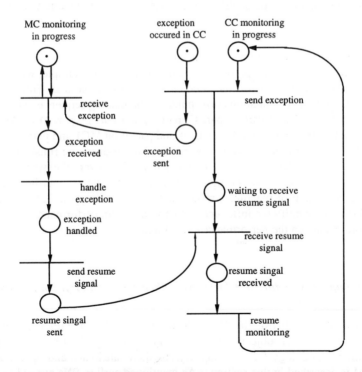

Figure 12.12 PNM of exception handling by main controller.

In this PNM, an exception is raised by a cell controller (CC) and handled by the main controller. This net clearly shows the conditions and activities involved in this kind of exception handling. Similarly, a PNM similar to this can be developed to model the exception handling by CC when an exception is raised by a PLC.

Formulating the final PNM by analyzing the performance of FMS

As discussed in Chapter 8, the final PNM can be formulated by analyzing it to evaluate the performance of FMS. Estimating the number of AGVs and selecting their routings are important since they affect both the performance and control of FMS. By analyzing the FMS performance and deciding the number of AGVs and their routings, appropriate data structures and operations for AGV objects can be selected in the control software. Another design issue that affects both performance and control of FMS is production lot size (PLS) and moving lot size (MLS). By determining the optimum PLS, signals for changing the setup of machines can be appropriately given during the control of FMS. In the FMS considered, there are two different combinations of PLSs for parts A, B, C, and D namely, 2, 1, 3, 2 and 1, 1, 1, and 1.

In the first combination parts A, B, C, and D are produced in the lot sizes equal to their exact requirement as in the bill of materials of product 1. In the second case, these parts are produced in unit lot sizes. For example, with respect to MC1 and MC2 and based on the process sequences of parts in FMS (shown in Table 8.3), the first combination corresponds to the loading sequence part A, part A, and part B on MC1 and part C, part C, part C, and part D, part D on MC2. The second combination corresponds to part A and part B on MC1 and part C and part D on MC2. It is clear that the setup time required to produce one finished product in the former case is less compared to the latter. However, the work in process inventory in the former may be more compared to the latter. Further, PLSs may affect the utilization of machines, robots, and AGVs. The quantitative analysis of PNM allows to quantify the influence of PLSs on system performance. See Chapter 8 for more details on the performance evaluation.

12.3.2 Complete Structure of Objects with Their Static and Dynamic Relations

After finalizing the PNM, the OMT diagram is combined with it to design the complete structure of object classes as shown in Fig. 12.13. A simple and systematic methodology for selecting objects, their attributes and operations from the PNM is described in this section. As mentioned earlier, PNs can aid to identify the potential objects and their corresponding data structures and operations. Places

12.3. Illustration of the Methodology with an FMS

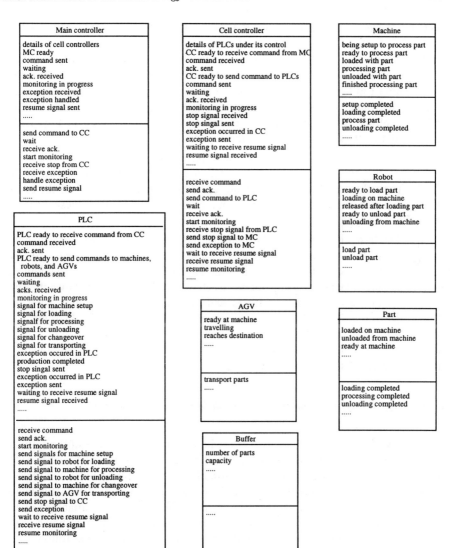

Figure 12.13 Class definitions of important object classes in FMSs.

in PNM aid to identify objects and their data structures, and transitions aid to find operations for objects. Generally objects are identified when developing the OMT diagram for an FMS. The attributes and operations of the objects can be formulated with aid of its PNM. For the FMS example, MC, CC, PLC, machine, AGV, robot, part, and buffer are selected as potential object classes from the OMT diagram. For convenience, those places corresponding to the objects are shown as concentric circles in Figs. 12.10 and 12.11. The places that represent the intermediate states of objects such as command_sent (for MC in Fig. 12.11) and loaded_with_part (PAM1 for Machine 1 in Fig. 12.10); and the places that model conditions of processes such as waiting and monitoring_in_progress (for MC in Fig. 12.11), and AGV_travelling_from_machine_1_to_machine_2 (AM12 for AGV in Fig. 12.10) aid in selecting the data structures for the objects.

In Figs. 12.10 and 12.11, this type of places are shown as normal circles and aid in selecting the data structures (also called as data attributes) as shown in Fig. 12.13. The values of these attributes are represented by the presence of the tokens in the corresponding places modeling them. Normally, the value of these attributes are of Boolean type.

The transitions in the PNM represent the activities or events corresponding to objects such as *send_command_to_CC* and *receive_acknowledgment* (in Fig. 12.10), *AGV_starts_from_machine* (in Fig. 12.11). Transitions corresponding to each object are selected as its operations as shown in Fig. 12.13. It shows the class definitions of important object classes and their data structures and operations. These classes are essential for development of object-oriented control software. A software engineer can understand the logic structure of the whole system by looking at these classes. Hence, these object classes are the fundamental building blocks of an object-oriented software system.

12.3.3 Reusability, Extendibility, and Modifiability of the Design

Reusability and extendibility are two of the important characteristics of the software generated by using OOD [Meyer, 1988; Rumbaugh *et al.*, 1991] which makes easily adaptable and maintainable software. Reusability is the ability to reuse a module or component developed for a given design, in a new design [Meyer, 1988; Rumbaugh *et al.*, 1991]. To illustrate the reusability, extendibility, and modifiability of the design, consider now that the designer decides modifying the FMS as shown in the Fig. 12.14. The modifications to the initial FMS can be summarized as follows:

1. Include a flexible assembly cell (FAC) consisting of two new machines, MC5 and MC6; two new robots, R5 and R6 for pick and place operations, and two part feeders,
2. Include a mobile robot with the additional tracks in FAC to transport material and parts among machines and robots, and
3. Change the operation management strategy of the FMS from push to pull. In other words, the machines process parts only when there is a demand.

12.3. Illustration of the Methodology with an FMS

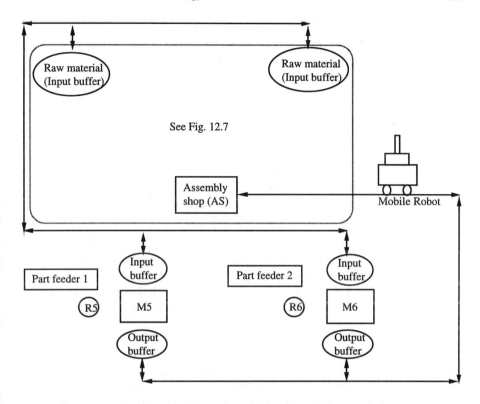

Figure 12.14 The expanded configuration of FMS.

The OMT diagram corresponding to the new FMS is shown in Fig. 12.15. It omits the OMT diagram portion that has not changed in Fig. 12.8. It is noticed that the earlier OMT diagram is reused by extending it to include the elements newly added to the FMS. The third new modification stated above changes the dynamic relations among objects in FMS and may affect the system performance. To accommodate this modification, we redefine the PNM as shown in Fig. 12.16. Observe that the same PNM (Fig. 12.10) used earlier to study the push paradigm is reused to study the pull paradigm with slightly modification. This modification is shown as the dotted arcs modeling the pull paradigm. The system performance can be evaluated again as discussed in Chapter 8.

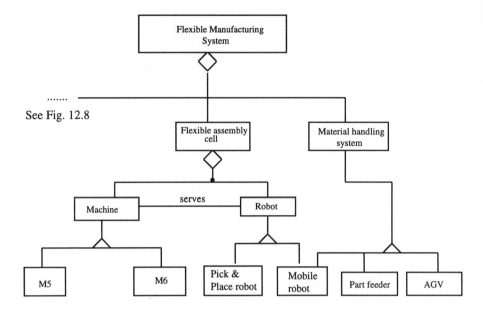

Figure 12.15 OMT diagram for the expanded FMS.

The control software developed earlier for the FMS need not be much changed due to the modularity of PNMs. Since the PNMs modeling the FMS corresponding to both push and pull paradigms are basically same, the data structures of objects need not be changed to include the new specifications of FMS. Hence, this section has briefly illustrated how the proposed design methodology supports reusability, extendibility, and modifiability concepts in developing control software.

12.4 Summary

Development of integrated FMS control software that can be used for planning, scheduling, monitoring, simulation, and control is difficult and hence attracting the growing attention of researchers and practitioners. To ease the task of developing FMS control software, a systematic design methodology is presented. It is the combination of OOD concepts, OMT diagrams, and PNs. An FMS example is used to illustrate the methodology. OMT diagram for the FMS is developed to find the objects and the static relationships among them. PNMs are formulated to study the performance of the system and to help identify the data structures and operations of FMS objects.

12.4. Summary

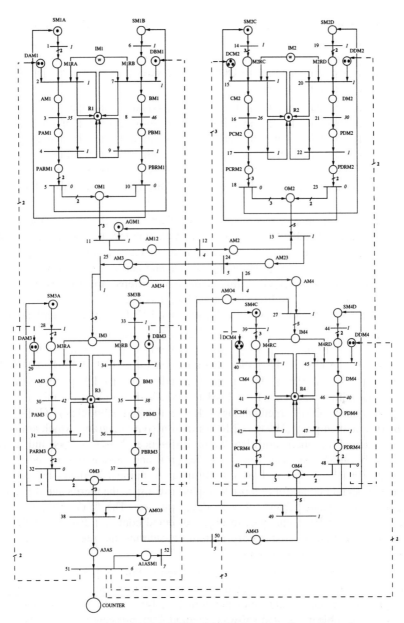

Figure 12.16. PNM of the FMS under pull paradigm.

NOTE:
1. Transition times are for FMS case
2. Initial marking is shown
3. $\overset{k\quad r}{\underset{}{\vert}}\;$ k is transition number, r is time units
4. PLS for A,B,C&D : 2,1,3 & 2
5. N = 1, n = 1

The reusability, extendibility, and modifiability of the control software system using this methodology are also illustrated by augmenting the original OMT and PNM to satisfy the new specifications of the FMS. The traditional methods for discrete event dynamic systems in OOD include state/state transition diagrams and event trace/interaction diagrams.

This chapter emphasized PNs as the dynamic model in OOD with the following advantages: 1) By adopting the bottom-up approach of PN modeling, PNs can support two important characteristics of software generated by OOD namely reusability and extendibility; 2) PN models offer a systematic method to identify the data structures and operations of objects in the software system; 3) PNs can be used as an integrated tool to both control the system and analyze the system performance.

The significance of this chapter is two fold. From the OOD point of view, this chapter has embedded PNs as a dynamic modeling tool in an OOD approach to achieve the objectives never before possible, e.g., explicit description of concurrency and synchronization among objects and performance analysis of FMS. From the practice point of view, this chapter has offered an effective systematic methodology to design modifiable, extendible, and reusable control software and helped further establish OOD, OMT diagrams, and PNs for industrial applications.

In the future, OMT diagrams and PNs needs to be extended to deal with such issues as real-time monitoring and fault tolerance, and communication among various objects in FMS. The authors are aware of the arrival of Unified Modeling Language (UML) which is the emerging standard for modeling object-oriented software modeling (Reference from UML). UML is the result of the mixture of the best ideas from OMT methodology, Rumbaugh's methodology (1994), and Jacobson's methodology (1994). UML contains activity diagrams which are similar to PNs. However, there are also lot of differences between PNs and activity diagrams because PNs have for more features and advantages as detailed in throughout this book. A benchmark study using traditional methods and proposed PN method should be useful to more formally investigate the merits and demerits of using PNs in object-oriented software development. The future studies include standardization of PN techniques in object-oriented control software development and implementation of a laboratory system considering the real time control and breakdowns of system components.

The class definitions of objects presented in this work need to be extended in order to address problems related to material requirement planning, computer aided process planning, and computer aided design. With such extended class definitions, the object-oriented software methodology presented in this chapter aids to develop a standard interface between business systems and control systems. The goal of such an interface is to reduce the risk, cost, and errors associated with sharing information among several types of computer systems.

12.4. Summary

Business systems consist of manufacturing/enterprise resource planning systems (MRP/ERP) and manufacturing execution systems (MES), used to determine resource planning and scheduling needs. Control systems include computer systems installed on the shop-floor. Currently, users have to specify and develop custom interfaces between business and control systems as part of an overall automation project. However, continual reinvention of the interface enormously increases the automation costs. By designing a standard interface in an object-oriented paradigm using Petri nets, the risk, cost, and errors in the automation project can be reduced. By using object-oriented concepts, the standard interface can be easily reused, modified, and extended for new automation projects. Also, by using Petri nets, an effective process integration between business and control systems can be achieved since Petri nets are ideal tools for process modeling. The results presented in this chapter represent a good beginning towards the above objectives.

CHAPTER 13

SCHEDULING USING PETRI NETS

13.1 Introduction

This chapter discusses how to use a Petri net model to accomplish the planning and scheduling of a flexible manufacturing system. The solution methodology is based on A* algorithm that is extensively used to search for a solution given a search space. The features of Petri net-based approaches are twofold. One is their easy handling of multiple lots for complex relations that may exist among jobs, routes, machines, and material handling devices. The other one is that the generated schedule is event-driven, deadlock-free, and optimal or near-optimal with respect to makespan. The drawback is their computation that could be huge if an inappropriate heuristic function is adopted. Two methods are often used to overcome the computational complexity. The first one is to find and use better heuristic-functions. The second one is to adopt hybrid heuristic search strategies. A hybrid search strategy combines the heuristic best-first (BF) strategy with the controlled backtracking (BT) strategy based on the execution of the Petri nets. If one can afford the computation complexities required by a pure best-first strategy, the pure best-first search can be used to locate an optimal schedule. Otherwise, the hybrid BF-BT or BT-BF combination can be implemented and a sub-optimal schedule is obtained given limited computation resources. Hence, the hybrid search scheme is controllable.

This chapter starts with a brief review on the scheduling methods presented in the next section. Section 13.3 presents a typical automated manufacturing cell and the detailed modeling processes using Petri nets. Various situations encountered in manufacturing systems are described and modeled using Petri nets. Section 13.4 describes four search algorithms: Best-first (BF), Back-tracking (BT), and two hybrid schemes, BF followed by BT, and BT followed by BF. Heuristic functions are also introduced. Section 13.5 discusses the application of best-first algorithm to scheduling of the automated manufacturing cell discussed in Section 13.3. Some comparisons with mathematical programming methods are made. Section 13.6 studies the hybrid search algorithms through scheduling of a semiconductor test facility.

13.2 A Brief Review

Scheduling problems arise when multiple kinds of job types are processed by multiple kinds of shared resources according to their technological precedence constraints. We need to determine the optimal input sequence of jobs and resource usage for a given job mix. The required ordering of operations within each job must be preserved. Production scheduling problems are very complex and have been proved to be NP-hard problems [French 1982]. Several major approaches to production planning and scheduling are as follows:

1. Heuristic dispatching rules which are widely used in practice. Good rules are obtained based on one's experience. They work but often not at the best. They can also be developed based on the system simulation models. The disadvantage lies in that most comprehensive models and results are difficult to develop and take tremendous computation. Moreover, simulation models are often too specific to particular situations and thus the obtained results cannot be very well generalized.
2. Mathematical programming methods which have been extensively studied by numerous researchers and can produce optimum results for some systems [Luh and Hoitomt, 1993; Chen, 1994]. However, only a few real applications exist in an industrial environment. The mathematical models have to ignore many practical constraints in order to solve these models efficiently. These practical constraints such as material handling capacity, complex resource sharing and routing, and sophisticated discrete-event dynamics are very difficult to be mathematically and concisely described. Furthermore, even if they are described, the algebraic equations and inequalities are very difficult to be understood by industrial engineers and management. The optimality will not hold if any parameters or structures change during an operational stage.
3. Computational intelligence based approaches that include expert or knowledge-based systems, genetic algorithms, and neural networks. Knowledge-based systems have difficulty in acquiring the efficient rules and knowledge. The

13.2. A Brief Review

results cannot be guaranteed the best. Both genetic algorithms and neural networks require considerable computation and also have formulation difficulties.

4. Other methods such as algebraic models and control theoretic methods are difficult to offer efficient solution methodologies. The methods based on CPM/PERT and queuing networks can provide efficient solution methodologies but cannot describe shared resources, synchronization, and lot sizes easily.

Petri net theory has been applied for modeling, performance analysis and discrete event control of manufacturing systems. Their advantages to represent the complex discrete event dynamics and all the important FMS characteristics have motivated several researchers to investigate their usage for planning and scheduling. Several features of the obtained schedules are:

1. They are event-driven. This facilitates the real-time implementation.
2. They are deadlock-free. Since the Petri net model of the system can be a detailed representation of all the operations and resource-sharing cases, a generated schedule is the one from the system's initial condition to the final desired one. It thus avoids any deadlock.
3. The completion time or makespan criterion is the optimization objective.

The drawbacks of this approach include:

1. A huge search space (reachability tree) is required for large complex systems. The speed depends heavily on a heuristic function selected. The study on the best heuristic functions is needed.
2. Most heuristic functions cannot guarantee the optimality of the obtained schedules. The question remains open on how good the resulting schedules are.
3. It remains difficult to use the criteria other than makespan in the search process. Other useful criteria include tardiness and due dates.

Previous work on the use of Petri nets for the scheduling and planning purpose is reported in [Shih and Sekiguchi, 1991; Hatono et al, 1991; Onaga et al., 1991, Shen *et al.*, 1992; Zhang, 1992; Lee and DiCesare, 1992-94; Sun et al., 1994; Chen and Jeng, 1995; Xiong, et al., 1995-1997]. Lee and DiCesare (1992-94) pioneered the research in using best-first search of the reachability tree of a Petri net to find the schedule with the objective of minimizing makespan. Various heuristic functions are investigated. The largest example includes five machines, three robots, and ten jobs each with varied number of processing steps from three to eight. The comparison with heuristic dispatching rules is made and the results show a significant advantage of Petri net approaches in reducing the overall makespan. Cyclic scheduling using Petri nets is investigated in [Onaga, et al., 1991; Proth and Minis, 1995]. Proth and Minis (1995) investigated both cyclic and non-cyclic planning and scheduling of manufacturing systems. Effective heuristic algorithms were proposed and applied to a three machine three job type system with a given job ratios. The goal is to maximize productivity and minimize the work-in-process. Li et al. (1995) investigated scheduling and re-scheduling of AGVs for flexible manufacturing systems. Timed Petri nets are used as a system model and effective heuristic dispatching rules are implemented whenever there is a conflict in using a machine or AGV. Furthermore, the rules are shown to be effective when one has to re-schedule the tasks when one or more AGVs malfunction. Xiong et al. (1994-1997) investigated fuzzy dispatching rules and hybrid heuristic search algorithms for scheduling of non-cyclic manufacturing systems using timed Petri nets. The largest system in [Xiong, 1997] contains eighteen resources with their quantities varying from 2 to 8, thirty jobs each with three test process steps. Hatono *et al.* (1991) employed the stochastic Petri nets to describe the uncertain events of stochastic behaviors in FMS, such as failure of machine tools, repair time, and processing time. They developed a rule base to resolve conflicts among the enabled transitions. The proposed method, however, cannot handle the routing flexibility and deadlock situation.

Shen *et al.* (1992) presented a Petri net-based branch and bound method for scheduling the activities of a robot manipulator. To cope with the complexity of the problem, they truncate the original Petri net into a number of smaller size subnets. Once the Petri net is truncated, the analysis is conducted on each subnet individually. However, due to the existence of the dependency among the subnets, the combination of local optimal schedules does not necessarily yield a global optimal or even near-optimal schedule for the original system. Zhou, Chiu and Xiong (1995) also employed a Petri net based branch and bound method to schedule flexible manufacturing systems. In their method, instead of randomly selecting one decision candidate from candidate sets (enabled transition sets in Petri net based models), they select the one based on heuristic dispatching rules such as SPT. The generated schedule is transformed into a marked graph for cycle time analysis.

Zuberek (1995; 1996) evaluated different fixed schedules for a flexible manufacturing cell using timed Petri nets. Uzma et al. (1995) explored the use of timed PN to investigate anticipatory scheduling for FMS. The anticipatory scheduling is compared with other fixed or optimized schedules under certain parameter conditions. The simulation results illustrate the robustness of the anticipatory scheduling policy when the system parameters are subject to change.

13.3 Petri Net Modeling for Scheduling

To schedule a manufacturing system, one has to associate the operational times with either transitions or places. In [Proth and Minis, 1995], the times are associated with transitions and the resources are modeled using places that connect the transitions (representing operations) via self-arcs. This chapter follows another strategy which is more often used [Lee and DiCesare, 1992-94; Xiong et al., 1993-96]. It uses a place to represent either a resource status or an operation. A transition represents either start or completion of an event or operation process, and the stop transition for one activity is the same as the start transition for the next activity following [Zhou and DiCesare 1993, Lee and DiCesare 1994]. Token(s) in a resource place indicates that the resource is available and no token indicates that it is not available. A token in an operation place represents that the operation is being executed; and no token shows none being performed. A certain time may elapse between the start and the end of an operation. A time delay is thus associated with the corresponding operation place.

A bottom-up method presented in Chapter 5 is used to synthesize the Petri net model of a system for scheduling. First, a system is partitioned into sub-systems according to the job types, then sub-models are constructed for each sub-system, and a complete net model for the entire system is obtained by merging Petri nets of the sub-systems through the places representing the shared resources.

To illustrate the Petri net modeling for scheduling, we use a simple example adopted from [Ramaswam and Joshi, 1995]. This automated manufacturing system as shown in Figure 13.1 has three machines, one robot and one part load/unload station.

The robot is responsible for handling parts between machines, loading from the load station (incoming conveyor) and unloading to the unload station (outgoing conveyor). There are four jobs as shown in Table 13.1. Each job follows a fixed route. The operations and transporting times are given in Table 13.2. In the table, $O_{i,j,k}$ stands for the j-th operation of the i-th job being performed by the k-th machine, L_i for loading of the i-th job from the load station, U_i for unloading of the i-th job to the unload station, and $R_{i,j}$ for transportation of the i-th job for its j-th operation via Robot.

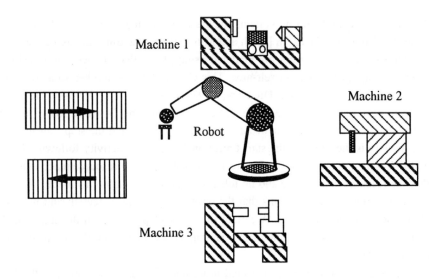

Figure 13.1 An automated manufacturing system

Table 13.1 Job requirements or routes

Operations/Jobs	J_1	J_2	J_3	J_4
1	M_1	M_2	M_1	M_3
2	M_2	M_1	M_2	M_2
3	M_3	M_3	M_3	M_1

13.3. Petri Net Modeling for Scheduling

Table 13.2 Machine operation and robot transportation times ($O_{i,j,k}$: the j-th operation of the i-th job being performed by the k-th machine, L_i: loading of the i-th job from the load station, U_i: unloading of the i-th job to the unload station, and $R_{i,j}$: transportation of the i-th job for its j-th operation via Robot)

Operation	Time Units	Transport	Time Units
$O_{1,1,1}$	40	L_1	5
$O_{1,2,2}$	100	$R_{1,2}$	3
$O_{1,3,3}$	36	$R_{1,3}$	5
$O_{2,1,2}$	65	U_1	4
$O_{2,2,1}$	45	L_2	5
$O_{2,3,3}$	98	$R_{2,2}$	3
$O_{3,1,1}$	212	$R_{2,3}$	6
$O_{3,2,2}$	73	U_2	4
$O_{3,3,3}$	32	L_3	6
$O_{4,1,3}$	35	$R_{3,2}$	7
$O_{4,2,2}$	65	$R_{3,3}$	4
$O_{4,3,1}$	55	U_3	5
		L_4	4
		$R_{4,2}$	3
		$R_{4,3}$	5
		U_4	5

13.3.1 Petri Net Model with Traditional Assumptions

The traditional job-shop scheduling often assumes that the material handling action can be ignored and unlimited intermediate storage is available. With such assumptions, we can obtain its Petri net model as follows.

First, consider job J1. According to the precedence relations specified in Table 13.1, we can order the following processes: M1 processes J1, M2 processes J1, and M3 processes J1. Between two processes, we have a place modeling the unlimited buffer space. Then we need an initially marked place to represent that initially one job of type J1 is to be processed. We need another place to keep the final product of J1. We also need three places to represent that all three machines are initially available. The resulting sub-model for J1 is pictured in Fig. 13.2. Similarly, we can derive the other three Petri net sub-models for jobs J2-4 as shown in Fig. 13.2. Since all four places, p_{17}, p_{28}, p_{37}, and p_{49} represent the availability of M1, they can be merged into a single place p_1 as shown in Fig. 13.3. By the same token, we merge p_{18}, p_{27}, p_{38}, and p_{48} into p_2, and p_{19}, p_{29}, p_{39}, and p_{47} into p_3. Therefore, we obtain a complete Petri net model for the system by merging all four sub-models through shared resources places modeling availability of Machines M1-3, as shown in Fig. 13.3.

Deterministic time delays are associated with the places, which model the machining processes. Based on Table 13.2, we associate delay times 40, 100, and 36 with places p_{11}, p_{13}, and p_{15} in Fig. 13.3, respectively for Job J1. Places p_{10}, p_{12}, p_{14}, p_{1-3} are associated with zero time delay. We can perform the same association for other sub-models for J2-J4 according to Table 13.2. Finally, we obtain a timed Petri net ready for heuristic search based scheduling algorithms to be discussed later.

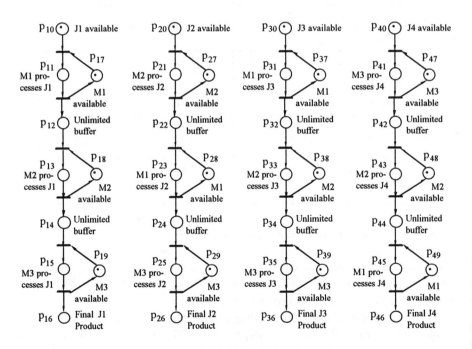

Figure 13.2 Four sub-models for J1-4.

13.3. Petri Net Modeling for Scheduling

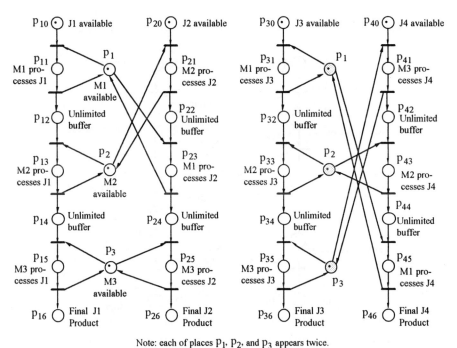

Note: each of places p_1, p_2, and p_3 appears twice.

Fig. 13.3 The complete Petri net model for an automated manufacturing system

13.3.2 Petri Net Model with Finite Buffer Size

In a practical manufacturing environment, the assumption of unlimited buffer space is unrealistic. For our automated manufacturing system example, the number of intermediate storage slots is limited and can be zero. Figure 13.4 shows the Petri net model for sub-systems job J1 and J2 assuming that no intermediate storage is provided. The models for J3 and J4 can be constructed in a similar way. Note that the present Petri net model is simplified by eliminating places modeling intermediate storage between two machining processes in the net model in Fig. 13.3.

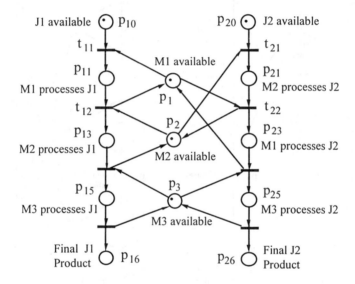

Figure 13.4 The Petri net model for the sub-systems J1 and J2 when no intermediate storage is provided.

13.3.3 Petri Net Model with Multiple Lot Size

Lot-size can be very easily represented in Petri net modeling. For example, if Job J1's lot size is k_1, we can model this conveniently by initially marking the place p_{10} in Figs. 13.3-4 with k_1 tokens. Finite buffer size is modeled as follows. First an initially marked place "availability of buffer slots" is created beside the existing place modeling unlimited buffer space in Fig. 13.3. Then an output and input arcs are introduced from the newly created place and connected to the input and output transitions of the existing place, respectively. Furthermore, the number of available

13.3. Petri Net Modeling for Scheduling

buffer slots is specified by the number of token in the initial marking of the newly created place. For example, as shown in Fig. 13.5, p_{12} is the existing place and p'_{12} is a newly created one. Output and input arcs are p'_{12}, t_{12}) and (t_{13}, p'_{12}). Input and output transitions of p_{12} are t_{12} and t_{13}. Two tokens in p'_{12} means two slots available. According to [Zhou and DiCesare, 1996], buffer spaces can also be shared among different processes. Petri net modeling of such a case is also convenient and manufacturing examples can be seen in [Zhou and DiCesare, 1996]. Figure 13.5 shows the Petri net model for sub-system J1. It shows for job J1 that the lot size is k_1, the number of intermediate buffer slots between M1 and M2 is 2, and the number between M2 and M3 is 1. The system with different scenarios of lot sizes and buffer sizes can be conveniently and visually modeled only by varying the number of available tokens of those corresponding places in the initial marking.

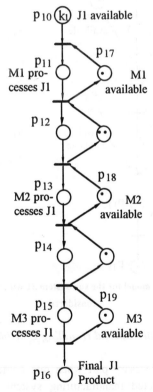

Figure 13.5 The Petri net model for the sub-system J1 with multiple lot and finite buffer size.

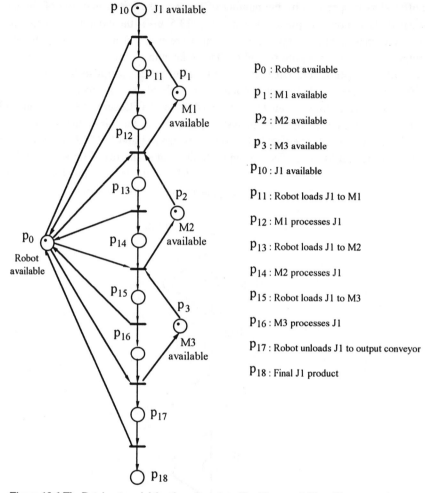

Figure 13.6 The Petri net model for the sub-system J1 with material handling operation considered.

13.3.4 Petri Net Model with Material Handling Considered

Material handling is a necessary part when one comes to implement a practical schedule. For the automated manufacturing system under discussion, a single material handler, i.e., the robot, is used for loading and unloading. Suppose that there is no intermediate buffer available for the processes, then we can construct the sub-model for Job J1 as shown in Fig. 13.6. In this model, the robot's handling time

13.3. Petri Net Modeling for Scheduling

can be specifically considered by associating the time delay with the places representing the loading, unloading, and transportation operations. Similarly, we can built up all the other three sub-models for J2-J4 one by one, and then merge all the four sub-models into a complete Petri net model.

13.3.5 Petri Net Model for Flexible Routes

Suppose that J1 can take either route M1-M2-M3 or route M2-M1-M3. Its Petri net model is constructed as shown in Fig. 13.7. Such flexible routes can be conveniently represented through choice structures. Thus the best schedules can be derived for cases that require both flexible routing and resource allocation to be considered at the same time. Note that the search space is however increased.

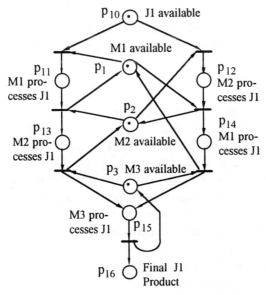

Figure 13.7 The Petri net model for the sub-system J1 with alternative routes.

13.4 Best-First, Backtracking, and Hybrid Search Algorithms

13.4.1 Best First Search and Backtracking Search

An event-driven schedule is searched in a timed Petri nets (TPN) framework to achieve minimum or near minimum makespan. This chapter employs deterministic timed Petri nets by associating fixed time delays with places. The transitions can fire with a zero duration that is consistent with the definition of non-timed Petri nets. In the Petri net model of a system, firing of an enabled transition changes the token distribution (marking). A sequence of firings results in a sequence of markings, and all possible behavior of the system can be completely tracked by the reachability graph of the net. The search space for the optimal event sequence is the reachability graph of the net, and the problem is to find a firing sequence of the transitions in the Petri net model from the initial marking to the final one. Combining a best-first graph search algorithm A* [Pearl 1984] and the Petri net execution develops a heuristic search algorithm for scheduling.

In the following algorithms, m_0 is an initial marking of a Petri net mode, m and m' are markings, f is a function of a marking to be discussed later, OPEN is a list of markings to be explored, and CLOSED a list of markings explored.

Algorithm 13.1 (Best-First):
1. Put the start node (initial marking) m_0 on OPEN.
2. If OPEN is empty, exit with failure.
3. Remove from OPEN and place on CLOSED a marking m whose f is the minimum.
4. If marking m is a goal node (final marking), exit successfully with the solution obtained by tracing back the pointers from marking m to marking m_0.
5. Otherwise find the enabled transitions of marking m, generate the successor markings for each enabled transition, and attach to them pointers back to m.
6. For every successor marking m' of marking m:
   (a) Calculate $f(m')$.
   (b) If m' is on neither OPEN nor CLOSED, add it to OPEN. Assign the newly computed $f(m')$ to marking m'.
   (c) If m' already resides on OPEN or CLOSED, compare the newly computed $f(m')$ with the value previously assigned to m'. If the old value is lower, discard the newly generated marking. If the new value is lower, substitute it for the old and direct its pointer along the current path. If the matching marking m' resides on CLOSED, move it back to OPEN.
7. Go to step 2.

13.4. Best-First, Backtracking, & Hybrid Search Algorithms

The function f (m) in the above algorithm is the sum of two terms, g(m) and h(m). f(m) is an estimate cost (makespan) from the initial marking to the final one along an optimal path which goes through the marking m. The first term, g(m), is the cost of a firing sequence from the initial marking to the current one. The second term, h(m) is an estimated cost of a firing sequence from current marking m to the final marking, called heuristic function. The following heuristic function is used:

$$h(m) = \max_i \{\xi_i(m), i = 1, 2, ..., N\}$$

where $\xi_i(m)$ is the sum of operation times of those remaining operations for all jobs which are planned to be processed on the ith machine when the current system state is represented by marking m. N is the total number of machines. The purpose of a heuristic function is to guide the search process in the most profitable direction by suggesting which transition to fire first.

For the above heuristic function, h(m) is a lower bound to all complete solutions descending from the current marking, i.e.,

$$h(m) \leq h^*(m), \forall m$$

where h*(m) is the optimal cost of paths going from the current marking m to the final one. Hence, h(m) is admissible, which guarantees for an optimal solution (Pearl 1984).

At each step of the best-first search process, we apply an appropriate heuristic function to each of the generated markings and then select the most promising marking among them. Next, we expand the chosen marking by firing all enabled transitions under it. If one of successor markings is a final marking, we can quit. If not, all those new markings are added to the set of markings generated so far. Again, the most promising marking is selected and the process continues. Additional heuristic functions are also investigated and used [Lee, 1993], including:

$$h_1(m) = 0$$
$$h_2(m) = -w \bullet dep(m)$$

where w is a weighting factor and dep(m) is the depth of the marking m in the reachability graph

$$h_3(m) = w \bullet A \bullet (E - dep(m))$$

where w is a weighting factor, A is the mean of all the possible operation times, E is the sum of the number of transition firings that each token in the initial places must nominally go through to reach the final places, and dep(m) is the depth of the marking m in the reachability graph.

Once the Petri net model of the system is constructed, given initial and final markings, an optimal schedule can be obtained using the above algorithm with an admissible heuristic function. But for a sizable multiple lot size scheduling problem, it is very difficult or impossible to find the optimal solution in a reasonable amount of time and memory space. Therefore, two directions emerge. The first one tries to find an effective heuristic function. This method can be very effective for certain classes of Petri nets but fail for general cases. It cannot in general predict whether a selected heuristic function can fulfil the need at the present time. It is a trial-error method in handling the time and space problem. Another direction is to combine different search strategies. For example, the heuristic best-first strategy can be combined with the controlled backtracking strategy. The backtracking method applies the last-in-first-out policy to node generation instead of node expansion. When a marking is first selected for exploration, only one of its enabled transitions is chosen to fire, and thus only one of its successor markings is generated. This newly generated marking is again submitted for exploration. When the generated marking meets some stopping criterion, the search process backtracks to the closest unexpanded marking which still has unfired enabled transitions. This method is controllable in terms of time and memory space by selecting different depths. Note that the depth at marking m is defined as the number of transitions in the path from a starting marking m' to m.

Algorithm 13.2 (Backtracking):
1. Put the start node (initial marking) m_0 on OPEN.
2. If OPEN is empty, exit with failure.
3. Examine the topmost marking from OPEN and call it *m*.
4. If the depth of *m* is equal to the depth-bound or if all enabled transitions under marking *m* have already been selected to fire, remove *m* from OPEN and go to step 2; otherwise continue.
5. Generate a new marking *m'* by firing an enabled transition not previously fired under marking *m*. Put *m'* on top of OPEN and provide a pointer back to *m*.
6. Mark *m* to indicate that the above transition has been selected to fire.
7. If marking *m'* is a goal node (final marking), exit successfully with the solution obtained by tracing back the pointers from marking *m'* to marking m_0.
8. If *m'* is a deadlock marking, remove it from OPEN.
9. Go to step 2.

The best-first search strategy examines, before each decision, the entire set of available alternative markings, those newly generated as well as all those suspended in the past. The backtracking search strategy is committed to maintaining in storage only a single path containing the set of alternative markings leading to the current marking. It proceeds forward heedlessly to find a feasible schedule without considering the optimality. Since only the markings on the current firing sequence are stored, it requires significantly less memory compared with the best-first search algorithm or Algorithm 13.1. These will be demonstrated through an example later.

13.4.2. Hybrid Heuristic Search Algorithms

The need to combine BF and BT strategies is a result of computational considerations. For multiple lot size scheduling problems, if we cannot afford the memory space and computation time required by a pure BF strategy, we can employ a BF-BT combination that cuts down the storage requirement and computation time at the expense of narrowing the evaluation scope, thus risking the miss of an optimal schedule

In the following Algorithm 13.3 [Xiong, et al., 1996], the heuristic best-first search strategy is applied at the top of reachability graph of a timed Petri net model and a backtracking search strategy at the bottom. We begin with BF search until depth-bound dep_0 is reached. Then a BT search is employed using the best present marking as a starting node. If it fails to find a solution, we return to get the second best marking on OPEN as a new root for a BT search, and so on. OPEN0 below is a list of markings to be explored during a new BT search starting at a marking m.

Algorithm 13.3 (Hybrid BF-BT):
1. Put the start node (initial marking) m_0 on OPEN.
2. If OPEN is empty, exit with failure.
3. Remove from OPEN and place on CLOSED a marking m whose f is the minimum.
4. If marking m is a goal node (final marking), exit successfully with the solution obtained by tracing back the pointers from marking m to marking m_0.
5. If the depth of marking m is greater than the depth-bound dep_0, go to Step 9; otherwise continue.
6. Find the enabled transitions of marking m, generate the successor markings for each enabled transition, and attach to them pointers back to m.
7. For every successor marking m' of marking m:
    (a) Calculate $f(m')$.
    (b) If m' is on neither OPEN nor CLOSED, add it to OPEN. Assign the newly computed $f(m')$ to marking m'.
    (c) If m' already resides on OPEN or CLOSED, compare the newly computed $f(m')$ with the value previously assigned to m'. If the old value is lower, discard the newly generated marking. If the new value is lower, substitute it for the old and direct its pointer along the current path. If the matching marking m' resides on CLOSED, move it back to OPEN.
8. Go to Step 2.

9. Take marking m as a root node for BT search, put it on OPEN0.
10. If OPEN0 is empty, go to Step 2.
11. Examine the topmost marking from OPEN0 and call it m'.
12. If all enabled transitions under marking m' have been selected to fire, remove it from   OPEN0 and go to Step 10.
13. Generate a successor marking m'' for one enabled transition not previously fired, calculate $g(m'')$, put m'' on top of OPEN0 and provide a pointer back to m'.
14. If marking m'' is a goal node (final marking), exit successfully with the solution obtained by tracing back the pointers from marking m'' to the initial marking m_0.
15. If m'' is a deadlock marking, remove it from OPEN0.
16. Go to Step 10.

An opposite approach is to start a backtracking search on the top of the reachability graph followed by heuristic best-first ending. This strategy is implemented in Algorithm 13.4. We begin BT until a depth-bound dep_0 is reached. Then we employ the heuristic BF search from the current marking until it returns the final marking. If the BF search fails to find a solution, we return to backtracking and again use BF upon reaching the depth-bound dep_0.

Algorithm 13.4 (Hybrid BT-BF)
1. Put the start node (initial marking) m_0 on OPEN0.
2. If OPEN0 is empty, exit with failure.
3. Examine the topmost marking from OPEN0 and call it m.
4. If all enabled transitions under marking m have already been selected to fire, remove m from OPEN0 and go to Step 2; otherwise continue.
5. If the depth of marking m is greater than the depth-bound dep_0, go to Step 10; otherwise continue.
6. Generate a new marking m' by firing an enabled transition not previously fired under marking m. Put m' on top of OPEN0 and provide a pointer back to m.
7. Mark m to indicate that the above transition has been fired.
8. If m' is a deadlock marking, remove it from OPEN0.
9. Go to step 2.
10. Take marking m from BT search as the start node m_0 and put it on OPEN.
11. If OPEN is empty, go back to Step 2 and return to backtracking search.
12. Remove from OPEN and place on CLOSED a marking m whose f is the minimum.
13. If marking m is a goal node (final marking), exit successfully with the solution obtained by tracing back the pointers from marking m to m_0. Otherwise continue.
14. Find the enabled transitions of marking m, generate the successor markings for each enabled transition, and attach to them pointers back to m.

15. For every successor marking m' of marking m:
    (a) Calculate $f(m')$.
    (b) If m' is on neither OPEN nor CLOSED, add it to OPEN. Assign the newly computed $f(m')$ to marking m'.
    (c) If m' already resides on OPEN or CLOSED, compare the newly computed $f(m')$ with the value previously assigned to m'. If the old value is lower, discard the newly generated marking. If the new value is lower, substitute it for the old and direct its pointer along the current path. If the matching marking m' resides on CLOSED, move it back to OPEN.
16. Go to step 2.

In both Algorithms 13.3-4, the heuristic function $h(m) = \max_i\{\xi_i(m), i = 1, 2, ..., N\}$ is used. $\xi_i(m)$ is the sum of operation times of those remaining operations for all jobs which are planned to be processed on the ith machine when the current system state is represented by marking m. N is the total number of machines. H(m) is a lower bound to all complete solutions descending from the current marking and thus admissible. This guarantees an optimal solution if a pure BF strategy is applied. The backtracking strategy is controllable through the depth-bound dep_0, i.e., if one can afford the memory space required by a pure BF strategy, only the pure BF search is employed, and thus an optimal schedule is obtained. Otherwise, a hybrid BF-BT or BT-BF combination can be implemented. This cuts down the storage requirement at the cost of narrowing the evaluation scope. i.e., risking the miss of the optimal one.

13.5 Scheduling An Automated Manufacturing System

Consider the automated manufacturing system discussed in Section 13.3. If we follow the traditional assumptions in which the material handling action is ignored and unlimited intermediate storage is available, we can build a Petri net model as shown in Fig. 13.3 and associate time delays with those places modeling processes. We can then employ Algorithm 13.1 to the model to obtain an optimal schedule that minimizes the makespan. In the algorithm, the admissible heuristic function
$$h(m) = \max_i\{\xi_i(m), i = 1, 2, ..., N.\}$$
is used where $\xi_i(m)$ is the sum of operation times of those remaining operations for all jobs which are planned to be processed on the ith machine when the current system state is represented by marking m. N is the total number of machines. The Gantt chart of the resulting optimal schedule is shown in Figure 13.8. This same

result can also be obtained by formulating it as a mathematical programming problem and solving it as demonstrated in [Ramaswam and Joshi, 1995].

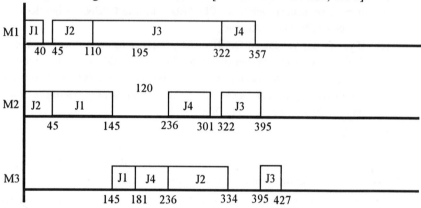

Figure 13.8 The optimal schedule without considering material handling and limited buffer

In a practical manufacturing environment, the assumption of unlimited buffer space is unrealistic. For an automated manufacturing cell, the number of intermediate storage slots between two machines is limited or even zero. The schedule shown in Fig. 13.8 will lead into a deadlock state if the actual system provides no intermediate storage. It becomes an infeasible schedule even though the constraints for precedence relations and processing times are satisfied. The Petri net model for the intersection of jobs J1 and J2 shown in Fig. 13.4 can be used to illustrate this situation. At the initial state all machines and jobs are available. According to the schedule shown in Fig. 13.8, at the time instant 0, both enabled transitions t_{11} and t_{21} fire, representing that job J1 starts its first operation on Machine 1 and J2 on Machine 2. Job J1 finishes its first operation on Machine 1 at time instant 40 and is then waiting for its second operation on Machine 2. Job J2 finishes its first operation on Machine 2 at time instant 45 and is waiting for its second operation on Machine 1. This circular waiting situation leads into a deadlock state in which neither transition t_{12} nor t_{22} is firable at the marking that marks p_{11}, p_{21}, and p_3 only in the J1 and J2 related Petri net model in Fig. 13.4. Note that a standstill may be local initially. It may be propagated leading to the entire system standstill or system deadlock. The example deadlock is in fact a system deadlock since M3 waits M2 to finish J1 too.

When the system provides no intermediate storage, we can construct Petri net sub-models for J3 and J4 as done for J1 and J2 (depicted in Figure 13.4) and merge the sub-models to get the complete one. Then, we apply Algorithm 13.1 to it

13.5. Scheduling an Automated Manufacturing System

to obtain an optimal deadlock-free schedule. The resulting deadlock-free schedule is shown in Figure 13.9 in the form of Gantt chart.

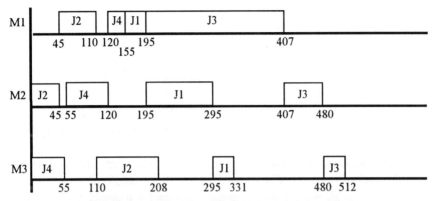

Figure 13.9 The optimal deadlock-free schedule for no buffer case

In the above cases, the system's buffer size is assumed infinite or zero and each job's lot size is 1. We often encounter the cases with other buffer sizes and multiple lot sizes in a practical manufacturing environment. The Petri net model for sub-system J1 is shown in Fig. 13.5 as such example. The system with different scenarios of lot sizes and buffer sizes can be conveniently and visually modeled by varying the available token of those corresponding places in the initial marking only.

Table 13.3 shows scheduling results for several different lot sizes, and the size of each intermediate buffers is set to 2 for the automated manufacturing system. By marking with zero token in the place representing availability of J3 jobs, we derive a schedule with J3's lot size being zero without changing the Petri net structure or model. Given a fixed job lot size, we can also find optimal schedules for varying buffer sizes. This helps designers find best buffer size to optimize a system. Figure 13.10 shows scheduling results for a fixed lot size (20, 20, 20, 20) with each buffer size varying from 1 to 8.

Table 13.3 The scheduling results for several different lot sizes

Lot	Size			Makespan
J1	J2	J3	J4	
2	2	0	2	455
5	4	6	3	1942
10	10	10	10	3638
20	20	20	20	7171

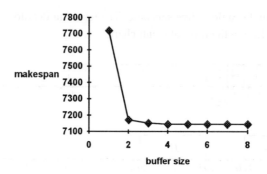

Figure 13.10 The optimal schedules for lot size (20, 20, 20, 20) and varying buffer size.

Mathematical programming techniques often have formulation difficulties for the features of varying buffer size and lot size. For example, in their pioneering work of using mathematical programming methods for deadlock-free scheduling [Ramaswamy and Joshi, 1996], the lot size of each job is limited to 1, and the proposed solution scheme for deadlock-free schedules is applicable to only the problems with m machines and $\lfloor m/2 \rfloor$ buffer slots where $\lfloor m/2 \rfloor$ represents the largest integer less than m/2.

To schedule the operations of material handling, Figure 13.6 shows the Petri net model for the sub-system J1 with the material handler, i.e., robot in this example, as a shared resource. The Petri net model for the whole system can be obtained by merging all four sub-models. Using Algorithm 13.1, we obtain the following optimal deadlock-free event sequences for each shared resource:

Machine 1: performs in sequence Operation 2 of J2, Operation 3 of J4, Operation 1 of J1, and Operation 2 of J3;

Machine 2: performs in sequence Operation 1 of J2, Operation 2 of J4, Operation 2 of J1, and Operation 2 of J3;

Machine 3: performs in sequence Operation 1 of J4, Operation 3 of J2, Operation 3 of J1, and Operation 3 of J3; and

Robot: transports in sequence J4 from load station to Machine 3, J2 from load station to Machine 2, J2 from Machine 2 to Machine 1, J4 from Machine 3 to Machine 2, J2 from Machine 1 to Machine 3, J4 from Machine 2 to Machine 1, J4 from Machine 1 to unload station, J1 from load station to Machine 1, J1 from Machine 1 to Machine 2, J3 from load station to Machine 1, J2 from Machine 3 to unload station, J1 from Machine 2 to Machine 3, J1 from Machine 3 to unload station, J3 from Machine 1 to Machine 2, J3 from Machine 2 to Machine 3, J3 from Machine 3 to unload

13.6. Modeling & Scheduling of Semiconductor Test Facility

station>.

The optimal deadlock-free schedule is shown in Fig. 13.11 in the form of a Gantt chart and the makespan is 560 time units. By employing Algorithm 13.1, the computation time is 0.13 CPU seconds to generate the schedule in Fig. 13.9 when only buffer availability is considered, 0.41 CPU seconds to generate the schedule in Fig. 13.11 when both material handling and buffer availability are considered in Sun Sparc 20. In [Ramaswamy and Joshi 1996] in which mathematical programming methods are used, the CPU time is increased from 0.71 seconds to 67.0 seconds in IBM ES/3090-600S for the above two schedules. It is clear that the use of Petri nets for optimal deadlock-free scheduling results in a significantly small variation in computation. This is not the case for mathematical programming case, however.

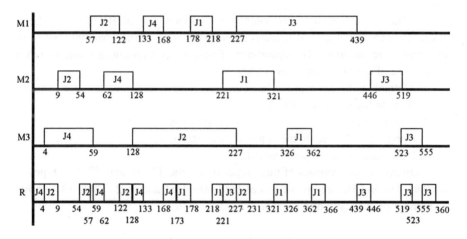

Figure 13.11 The optimal deadlock-free schedule including the operations of material handling

13.6 Modeling and Scheduling of a Semiconductor Test Facility

A job-shop system is a specific type of production systems composed of a certain set of machines and a variety of jobs that must be produced using the machines. The manufacturing process of each job is specified as a sequence of machines to visit, i.e., as a routing into the system. Any routing is allowable but is defined uniquely for each job. Furthermore, the time spent by the jobs on the machines is assumed to be fixed and deterministic. Finally, the sequencing of the jobs on the machines is also assumed to be given, as well as the order with which jobs are loaded into the system (input sequencing).

There are four main stages in a typical Integrated Circuit (IC) manufacturing process: wafer fabrication, wafer sort, assembly cycle, and final test [Chen. 1994]. The first stage of IC production is called wafer fabrication. In wafer fabrication, ICs are manufactured on a silicon or gallium arsenide wafer using photolithography, etching, diffusion, and ion implantation processes. In the next stage, wafer sort, the individual circuits (dice) on a wafer are tested for functionality by means of electrical probes. Dice that fail to meet specifications are marked with an ink dot. The wafer then goes to assembly cycle, where the wafer is sawed; the defective dice are discarded; the good dice are bounded to the lead frames; the wires are bounded and then encapsulations are followed. After the assembly cycle, each IC ship is subjected to final test to determine whether or not it is operating at the required specifications.

This section focuses on scheduling problems for an IC final test floor. Normally, a task for final test requires a combination of a tester, handler, and some other hardware facilities. The operations of setting up a testing machine to test a certain type of products consist of [Uzsoy, *et al*. 1991]:

1) Obtaining the appropriate handler, load board, and contacts and bringing them to the tester concerned.
2) Connecting the hander, contacts and load boards to the tester.
3) Bringing the hander to the required temperature.
4) Downloading the required software.

Our example system consists of three types of testers, T1, T2 and T3, two types of handlers, H1 and H2, and two types of hardware, Ha1 and Ha2. Table 13.4 shows the number of each type of resources.

Table 13.4 The number of each type of resources

Facility	Type	Quantity
Tester	T1, T2, T3	1
Handler	H1	1
Handler	H2	2
Hardware	Ha1	2
Hardware	Ha2	1

Table 13.5 shows the possible resource combinations for IC final test. Each combination consists of a workcenter and looks as a single machine for scheduling. There are 4 jobs to be scheduled. Table 13.6 shows the job requirements.

13.6. Modeling & Scheduling of Semiconductor Test Facility

Table 13.5 Workcenters for final test

Workcenter	Resource Combination
M_1	T1+H1+Ha1
M_2	T2+H2+Ha2
M_3	T3+H2+Ha1

Table 13.6 Job Requirements

Operations/Jobs	J_1	J_2	J_3	J_4
1	$(M_1,2)$	$(M_3,4)$	$(M_1,3)$	$(M_2,3)$
2	$(M_2,3)$	$(M_1,2)$	$(M_3,5)$	$(M_3,4)$
3	$(M_3,4)$	$(M_2,2)$	$(M_2,3)$	$(M_1,3)$

Note: (M_i, K) means the corresponding operation takes Machine M_i K time units.

Figure 13.12 shows the Petri net model for this system. The three test stages of Job i are modeled by p_{i1}, t_{i1}, Ö, t_{i6}, and p_{i7} for i = 1, 2, 3 and 4, respectively. The places and transitions are interpreted in the same figure. The modeling is briefed as follows. First, model a Petri net model for each job based on their sequence and the use of machines. Then merge these sub-models to obtain a complete Petri net model for the system through shared resource places which model the availability of machines. The lot size is represented by the number of tokens in places p_{11}-p_{41}. For example in Figure 13.12, the lot sizes for jobs J1-4 are 2, 1, 2, and 3, respectively. It should be noted that the system is basically a job-shop type of systems based on its requirements or its Petri net model. Unlimited buffer space between two process steps is assumed for this test facility.

Several different lot sizes of each job in the example are tested and makespans, numbers of generated markings and CPU times in Sun SPARC 20 are shown in Table 13.7 by employing Algorithms 13.1 (Best-First) and 13.2 (Back-Tracking).

Table 13.7 Scheduling results for the IC test facility

Lot sizes				Makespan		Number of markings		CPU time (Sec.)	
J_1	J_2	J_3	J_4	BF	BT	BF	BT	BF	BT
1	1	1	1	17	21	155	25	0.16	0.06
2	2	1	1	25	33	501	37	0.56	0.1
5	5	2	2	58	105	3437	85	14	0.16
8	8	4	4	100	198	9438	145	112	0.23
10	10	6	6	134	274	23092	193	720	0.38

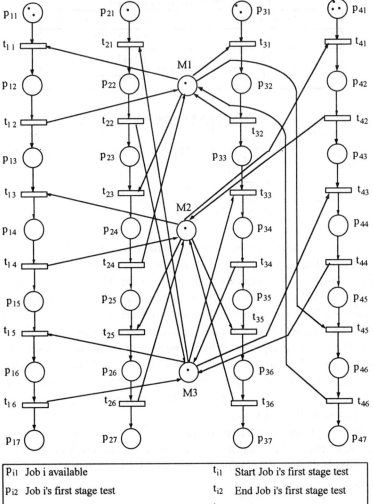

p_{i1}	Job i available	t_{i1}	Start Job i's first stage test
p_{i2}	Job i's first stage test	t_{i2}	End Job i's first stage test
p_{i3}	Waiting for Job i's second stage test	t_{i3}	Start Job i's second stage test
p_{i4}	Job i's second stage test	t_{i4}	End Job i's second stage test
p_{i5}	Waiting for Job i's third stage test	t_{i5}	Start Job i's third stage test
p_{i6}	Job i's third stage test	t_{i6}	End Job i's third stage test
p_{i7}	Job i's test complete		$i = 1, 2, 3,$ and 4
M_j	Machine j is available for $j = 1, 2,$ and 3		

Fig. 13.12 The complete Petri net model for the semiconductor test facility

13.6. Modeling & Scheduling of Semiconductor Test Facility

From Table 13.7, we see that Algorithm 13.1 finds the optimal solutions at the expense of greater computation complexity, while Algorithm 13.2 reduces the computation complexity at the expense of optimality. For many practical scheduling problems, it is desired to get a good solution (even not optimal) in a reasonable amount of time and storage. This suggests a combination of best-first search and backtracking search for a trade-off.

For the above example, we make a comparison between two hybrid search strategies: Algorithm 13.3 (BF followed by BT) and Algorithm 13.4 (BT followed by BF). We set the different depth bounds to find out the relations between the optimality and computation complexity.

The three sets of lot size (5, 5, 2, 2), (8, 8, 4, 4) and (10, 10, 6, 6) are tested. We employ both Algorithms 13.3 and 13.4. The scheduling results of makespan, number of generated markings and computation time are shown in Table 13.8-10 for the above lot sizes (5, 5, 2, 2), (8, 8, 4, 4) and (10, 10, 6, 6), respectively. The optimal makespans for different cases obtained from pure BF search in Table 13.7 are also shown in these tables.

Table 13.8 Scheduling results of the Example for lot size (5, 5, 2, 2)

Depth for BF search	Makespan		Number of markings		CPU time (Sec)		Optimal makespan
	BF-BT	BT-BF	BF-BT	BT-BF	BF-BT	BT-BF	pure BF
20	94	88	571	248	0.65	0.38	58
40	85	80	1607	484	4	0.8	58
50	79	70	2132	1247	6	3.6	58
60	74	64	2775	1520	8	6.5	58
80	64	62	3308	1687	11	7	58

Table 13.9 Scheduling results for lot size (8, 8, 4, 4)

Depth for BF search	Makespan		Number of markings		CPU time (Sec)		Optimal makespan
	BF-BT	BT-BF	BF-BT	BT-BF	BF-BT	BT-BF	pure BF
40	168	163	3888	585	24	1.4	100
60	154	140	5234	1590	38	7	100
80	140	121	7699	2873	49	18	100
100	127	112	8819	4545	90	36	100
120	108	104	9233	8045	104	76	100

Table 13.10 Scheduling results of the Example for lot size (10, 10, 6, 6)

Depth for BF search	Makespan		Number of markings		CPU time (Sec)		Optimal makespan
	BF-BT	BT-BF	BF-BT	BT-BF	BF-BT	BT-BF	pure BF
80	206	209	6281	1254	64	5	134
100	198	181	12341	2315	240	16	134
120	180	162	16602	8495	480	139	134
140	169	150	20155	11368	540	390	134
160	153	148	21797	18875	660	560	134

Both Algorithms 13.3 (BF-BT) and 13.4 (BT-BF) cut down the computation complexity by narrowing the evaluation scope at the expense of losing the optimality. The relations of computation complexity (number of generated markings and computation time) reduced versus optimality lost are shown in Figs. 13.13-15 for three different sets of lot size (5, 5, 2, 2), (8, 8, 4, 4) and (10, 10, 6, 6) respectively. In these figures, the percentage of optimality lost, which is the comparison of the makespan, is defined as

$$Percentage\ of\ Optimality\ Lost = \frac{Makespan(Hybrid) - Makespan(BF)}{Makespan(BF)} \bullet 100\%$$

Where *Makespan(Hybrid)* means the makespan using a hybrid search method which is either BF-BT or BT-BF and *Makespan(BF)* is the makespan using best-first (BF) search method with the previously discussed admissible heuristic function.

The computation complexity is characterized by two indices, the storage and the CPU time. The percentage of storage (generated markings) reduced is defined as

$$Percentage\ of\ Storage\ Reduced = \frac{Storage(BF) - Storage(Hybrid)}{Storage(BF)} \bullet 100\%$$

Where *Storage(BF)* is the number of markings generated or searched when the BF method is used and *Storage(Hybrid)* is the number of markings generated when a hybrid method is used. The percentage of computation time reduced is defined as

$$Percentage\ of\ Time\ Reduced = \frac{CPU(BF) - CPU(Hybrid)}{CPU(BF)} \bullet 100\%$$

Where *CPU(BF)* is the CPU time a BF method takes and *CPU(Hybrid)* is the CPU time a hybrid method takes. Again, a hybrid method can be either a BF-BT or a BT-BF method.

13.6. Modeling & Scheduling of Semiconductor Test Facility

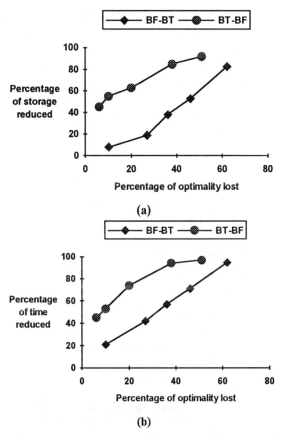

Figure 13.13 (a) Percentage of storage and (b) Percentage of computation time reduced versus percentage of optimality lost for lot size (5, 5, 2, 2)

From the test results the following conclusions are drawn. The hybrid heuristic search which employs the heuristic best-first search at the bottom of the Petri net reachability graph (Algorithm 13.4) performs much better than the one which employs the heuristic best-first search at the top of the Petri net reachability graph (Algorithm 13.3). This is due to two reasons. One is that the performance of heuristic best-first search is at its best when its guiding heuristic is more informed, and this usually happens at the bottom of the search graph [Pear, 1984]. Thus BT-BF search greatly reduces the computation complexity comparing with BF-BT search which employs the heuristic best-first search at the top of the search graph. Another reason is that there are fewer firing transitions for the markings at the bottom of

Petri net reachability graph than at the top. This is because at the late stages of a scheduling task, the reduced number of remaining operations reduces the number of choices. Hence, the number of alternatives considered in each decision for BT-BF search is less than the one for BF-BT search. However, the important decisions with respect to the quality of a schedule may happen at the early stages of the scheduling activity, this increases the likelihood of missing the critical candidates for BT-BF search which employs backtracking search instead of best-first search at the early stage.

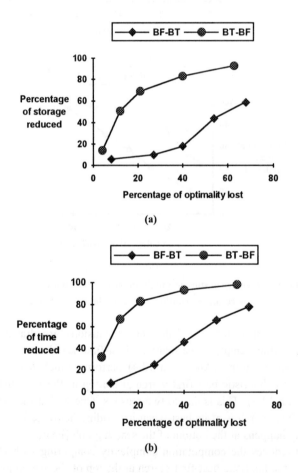

Figure 13.14 (a) Percentage of storage and (b) Percentage of computation time reduced versus percentage of optimality lost for lot size (8, 8, 4, 4)

13.7. Summary

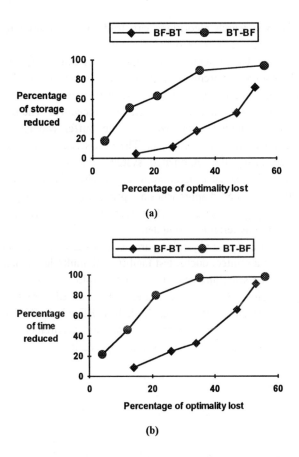

Figure 13.15 (a) Percentage of storage and (b) Percentage of computation time reduced versus percentage of optimality lost for lot size (10, 10, 6, 6)

13.7 Summary

Timed Petri nets provide an efficient method for representing concurrent activities, shared resources, precedence constraints and multiple lot sizes. This chapter presents Timed Petri nets and heuristic search methods for scheduling of flexible manufacturing systems. The emphasis is put on the best-first algorithms and two hybrid search algorithms which combines the best-first and back-tracking search algorithms. The best-search algorithms can lead to the optimal result when an admissible heuristic function is used and computational resources are sufficient for sizable systems. The back-tracking search algorithms can significantly reduce the computational resource requirements at the cost of narrowing the search space thus increasing risk in missing the optimal solutions. A typical automated manufacturing system is used to illustrate the Petri net modeling procedures. The easy handling of precedence relations, multiple lot size, variable buffer capacities, and routing flexibility is demonstrated with Petri nets. The best-search algorithm is developed and used to scheduling this automated system. The computational results are compared with mathematical programming approaches. They show that the Petri net and heuristic search based approaches are promising in deriving a deadlock-free and event-driven schedule for a manufacturing system with variable buffer size, multiple lot-size, and material handling details.

Two hybrid heuristic algorithms are used to search for an optimal or near-optimal schedule of a semiconductor test facility with multiple lot sizes for each job type considered. The searching scheme is controllable, i.e., if one can afford the memory space required by a pure BF strategy, the pure BF search can be used to locate an optimal schedule. Otherwise, the hybrid BF-BT or BT-BF combination can be implemented, which can cut down the storage requirement at the cost of a smaller evaluation scope, thereby the solution's optimality.

CHAPTER 14

PETRI NETS AND FUTURE RESEARCH

Many novel applications of Petri nets continue to emerge. The success of Petri nets and related technologies can be greatly achieved only when more industrial engineers and designers use them together with other techniques in their system development and operation. More research and development efforts in the following areas are still needed.

14.1 CASE Tool Environment

There are several PN CASE tools available that are listed in the web sites mentioned at the end of the Chapter 7. However, the widely spread industrial use of PNs is limited. This fact is even truer in North American and Asian countries than in European countries, perhaps because PNs were initially originated and acknowledged in European countries. Even though the limited use of PNs in industry can be attributed to several reasons, they can be broadly classified into two major categories as follows:

- Limited awareness of PNs in industry
- Limited availability of industrial-strength PN CASE tools

When one closely observes the above two categories it can be concluded that they are directly related to each other. In other words, if the awareness of PNs in industry were increased, the availability of industrial-strength PN CASE tools would be also increased. Similarly, if there is an increase in the number of industrial-strength CASE tools in the market, the chances of PN usage in industry are more, which in turn will increase the awareness of PNs in industry.

While it is very difficult to decide which one of the above two categories is more important, the following can be stated:

There are several academic universities throughout the world carrying the research on PNs by discovering new types of PNs, new application areas of PNs, new algorithms to minimize the execution speed and memory to store PNs, etc. However, just as any technology, the success of PNs is better judged when it is widely used in industry. It is needless to emphasize the role of CASE tools to increase the application of a technology. Nevertheless, there are only few industrial-strength PN CASE tools available in the market. The following are only some of the criteria for assessing whether a CASE tool is an industry-strength CASE tool or not. Note that the following criteria are not much important for the PN research community as such but they are very much crucial for the PN CASE tool vendors who contribute to the success and widely spread industrial use of CASE tools.

- **The simplicity and usability of the CASE tool**

    The CASE tool should be easily usable and understandable by people who do not know much of the details of PNs. Some of the CASE tools available impose extra constraints on using PNs by defining new and advanced concepts of PNs that are even not widely known in the PN community. The PN CASE tool should be developed with a novice PN user in mind but not an advanced user who knows the intricate details of PN theory. Our experience suggests that timed PNs with token game simulation capabilities are quite powerful and enough for most of the industrial applications that do not involve in directly controlling the modeled systems, but just need to capture the coordination, synchronization, etc. among the resources, events, and states related to the system under investigation. Special attention needs to be given to printing, zooming, and unzooming abilities of PN drawing.

- **The cost of the CASE tool**

    The cost of the CASE tool is a very important factor that decides the number of copies that an industry can buy from the vendor of the CASE tool. Most of the available PN CASE tools are very costly. This may again be due to the fact that there is not enough competition among the PN CASE tool vendors. Since there are only few industrial-strength CASE tools, the cost of these CASE tools are relatively higher than the CASE tools of any other similar technology.

- **The ability of the CASE tool to be used along with other CASE tools**

    In the current information age every application program that runs on a computer is related to another application directly or indirectly. For example, one may use one application for word processing and another application for drawing graphics. However, the material typed in the word processing application can be pasted and modified in the application that is used to draw graphics and vice-versa. This is what is

14.1. Case Tool Environment

called as "component technology" in the area of object-oriented technology. An example of the above inter-operability of two applications is Microsoft Word and Microsoft PowerPoint. In the current age where information dominates the world we live in, the compatibility of an application to work with other applications is very crucial for the widely spread use of an application. Hence, PN CASE tools should be easily compatible to work with other general and normally used desktop applications. In other words, PN diagrams drawn in a PN CASE tool should be easily exported to other desktop applications. Similarly, PN CASE tools should also support the importing of content-body (i.e., word processing matter, graphics, etc.) developed in other desktop applications. This compatibility of PN CASE tools with other CASE tools will certainly help to treat PNs as an integral part of the other desktop tools and thereby promotes the use of PNs in preparing reports for management, labor, etc. who use PNs in their normal business communications.

Languages such as Java need to be used to develop PN CASE tools for improving the compatibility of running the CASE tool on any operating system. Currently most of the CASE tools run either on Unix or on Windows operating systems. There is also an additional cost to buy CASE tool that runs on more than one operating system. By using languages such as Java, a PN CASE tool can be executed irrespective of the operating system and hardware, thereby reducing the cost of the CASE tool.

- **The availability of technical support and maintenance by the CASE tool vendor**

  When a CASE tool is sold to an industry, the vendor should provide the necessary technical support for users of the CASE tool. One of the reasons for early rejection of the CASE tools by users in the industry is the unavailability of knowledgeable and courteous technical support. Furthermore, the CASE tool sold should be maintained in the sense that whenever a new upgrade is developed (for example, to run the CASE tool on a new operating system, to fix the previous errors in the software, etc.), it should be automatically notified to the user community of the CASE tool so that they are given the most recent and up-to-date version of the CASE tool. Some CASE tool vendors provide on-site consulting and training help to users of the CASE tool during the initial usage period of the CASE tool.

Given the above criteria for an industrial-strength PN CASE tool, more CASE tool vendors need to come forward to develop and offer PN CASE tools to the industry. Research studies that do benchmark studies on the commercially available PN CASE tools are also highly needed to compare, contrast, and evaluate different PN CASE tools. Such studies will not only contribute to improve the quality, usability, and applicability of PN CASE tools in the industry but also highlight the research opportunities to better develop PN CASE tools.

Such studies may need to follow the following three steps:

1. Classify the available PN CASE tools according to the types of PNs supported (e.g., simple PNs, timed PNs, colored PNs, etc.), type of application areas of PNs (e.g., manufacturing, information systems, telecommunication, control, hardware-software co-design, etc.), type of hardware on which the CASE tool runs (e.g., IBM compatible personal computer, Unix based workstation, etc.), type of operating systems on which the CASE tool runs (e.g., Windows NT, Solaris, HP Unix, etc.), the level of user knowledge (e.g., novice PN user, advanced PN user, etc.), scope of the PN usage (e.g., modeling and analysis only, modeling and simulation only, modeling, simulation, and control only, etc.);

2. Develop some criteria for application of PNs in a given application area such as manufacturing, information systems, telecommunications, control, hardware-software co-design, etc. Note that this criterion is in addition to the general industrial-strength PN CASE tool criteria detailed in the earlier paragraphs. This is because the criteria to be satisfied in one particular application area may or may not be considered important in other application areas.

    For example, the generation of relational database tables to store the data modeled by PN models is very important in an industrial application that belongs to the application area of information systems. However, the generation of database tables from PN models that are intended to check the absence of deadlocks in a communication protocol is not much important. Note that the PN models that model the communication protocol come under the application area of telecommunications. In this example described for the communication protocol, automatic generation of reachability tree from the PN models is much more important than that of the database tables from the PN models. Of course, if there is a CASE tool that satisfies all the needs of all the application areas, it is ideal and desirable. However, when designed properly such a CASE tool may contribute to increase the complexity of the usage of the CASE tool and violates one of the criteria listed earlier - *the simplicity and usability of the CASE tool.* This potential danger can be resolved if the CASE tool vendor develops a CASE tool that comes along with different modules or add-on packages such that each module or package is unique to each application area. The bottom line is that each user community of each application area has a certain set of specific needs and preferences that are different from that of other user communities belonging to other application areas. Hence, a CASE tool vendor needs to design the CASE tools keeping in mind the potential danger listed above; and

3. Use the results of the above two steps to propose improvements and suggestions for increasing the industrial applicability of the PN CASE tools.

14.2 Scheduling Large Production Systems

Many industrial problems are sizable. For example, a semiconductor manufacturing system often contains over a hundred machines. Multiple machines are often installed to deal with a critical process of wafers thus one can still accomplish some "urgent" jobs in case that some of them malfunction. These "urgent" jobs carry priority over the others often due to their higher values or more importance. Many machines can take from one to dozens of wafers at a time and spend almost the same time and resource to run one or more through. Thus a relatively more expensive machine needs to look ahead to check if there are forthcoming processes which need to be undertaken. Many wafers are required to go through the same process for a number of times. From the beginning to the end, wafers need to go through hundreds of processes or steps. Any error in this sequence may generate serious consequences, e.g., scraping the batch of wafers and thus implying a great amount of financial loss. In today's factory, each batch of wafers has a smart card that records its sequence, the past history, current status, and future plan. Each machine reads the card before its processing and writes to the card when it finishes the batch. Hence, the errors are minimized. Production managers are in charge of daily schedules which could involve only a single product type (a single batch of wafers that go through the same processes) or several product types. They often face rush jobs that require rescheduling of their current resources in order to get these rush jobs done first. For those most critical or expensive resources, they have to predict if there are forthcoming batches requiring them to wait. If so and the waiting time is reasonable, the batch has to wait for the other forthcoming batches. Due to the complexity involving in the above scheduling environment, a semiconductor manufacturing facility is often operated based on the experience of production managers and process engineers. The challenges are how to optimize such systems' performance in terms of both productivity improvement and cost saving. A scheduling algorithm designed based on a Petri net frame seems feasible but needs to resolve the following research issues:

1. Modeling Issues: The use of Petri nets to represent such a system is theoretically and practically convenient. Each job type has its processes defined or precedence relation given. Each process's time is also known

and fixed. Unlimited buffer capacity of each machine can be assumed if the transportation between machines is manually performed, which is common in many semiconductor manufacturing facilities. Note that each machine, however, often has its own automated wafer handling devices. The challenge comes from the size of the net model. A simplified representation with the Petri net as an underlying mathematical tool could be more appropriate. Special modules should be created for this huge net's graphical representation in computer monitors. Thus hundreds of processes and machines can all be well represented. Tokens can be still used to represent batches of wafers. They model smart cards that carry the information regarding the batch. System decomposition and composition techniques need to be developed for large systems to facilitate modeling and scheduling.

2. Solution Methodologies: The feasibility of using heuristic search algorithms has to be more carefully investigated. It seems clear that a brute-force approach will not work. Heuristic functions have to be synthesized based on such systems' characteristics. On-line rescheduling has to be generated for rush jobs as well as the cases of unexpected breakdown machines. For a small system, a same type machine's failure requires us to remove one token from the resource place that models the availability of the same type machines. We then run the model to get another good schedule. When the machine is recovered, we add a token back and then run the net model based on the current state to get another schedule. The question is how quickly the algorithm can generate a good solution. The combination of the other rule-based techniques with the Petri net approaches may prove effective in handling either rush jobs or the machine status change cases.

The anticipatory scheduling policy [Uzma *et al.*, 1995] is another effective way to handle such large system's scheduling problem. When the system's performance and external requirements change, such a policy s often more robust compared with fixed scheduling policy that would be optimized under certain parameter conditions. The systematic applications of such a policy to large systems need to be explored further. The benchmark studies of the different scheduling methodologies on industrial-scale systems are also needed.

3. Integration with the existing information systems: The scheduling algorithms have to be integrated with the existing information systems that record all the information related to the machine status, job progress, and other information on job in the queue.

14.3 Petri Nets and Supervisory Control Theory

In a Ramage and Wohnam's framework for supervisory control [Ramage and Wohnam, 1987], a discrete-event system is modeled as a finite automaton or state machine and events are divided into two classes: controllable and uncontrollable where the former can be enabled or disabled. The application of their framework to Petri nets has already produced many interesting results and methods [Holloway and Krogh, 1990; Giua, 1994; Alpan, 1997; Holloway et al., 1997]. Petri nets are used as discrete-event models instead of finite sate machines. Petri net languages and supervisory design and validation methods are discussed for special classes of Petri net models [Guia, 1994]. Thus supervisory control theory in [Ramage and Wohnam, 1987] is extended. In [Alpan, 1997], a system is described as an uncontrolled plant in Petri nets and then a supervisory controller is synthesized to satisfy the controlled system's requirements and specifications. Thus two design steps are:

1. Design a Petri net model for the uncontrolled plant, and
2. Synthesize the controller based on the specifications.

Instead of the synthesis of a Petri net model for control directly from a plant [Zhou and DiCesare, 1993], this approach avoids the reconstruction of a plant model when only control requirements and specifications vary. When a plant is changed, both plant model and supervisory controller often need modifications. Three challenging issues are:

1. Compact and Petri net-based supervisory controllers: The resulting controllers based on the known procedures often result in a large size and unmanageable supervisory controller which is specified through an automaton [Alpan, 1997]]. Thus state explosion problems are encountered for sizable systems. The question is whether one can generate a procedure that can generate a Petri net supervisory controller that has much smaller structure through places with multiple tokens, for example. While reachability tree generation algorithm can convert a Petri net into a huge automaton, the reversion, i.e., converting a state machine into a compact Petri net, is a very difficult, if not solvable, problem.

2. Expand the control specification including timing requirements: Some preliminary work was performed in [Alpan, 1997] for some special Petri nets. The work is far from the completion for general Petri nets. Instead of focusing on general case Petri nets, special classes of Petri nets discussed in this book are very useful in practice. An example can be seen in [Suzuki et al., 1994] in which supervisory control theory was applied to assembly Petri nets although time delays were not included. The work with timing consideration also becomes related to the planning and scheduling research. The derivation and integration of

scheduling algorithms and supervisory control algorithms should be a very interesting research area.

3. Application to Sizable Systems: Due to the state explosion problems encountered in the modeling and design stages, an automaton-based supervisory control theory has not reached wide industrial applications yet. Therefore, development of industrial applications for sizable systems becomes very important in motivating the investigation of Petri nets and supervisory control theory. Easy-to-use procedures and CAD tools are also needed.

14.4 Application to Multi-lifecycle Engineering Research

To remain competitive in today's market, manufacturers have to produce at constant quality, in required quantities, at the right time, and in a flexible and agile way. To keep the environment clean, they must also extend the concept of Design For Lifecycle to the concept of Design For Multi-lifecycles. Designers have to consider the design issues related to clean manufacturing processes, waste-stream processing, and the reuse of obsolete or malfunctioning products [Zhou et al., 1996].

The past product lifecycle is illustrated in Fig. 14.1(a). Engineering materials are refined from natural resources. Then they are manufactured into component parts, assembled into sub-systems, and finally to systems or consumer products. These products are consumed by users and obsolete or malfunctioning products are disposed as waste that is eventually landfilled. Waste streams exist at each stage. The reuse and recycle are not stressed. The current research and industrial practice in the area of green production, green products, design for environment, and sustainable development emphasizes a multi-lifecycle engineering approach. At each state, its reusing or recycling into the usable material, components and systems as shown in Fig. 14.1(b) minimize the waste stream. Furthermore, the materials or components resulting from a process are used as the next level product's inputs if they are no longer reusable for the present product. For example, the glass of CRT in a computer system may be used as building material. Thus lifecycles of the materials are extended while their dumping into the earth is avoided.

Petri nets can be used into this area of multi-lifecycle engineering in the following aspects:

1. Modeling and representation tools for the detailed components and unit-processes. The focus is on the physical material flow, energy flow, and information flow. The environmental and economical effects as well as other criteria can be evaluated from the models. The detailed procedure and their application scope need to be further investigated.

14.4. Application to Multi-Lifecycle Engineering Research

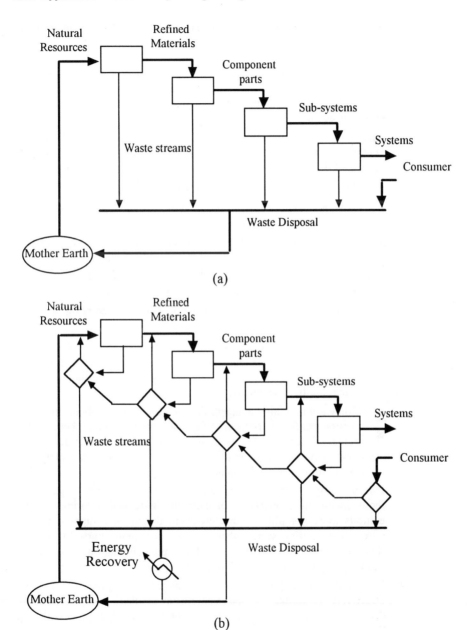

Figure 14.1 (a) Past approach and Design for Lifecycle and (b) Current approach and Multi-lifecycle Engineering.

2. Resource and process optimization can be performed based on the Petri net models in terms of both completion time as well as environmental impact. Incorporation of the environmental data (such as energy use, waste material generation, etc.) into such a model and its evaluation are not a trivial task due to the complexity involved in the product design and manufacturing, usage, and post-live treatment. More research is clearly needed.
3. Discrete event modeling and simulation of manufacturing/ demanufacturing systems. Many manufacturers have to take a varying portion of their products back because major categories as follows:
   - The products may not satisfy a customer,
   - They are no longer working properly due to a customer's inappropriate usage; and
   - They contain design and manufacturing flaws or present certain danger to, e.g., health or environment.

   These products have to be examined to face several choices:
   - If there is nothing wrong with them, the manufacturer just needs to perform minor reconditioning work (such as painting and re-packaging);
   - If any faults are found with certain components, replace these components with good ones, test them, and ship them new products. To perform this, they have to be disassembled in order to extract certain faulty components;
   - If they are partially damaged, disassemble them to obtain the good components for their reuse, to facilitate raw material recycling back to the fabrication of new components and to extract hazardous components out of the systems to ensure environmental safety and avoid future costly environmental liability.

   Due to many uncertainty associated with the quantity and quality of these incoming products, a manufacturing/demanufacturing system exhibits more complicated discrete-event dynamics compared with a pure manufacturing system that takes homogeneous parts to produce final products. Hence, its modeling and simulation is a more challenging problem that is worth much exploration.
4. Modeling and adaptive planning of disassembly processes can be effectively performed in the framework of Petri nets as partially demonstrated in [Zussman and Zhou, 1997]. A principal process during the demanufacturing of worn-out or malfunctioning products is disassembly that enables the dumping, cleaning, repair or replacement of components as desired. Products subjected to disassembly exhibit uncertainty in the product structure and component conditions. Hence

the termination goal (the level of disassembly) is subject to change, and the disassembly plan must be adapted. Disassembly Petri nets presented in this book has been augmented for the modeling and adaptive planning of disassembly processes in [Zussman and Zhou, 1997]. Planning and execution algorithms should be developed to guarantee the optimal demanufacturing value when each node's utility function (benefit of a subassembly or part and cost of a disassembly operation) remains deterministic, and the best average demanufacturing value when the disassembly operation success rates vary. Integration of DPNs with Bayesian networks to model uncertainty in product structures and component conditions shall result in a more powerful demanufacturing system [Zussman *et al.*, 1995]. The applications of Petri nets to modeling and analysis of sequential and concurrent disassembly systems are reported in [Thomas *et al.*, 1996]. More investigation and applications to industrial-scale systems along this line need to performed

The progress in the above fields should facilitate the design and operation of clean and agile computer-integrated manufacturing systems.

14.5 Benchmark Studies and Comparisons

Many methods can be used in the development of complex systems. A set of benchmark studies should be created and used to make comparisons of different methods. Among these methods are Ladder Logic Diagrams (LLDs), Sequential Function Chart (SFC), Petri nets, state machines, and structure text. Current industrial automated manufacturing systems work with a number of inputs and outputs varying from perhaps ten to several thousands and have communication capacity. Traditional discrete-event control design approaches such as LLDs prove entirely ineffective on large-scale problems. SFC and Petri net-based methodologies become increasingly important to handle them. SFC is an international standard based on GRAFCET language introduced in France in 1977. Chapter 11 of this book compares LLDs and Petri nets through a discrete-event manufacturing system. It represents, however, just a portion of the research effort to perform benchmark studies. The currently used discrete event control methods are reviewed in [Zhou and Twiss, 1996]. The comparison is made between LLDs and Petri net design methods for an industrial-size control problem [Zhou and Twiss, 1995; 1998]. More comparisons are needed in terms of their design efficiency, cost, adaptability, and maintainability on a flexible manufacturing system. The comparisons and benchmark studies should be extended to other methods.

The second important research issue is design recovery from old or unstructured design in general-purpose languages and LLDs to more advanced methods such as

SFCs and Petri nets. This is motivated by the practical needs to have the easier understanding, modification and maintenance of existing programs and to achieve higher system integration. The research tasks include defining mathematical equivalence between two discrete-event controllers and converting specifications and LLD programs into SFCs and Petri nets. Such research shall advance the knowledge, tools, and methodologies for designing and optimizing discrete-event controllers in large industrial automation projects.

Another relevant area is the benchmark studies of various deadlock avoidance or prevention policies for FMS. The present studies focus on the logical aspects of the discrete-event control of FMS. Introduction of deterministic as well as stochastic times into such systems poses more difficult problems in order to optimize performance of FMS. Realistic examples arising from industrial practice should be created and provided for such benchmark studies, in which operational parameters should be allowed to change to be a fixed value or following certain stochastic distributions. Several promising deadlock avoidance approaches based on Petri nets and other formalisms are referred to [Banaszak and Krogh, 1990; Ezpeleta et al., 1995; Hsieh and Chang, 1994; Barkaoui and Ben Abdallah, 1995; Fanti et al., 1997; Reveliotis et al., 1997].

BIBLIOGRAPHY

Adamides, E. D., E. C. Yamalidou, and D. Bonvin "A Systemic Framework for the Recovery of Flexible Production Systems," *International Journal of Production Research*, Vol. 34, No. 7, 1996, pp. 1875-1893.

Ajmone Marsan, M., G. Balbo, and G. Conte, "A class of generalized stochatic Petri nets for performance analysis of multiprocessor systems," *ACM TOCS*, Vol. 2, No. 2, May 1984, pp. 93-122.

Ajmone Marsan, M. and G. Chiola, "On Petri nets with deterministic and exponentially distributed firing times,"*Lecture Notes in Computer Science*, No. 254, 1986, pp. 132-145.

Ajmone Marsan, M., G. Balbo, G. Conte, S. Donatelli, and G. Franceschinis, *Modeling with Generalized Stochastic Petri Nets*, John Wiley & Sons, New York, 1995.

Akella, P. and F. DiCesare, "Closed Form Analysis with Petri Nets," *Proc. of IEEE Int. Conf. on Systems, Man, and Cybernetics*, Chicago, IL, 1992, pp. 643-649.

Alanche, D., D. F. Benzakour, P. Gillid, P. Rodriguel, and R. Valette, "PSI: A Petri Net Based Simulator for Flexible Manufacturing Systems," *Advances in Petri Nets*, Lecture notes in computer science, 188, Springer Verlag, 1984, pp. 1-14.

Al-Jaar, R. Y., and A. Desrochers, "Performance Evaluation of Automated Manufacturing Systems Using Generalized Stochastic Petri nets," *IEEE Trans. on Robotics & Automation*, Vol. 6, No. 6, 1990, pp. 621-639.

Alla, H., P. Ladet, J. Martinez, and M. Silva, "Modeling and validation of complex systems by colored Petri nets: application to a flexible manufacturing system," *Advances in Petri Nets 1984*, G. Rozenberg, H. Genrich, and G. Roucairol (Eds.), Springer-Verlag, 1985, pp. 15-31.

Alpan, G., *Design and Analysis of Supervisory Controllers for Discrete Event Dynamic Systems*, Doctoral Dissertation, Industrial and Systems Engineering Department, Rutgers University, 1997.

Anthony, I. R., "Flexible Manufacturing Systems: Issues and Implementation," *Industrial Management*, Vol. 33, No. 4, July/Aug 1991, pp. 7-11.

Archetti, F., and A. Sciomachen, "Development, Analysis and Simulation of Petri Net Models: An application to AGV systems," *OR Models in FMSs*, 1987.

Attaran, M. "Flexible Manufacturing Systems: Implementing an Automated Factory," *Information Systems Management*, Vol. 9, No. 2, 1992.

AT&T QNA Software, *AT&T Software Solutions Group*, NJ, 1992.

AT&T, *Flexible Work Station: User Guide and Manual*, Version 1.01, 1989.

Azzopardi, D., D. J. Holding and G. Genovese, "Object Oriented Petri net synthesis for modelling a semiconductor testing plant," *Proc. IMACS-IEEE/SMC Int'l Conf. on Computational Engineering in Systems Applications*, Lille, France, 1996, pp. 374-378.

Baker, K. R., *Introduction to Sequencing and Scheduling*, New York: Wiley, 1974.

Balph, T., and J. Vittera, "Architectural Considerations in the Development of an IEEE 802.4 Token Bus Chip Set," *IEEE 4th Annual International Conference on Computers and Communications*, Phoenix, AZ, March 1985, pp. 439-443.

Banaszak, Z. A. and B. H. Krogh, "Deadlock avoidance in flexible manufacturing systems with concurrently competing process flows," *IEEE Trans. Robotics and Automation*, 6(6), 1990, pp. 724-734.

Barad, M. and D. Sipper, "Flexibility in Manufacturing Systems: Definitions and Petri net Modeling," *International Journal of Production Research*, Vol. 26, No. 2, 1988, pp. 237-248.

Barkaoui, K. and I. Ben Abdallah, "A deadlock prevention method for a class of FMS," *Proc. the IEEE Int'l Conf. on Systems, Man and Cybernetics*, Vancouver, Canada, October 1995, pp. 4119-4124.

Basnet, C., P. A. Farrington, and D. B. Pratt, "Experiences in Developing an Object-Oriented Modeling Environment for Manufacturing Systems," *Proceedings of the 1990 Winter Simulation Conference*, IEEE, Piscataway, NJ, 1990, pp. 477-481.

Beck, C. L. and B. H. Krogh, "Models for simulation and discrete control of manufacturing systems," *Proc. of IEEE Int. Conf. on Robotics and Automation*, San Francisco, CA, 1986, pp. 305-310.

Berchi, R., and G. Frosi, "Design of the Material Handling System for a Manufacturing Line," *AIRO Workshop on Coordination Management by means of Petri Nets*, Modena, Italy, April 1988.

Best, E., "Structural theory of Petri nets: the free choice hiatus," *in Lecture Notes in Computer Science*, Spinger-Verlag, Vol., 254, 1987, pp. 168-206.

Bischak, D. P., and K. B. Stevens Jr., "An Evaluation of the Tandem Configuration Automated Guided Vehicle System," *Production Planning and Control*, Vol. 6, No. 5, 1995, pp. 438-444.

Black, J. T., "Cellular Manufacturing Systems Reduce Setup Time, Make Small Lot Production Economical," *Industrial Engineering*, Vol. 15, No. 11, 1983, pp. 39.

Bluementhal, M., and J. Dray, "The Automated Factory: Vision and Reality," *Technology Review*, Vol. 88, No. 1, January 1985, pp. 28-37.

Booch, G., *Object-oriented Analysis and Design with Applications*, The Benjamin Cummings Publishing Company Inc., 1994.

Boucher, T. O., M. A. Jafari, and G. A. Meredith, "Petri Net Control of an Automated Manufacturing Cell," *Computers in Industrial Eng.*, Vol. 17, No. 1-4, 1989, pp. 459-463.

Boucher, T. O., *Computer Automation in Manufacturing*, Chapman & Hall, London, 1996.

Bozer, Y. A., and M. M. Srinivasan, "Tandem Configurations for Automated Guided Vehicle Systems Offer Simplicity and Flexibility," *Industrial Engineering*, Vol. 21, No. 1, 1989, pp. 23-27.

Bruno, G. and P. Biglia, "Performance Evaluation and Validation of Tool Handling in Flexible Manufacturing Systems Using Petri Nets," *Proc. of 1985 IEEE Int. Workshop on Timed Petri Nets*, 1985, pp. 64-71.

Bruno, G., and G. Marchetto, "Process Translatable Petri Nets for the Rapid Prototyping of Process Control Systems," *IEEE Trans. on Software Eng.*, Vol. 12, No. 2, 1986, pp. 346-356.

Buffa, E. S., *Meeting the Competitive Challenge: Manufacturing Strategies for U.S. Companies*, Richard Irwin, Illinois, 1984, pp. 83.

Buzacott, J. A., "A Perspective on New Paradigms in Manufacturing," *Journal of Manufacturing Systems*, Vol. 14, No. 2, 1995, pp. 118-125.

Buzacott, J. A., and D. D. Yao, "Flexible Manufacturing Systems: A Review of Analytical Models," *Management Science*, Vol. 32, No. 7, July 1986, pp. 890-905.

Caloini, A., G. Magnani, and M. Pezze, "A technique for designing robotic control systems based on Petri nets," *IEEE Trans. on Control Systems Technology*, 6(1), 1998, pp. 72-87.

Campos, J., G. Chiola, J. M. Colom, and M. Silva, "Properties and performance bounds for timed marked graphs," *IEEE Trans. Circuit and Systems I*, Vol. 39, 1992, pp. 386-401.

Camurri, A. and A. Coglio, "A Petri net-based architecture for plant simulation," in *Proc. of 6th IEEE Int. Conf. on Emerging Technologies and Factory Automation*, Los Angeles, CA, September 9-12, 1997, pp. 397-402.

Cao, T. and A. C. Sanderson, "Task decomposition and analysis of robotic assembly task plans using Petri nets," *IEEE Trans. on Industrial Electronics*, 41(6), 1992, pp. 620-631.

Cao, T. and A. C. Sanderson, "Task Sequence Planning Using Fuzzy Petri Net," *IEEE Trans. on Systems, Man. and Cybernetics*, 25(5), 1995, pp. 755 - 768

Carlier, J. and P. Chretienne, "Timed Petri net schedules," *Advances in Petri Nets*, G. Rozenberg (ed.), Berlin: Springer Verlag, 1988, pp. 62-64.

Carlier, J. and E. Pinson, "An algorithm for solving the job-shop problem," *Management Science*, Vol. 35, No. 2, 1989, pp. 164-176.

Cassandras, C. G., *Discrete Event Systems*, Irwin, Homewood, IL, 1993.

Cecil, J. A., K. Srihari, and E. R. Emerson, "A Review of Petri-net Applications in Manufacturing," *Int. J. of Advanced Manufacturing Technology*, Vol. 7, No. 3, 1992, pp. 168-177.

Chaar, J. K., D. Teichroew, and R. A. Voltz, "Developing Manufacturing Control Software: a Survey and Critique, " *Int'l J. of Flexible Manufacturing Systems*, Vol. 5, No. 2, 1993, pp. 53-88.

Chaar, J. K., D. Teichroew, and R. A. Voltz, "Real-time Software Methodologies: Are They Suitable for Developing Manufacturing Control Software?" *Int'l J. of Flexible Manufacturing Systems*, Vol. 5, No. 2, 1993, pp. 95-128.

Chaar, J. K., R. A. Voltz, and E. S. Davidson, "An Integrated Approach to Developing Manufacturing Control Software," *Proceedings of the 1991 IEEE International Conference on Robotics and Automation*, Sacramento, CA, 1991.

Chan, C. C., and H. P. Wang, "Design and Development of a Stochastic High-level Petri net System for FMS Performance Evaluation," *Int'l J. of Production Research*, Vol. 31, No. 10, 1993, pp. 2415-2440.

Chang, Y. C., R. S. Sullivan, U. Bagchi, and J. R. Wilson, "Experimental Investigation of Real-time Scheduling in Flexible Manufacturing System," *Annals of Operations Research*, Vol. 3, 1985, pp. 355-377.

Chao, D. Y., (also named Yaw, Y), *Analysis and Synthesis of Distributed Systems and Protocols*. Doctoral Dissertation, Department of Electrical Engineering and Computer Science, University of California, Berkeley, 1987.

Chao, D. Y., M. C. Zhou and D. T. Wang, "Extending knitting technique to Petri net synthesis of automated manufacturing systems," *The Computer Journal*, Oxford University Press, 37(1), 1994, pp. 67-76.

Chen, P., S. Bruell, and G. Balbo, "Alternative methods for incorporating non-exponential distribution into stochastic timed Petri nets," *Proc. of the 3rd Int. Workshop on Timed Petri Nets and Performance Models*, IEEE-CS Press, Kyoto, Japan, 1989, pp. 187-192.

Chen, S. C. and M. D. Jeng, "FMS Scheduling using backtracking-free heuristic search based on Petri net state equations," *Proc. of IEEE Int'l Conf. on Systems, Man, and Cybernetics*, Vancouver, Canada, 1995, pp. 2153-2158.

Chen, T. R., *Scheduling for IC Sort and Test Facilities via Lagrangian Relaxation*, Ph.D. Dissertation, University of California at Davis, CA, 1994.

Chiola, G., " A software package for the analysis oof generalized stochastic Petri net models," *Proc. IEEE Int. Workshop on Timed Petri Nets*, Torino, Italy, July 1985, pp. 136-143.

Chiola, G., "A graphic Petri net tool for performance analysis," *Proc. of Int. Workshop on Modeling Techniques and Performance Evaluation*, France, 1987, pp. 323-333.

Chocron, D., and E. Cerny, "A Petri net Based Industrial Sequencer," *Proc. of IEEE Int. Conf. and Exhibit. on Industrial Control & Instrumentation*, 1980, pp. 18-22.

Christofides, N., A. P. Toth, and C. Sandi, *Combinatorial Optimization*. New York: John Wiley & Sons, 1978.

Chu, F. and X. Xie, "Deadlock analysis of Petri nets using siphons and mathematical programming," *IEEE Trans. Robotics Automat.*, vol. 13, no. 6, 1997, pp. 793-804.

Ciardo, G., *Manual for the SPNP Package*, Duke University, February, 1989.

Coffman, E. G., M. J. Elphick and A. Shoshani, "System deadlocks," *Computer Surveys*, Vol. 3, 1971, pp. 67-78.

Commoner, F., A. W. Holt, S. Even, and A. Pnueli, "Marked directed graphs," *J. Comp. Syst. Sci.*, 5, Oct. 1971, pp. 511-532.

Conway, R., W. L. Maxwell, and L. W. Miller, *Theory of Scheduling*. Ontario: Addison-Wesley, 1967.

Cooper, L. and D. Steinberg, *Introduction to Methods of Optimization*. Philadelphia: Saunders, 1970.

Courvoisier, M., R. Valette, A. Sahraoui and M. Combacau, "Specification and Implementation Techniques for Multilevel Control and Monitoring of FMS," *Computer Applications in Production and Engineering*, F. Kumara and A. Rolstadas (Eds.), Elsevier Science Publishers, IFIP, 1989, pp. 509-516.

Crockett, D., A. A. Desrochers, F. DiCesare, and T. Ward, "Implementation of a Petri Net Controller for a Machining Workstation," *Proc. of the IEEE Int. Conf. on Robotics and Automation*, Raleigh, NC, 1987, pp. 1861-1867.

Cumani, A., "ESP-a package for evaluation of stochastic Petri nets with phase-type distributed transition times," *Proc. of 1985 IEEE Int. Workshop on Timed Petri nets*, Torino, Italy, IEEE Computer Society, 1985, pp. 144-151.

Cumings, S., "Developing Integrated Tooling Systems: A Case Study at Garrett Turbine Engine Company," *Proc. of Fall Industrial Engineering Conf.*, Boston, MA, 1986.

Dallas, D. B., "The Impact of FMS," *Production*, Vol. 9, No. 10, October 1984, pp. 33-38.

Darrow, W. P., "International Comparison of Flexible Manufacturing System Technology," *Interfaces*, Vol. 17, November-December 1987:88.

David, R., and H. Alla, *Petri Nets and Grafcet*, Prentice Hall, New York, 1992.

David, R. and H. Alla, "Petri nets for modeling of dynamic systems - a survey," *Automatica*, Vol. 30, No. 2, 1994, pp. 175-202.

Davis, W. J., R. H. F. Jackson, and A. T. Jones, "Real time Optimization in the Automated Manufacturing Research Facility," *Progress in Material Handling and Logistics*, J. A. White and I. W. Dence (Eds.), Springer Verlag, 1989.

Desrochers, A. A., *Modeling and Control of Automated Manufacturing Systems*, IEEE Computer Society Press, 1989.

Desrochers, A. A. and R.Y. Al-Jaar, *Applications of Petri Nets in Manufacturing Systems*, IEEE Press, 1995.

Dhar, U. R., "Overview of Models and DSS in Planning and Scheduling of FMS," *Int'l J. of Production Economics*, Vol. 25, No. 1-3, Dec. 1991, pp. 121-127.

DiCesare, F. and A. A. Desrochers, "Modeling, Control, and Performance Analysis of Automated Manufacturing Systems using Petri Nets," *Control and Dynamic Systems*, Vol. 47, C. T. Leondes (Ed.), Academic Press, 1991, pp.121-172.

DiCesare, F., G. Harhalakis, J.-M. Proth, M. Silva and F. Vernadat (1993). *Practice of Petri Nets in Manufacturing*, Chapman and Hall, London.

Di Mascolo, M., Y. Frein, Y. Dallery, and R. David, "A unified modeling of Kanban systems using Petri nets," *Int. J. Flexible Manufacturing Systems*, 3, 1991, pp. 275-307.

Dimitrov, P., "The Impact of Flexible Manufacturing Systems (FMS) on Inventories," *Engineering Costs and Production Economics*, Vol. 19, Nos.1-3, May 1990, pp. 165-174.

D'Souza, K. and K. S. Khator, "A Petri Net Approach for Modeling Controls of a Computer-Integrated Assembly Cell," *International Journal of Computer Integrated Manufacturing*, Vol. 6, No. 5, 1993, pp. 302-310.

D'Souza, K. and A. Kelwyn, "A Control Model for Detecting Deadlocks in an Automated Machining Cell," *Computers and Industrial Engineering*, Vol. 26, No. 1, 1994, pp. 133-139.

Dubois, D. and K. Stecke, "Using Petri nets to represent production processes," *Proc. of the 22nd IEEE Conf. on Decision and Control*, San Antonio, TX, 1983, pp. 1062-1067.

Duffie, N. A., R. Chitturi, and J. Mou, "Fault-tolerant Heterarchical Control of Heterogeneous Manufacturing System Entities," *Journal of Manufacturing Systems*, Vol. 7, No. 4, 1988, pp. 315-328.

Dugan, J. B., A. Bobbio, A. Ciardo, and K. S. Trivedi "The design of a unified package for the solution of stochastic Petri net models," *Proc. of 1985 IEEE Int. Workshop on Timed Petri nets*, Torino, Italy, IEEE Computer Society, 1985, pp. 6-13.

Egbelu, P. J. and J. M. A. Tanchoco, "Characterization of automatic guided vehicle dispatching rules," *Int'l J. of Production Research*, 22(2), 1984, pp. 359-374.

Egbelu, P. J., and J. M. A. Tanchoco, "Potentials for bi-directional guide path for automated guided vehicle based systems," *Int'l J. of Production Research*, 24(6), 1986, pp. 1075-1097.

Esparza, J. and M. Silva, "Top-down synthesis of live and bounded free choice Petri nets," Technical Report, GISI-RR-90.01, Universidad De Zaragoza, 1990.

Ezpeleta, J., J. M. Colom, and J. Martinez, "A Petri net based deadlock prevention policy for flexible manufacturing systems," *IEEE Trans. Robotics Automat*, Vol. 11, 1995, pp. 793-804.

Falcione, A., and B. H. Krogh, "Design Recovery for Relay Ladder Logic," *IEEE Control Systems Magazine*, April 1993, pp. 90-98.

Fanti, M. P., B. Maione, S. Mascolo, and B. Turchiano, "Event-based feedback control for deadlock avoidance in flexible production systems," *IEEE Trans. Robotics and Automation*, Vol. 13, 1997, pp. 347-363.

Fernandez, E. B., and C. P. Han, "Object-oriented Design of Flexible Manufacturing Systems," *Proceedings of the 6th Annual Conference on Recent Advances in Robotics*, Gainesville, FL, 1993.

Ferrarini, L., "An Incremental Approach to Logic Controller Design with Petri Nets," *IEEE Trans. on Systems, Man, and Cybernetics*, 22(3), 1992, pp. 461-473.

Ferrarini, L., M. Narduzzi, and M. Tassan-Solet, "A new approach to modular liveness analysis conceived for large logic controllers design," *IEEE Trans. on Robotics and Automation* 10(2), 1994, pp. 169-184.

Ferrarini, L., "Computer aided design of logic controllers with Petri nets," in *Petri Nets in Flexible and Agile Automation*, M. C. Zhou (Ed.), Kluwer Academic, Boston, MA, 1995, pp. 71-92.

Florin, G. and S. Natkin, "Evaluation based upon stochastic Petri nets of the maximum throughput of a full duplex protocol," *Informatik Fachberichte*, C. Girault and W. Reisig (Eds.), Springer-Verlag: NY, 1982.

Freedman, P., "Time, Petri nets, and robotics," *IEEE Trans. Robotics Automation*, 7(4), 1991, pp. 417-433.

French, R. L., "Management Looking At CIM Must Deal Effectively With These Issues And Realities," *Industrial Engineering*, Vol. 16 August 1984, pp. 70.

French, S., *Sequencing and Scheduling: An Introduction to the Mathematics of the Job-Shop*. New York: Wiley, NY, 1982.

Garg, K., "An Approach to Performance Specification of Communication Protocols Using Timed Petri nets," *IEEE Trans. on Software Engineering*, Vol. 11, No. 10, 1985, pp. 1216-1255.

Gaymon, D. J., "Computers in the Tool Crib," *Manufacturing Engineering*, Vol. 103, No. 9, September 1986, pp. 41-44.

Gaymon, D. J., "Meeting Production Needs with Tool Management," *Manufacturing Engineering*, September 1987, Vol. 104, No. 9, pp. 41-47.

Gentina, J. C. and D. Corbeel, "Colored adaptive structured Petri net: a tool for the automated synthesis of hierarchical control of flexible manufacturing systems," *Proc. of IEEE Int. Conf. Robotics & Automat.*, Raleigh, NC, 1987, pp. 1166-1173.

Gershwin, B. S. and O. Berman, "Analysis of Transfer Lines Consisting of Two Unreliable Machines With Random Processing Times and Finite Storage Buffers," *AIIE Transactions*, Vol. 13, No. 1, 1981, pp. 2-11.

Ghosh, B. K., "Design and Performance Analysis Models of Computer Networks in CIM Systems," *Computers in Industry*, Vol. 12, 1989, pp. 141-152.

Gilbert, J. P. and P. J. Winter, "Flexible Manufacturing Systems: Technology and Advantages," *Production and Inventory Management*, 27(4), 1986, pp. 53.

Glassey, C. R. and Y. Hong, "Analysis of Behavior of an Unreliable n-stage Transfer Line With (n-1) Inter-stage Storage Buffers," *Int'l J. of Production Research*, Vol. 31, No. 3, 1993, pp. 519-530.

Glassey, C. R. and S. Adiga, "Berkeley Library of Objects for Control and Simulation of Manufacturing (BLOCS/M)," *Applications of Object-Oriented Programming*, L. J. Pinson, and R. S. Wiener(editors), Addison-Wesley, 1990, pp. 1-26.

Goldhar, J. D., "What Flexible Automation Means to Your Business", *Modern Material Handling*, Vol. 39, September 7, 1984, pp. 63-65.

Graham, J. H., S. M. Alexander, and W. Y. Lee, "Object-oriented Software for Diagnosis of Manufacturing Systems," *Proc. of the 1991 IEEE Int. Conf. on Robotics and Automation*, Sacramento, CA, 1991, pp. 1966-1971.

Gray, A. E., and E. K. Stecke "Tool Management in Automated Manufacturing: Operational Issues and Decision Problems," *Working chapter series, CMOM 88-03*, Williame Simon Graduate School of Business Administration, Univ. of Rochester, November 1988.

Groenevelt, H., L. Pintelon, and A. Seidmann, "Production Batching With Machine Breakdowns and Safety Stocks," *Operations Research*, 40(5), 1992, pp. 959-971.

Groover, M. P., *Automation, Production Systems, and Computer-Aided Manufacturing*, Prentice-Hall, Englewood, NJ, 1980.

Guha, R., D. Lange and J. Basiouni, "Software Specification and Design Using Petri nets," *Proceedings of the 4th International Workshop on Software Specification and Design*, 1987, pp. 225-230.

Giua, A. and F. DiCesare, "GRAFCET and Petri Nets in Manufacturing Systems," *Intelligent Manufacturing: Programming Environments for CIM*, W.A. Gruver and J. Boudreaux (Eds.), Springer-Verlag, London, 1993, pp. 153-176.

Giua, A. and F. DiCesare, "Blocking and Controllability of Petri Nets in Supervisory Control," *IEEE Trans. Automatic Control*, 39(4), 1994, pp. 818-823.

Giua, A. and F. DiCesare, "Petri Net Incidence Matrix Analysis for Supervisory Control," *IEEE Trans. Robotics and Automation*, 10(2), 1994, pp. 185-195.

Guo, D. L., F. DiCesare, and M. C. Zhou, "A Moment Generating Function Based Approach for Evaluating Extended Stochastic Petri Nets," *IEEE Trans. Automatic Control*, 38(2), 1993, pp. 321-327.

Gupta, D. and J. A. Buzacott, "A Framework for Understanding Flexibility of Manufacturing Systems," *J. of Manufacturing Systems*, 8(2), 1989, pp. 89-95.

Haidar, B., E. B. Fernandez, and T. B. Horton, "An Object-oriented Methodology for the Design of Control Software for Flexible Manufacturing Systems," *Proceedings of the 2nd Workshop on Parallel and Distributed Real-time Systems*," Cancus, Mexico, April 1994.

Hanisch, H.-M., "Analysis of place/transition nets with timed arcs and its application to batch process control," in *Applications and Theory of Petri Nets 1993*, Vol. 691, M. A. Marsan (Ed.), Springer Verlag, 1994, pp. 282-299.

Hanisch, H.-M., S. Koelbel, and M. Rausch, "A Modular Modeling, Controller Synthesis and Control Code generation Framework,," in *Preprints of 13th IFAC World Congress*, San Francisco, CA, July 1996, Vol. J, pp. 495-500.

Hanisch, H.-M., J. Thieme, A. Lueder, and O. Wienhold, "Modeling of PLC Behavior by Means of Timed Net Condition/Event Systems," in *Proc. of 6th IEEE Int. Conf. on Emerging Technologies and Factory Automation*, Los Angeles, CA, September 9-12, 1997, pp. 391-396.

Harrington, J., *Computer Integrated Manufacturing*, Industrial Press, New York, 1973.

Harry, B. and H. Malcolm, and K. Koos, "FMS Implementation Management: Promise and Performance," *Int'l J. of Operations and Production Management*, Vol. 10, No. 1, 1990, pp. 5-20.

Harvey, R. E., "Factory 2000," *Iron Age*, Vol. 227, June 1984, pp. 72-76.

Hatono, I., K. Yamagata and H. Tamura, "Modeling and on-line scheduling of flexible manufacturing systems using stochastic Petri nets," *IEEE Trans. on Software Engineering*, Vol. 17, No. 2, 1991, pp. 126-132.

Hays, R. H., and S. C. Wheelright, *Restoring Our Competitive Edge: Competing Through Manufacturing*, John Wiley, New York, 1984, pp. 192.

Heywood, P., "Four Generations of FMS," *American Machinist*, Vol. 32, No. 3, March 1988, pp. 62-63.

Hillion, H. P., and J. M. Proth, "Performance Evaluation of Job-Shop Systems Using Timed Event-Graphs," *IEEE Trans. on Automat. Contr.*, 34(1), 1989, pp. 149-154.

Hodgson, T. J., R. E. King, S. K. Manteih, and S.R. Schultz, "Developing Control Rules for an AGVS Using Markov Decision Processes," *Material Flow*, Vol. 4, 1987, pp. 85-96.

Holloway, L. E. and B. H. Krogh, "Synthesis of feedback control logic for a class of controlled Petri nets," *IEEE Trans. on Automat. Contr.*, 35(5), 1990, pp. 514-523.

Holloway, L. E., B. H. Krogh and A. Giua, "A survey of Petri net methods for controlled discrete event systems," *Discrete Event Dynamic Systems: Theory and Applications*, 7(2), April 1997, pp. 151-190.

Hsieh, D. Y. and S. C. Chang, "Dispatching-driven deadlock avoidance controller synthesis for flexible manufacturing systems," *IEEE Trans. Robotics and Automation*, 10(2), 1994, pp. 196-209.

Hsieh, S., and Y. Shih, "The Development of an AGVS Model by Union of the Modularised Floor-path nets," *Int'l J. of Advanced Manufacturing Technology*, Vol. 9, 1994, pp. 20-34.

Hsu, C. L., "Flexible Manufacturing System Controller Software Development by Object-oriented Programming," *Proc. of the Second Int. Conf. on Automation Technology*, Taipei, Taiwan, 1992, pp. 53-59.

Huang, H., and P. Chang, "Specification, modeling, and control of a flexible manufacturing cell," *Int'l J. of Production Research*, 30(11), 1992, pp. 2515-2543.

Huang, P. Y, "Analysis of the Necessary Conditions for Implementing JIT Production," *Zero Inventory Philosophy and Practice, Seminar Proceedings*, St. Louis, MO, 1984, pp. 24-29.

Huang, P. Y., and M. Sakurai, "Factory Automation: the Japanese Experience," *IEEE Trans. on Engineering Management*, Vol. 37, No. 2, 1990, pp. 103-108.

Hughes, T. and D. Hegland, "Flexible Manufacturing: The Way to the Winner's Circle," *Production Engineering*, Vol. 30, No. 9, September 1983, pp. 55.

Inman, A. R., "Flexible Manufacturing Systems: Issues and Implementation," *Industrial Management*, Vol. 33, No. 4, July/August 1991, pp.7-11.

Ismael, D. Jr., "Back to Basics: Just What is Involved in Implementing a Flexible Manufacturing System?," *Industrial Engineering*, Vol. 23, No. 4, 1991, pp. 43-44.

Jackovic, M., M. Vukobratovic, and Z. Ognjanovic, "An Appproach to the Modeling of the Highest Control Level of Flexible Manufacturing Cell," *Robotica*, Vol. 8, 1990, pp. 12-130.

Jacobson, I., et. al., *Object-Oriented Software Engineering - A Use Case Driven Approach*, Addison-Wesley, 1995.

Jafari, M. A., "An Architecture for a Shop-Floor Controller Using Colored Petri Nets," *Journal of Manufacturing Systems*, Vol. 4, No. 4, 1992, pp. 159-181.

Jafari, M. A. and T. O. Boucher, "A Rule Based System for Generating a Ladder Logic Control Program from a High Level System Specification," *Journal of Intelligent Manufacturing*, Vol. 5, No. 2, 1994, pp. 103-120.

Jain, S., "Basis for Development of a Generic FMS Simulator," *Proceedings of the 2nd ORSA/TIMS Conference on FMS: Operations Research Models and Applications*, Eds. K.E. Stecke, and R. Suri, 1986, pp. 393-403.

Jari, M., "The Success of FMS investments: Case studies from Small Industrial Economies," *Int'l J. of Technology Management*, 6(3), 1991, pp. 277-291.

Jeng, M.-D, S. W. Chou and C. L. Chung, "Performance evaluation of an IC fabrication system using Petri nets," *Proc. 1997 IEEE Int. Conf. Systems Man Cybern.*, Orlando, Florida, October 1997, pp. 269-274.

Jeng, M.-D, and F. DiCesare, "A Review of Synthesis Techniques for Petri Nets with Applications to Automated Manufacturing Systems," *IEEE Trans. Systems, Man, and Cybernetics*, 23(1), 1993, pp. 301-312.

Jeng, M.-D. and F. DiCesare, "Synthesis using resource control nets for modeling shared-resource systems," *IEEE Trans. Robotics Automat.*, 11, 1995, pp. 317-327.

Jeng, M.-D., C. Feng and S. W. Chou, "Modeling and analysis of the photo area in an IC manufacturing system," *Proc. IEEE Int. Conf. Systems Man Cybern.*, Beijing, China, October 1996, pp. 2973-2977.

Jeng, M.-D. and C. S. Lin, "Petri nets for the formulation of aperiodic scheduling problems in FMSs," *Proc. of 6th IEEE Int. Conf. on Emerging Technologies and Factory Automation*, Los Angeles, CA, September 9-12, 1997, pp. 375-380.

Johnson, M. E., L. Thompson, and R. Fontaine, "An Integrated Simulation and Shop-floor Control System," *Manufacturing Review*, 5(3), 1992, pp. 158-165.

Jones, A. T., and C. R. McLean, "A Proposed Hierarchical Control Model for Automated Manufacturing Systems," *J. of Manufacturing Systems*, 5, 1986, pp. 15-25.

Jothishankar, M. C. and H. P. Wang, "Determination of optimal number of Kanbans using stochastic Petri nets," *J. of Manufacturing Systems*, 11, 1992, pp. 449-461.

Jukka, R., and T. Iouri, "Economics and Success Factors of Flexible Manufacturing Systems: The Conventional Explanation Revisited," *Int'l J. of Flexible Manufacturing Systems*, Vol. 3, No. 2, 1990, pp. 169-190.

Jungnitz, H. and A. Desrochers, "Flow equivalent nets for the performance analysis of generalized stochastic Petri nets," *Proc. of IEEE Int. Conf. on Robotics and Automation*, Sacramento, CA, pp. 122-127.

Kaighobadi, M., and K. Venkatesh, "Investigating the Performance of Push and Pull Systems in Flexible Automation Environment Using Petri nets," *Proc.s of 1992 Decision Sciences Institute's Annual Meeting*, San Francisco, CA, 1992, pp. 1253-1255.

Kaighobadi, M. and K. Venkatesh, "Flexible Manufacturing Systems: an Overview," *Int'l J. of Operations and Production Management*, 14(4), 1994, pp. 26-49.

Kakati, M. and U.R. Dhar, "Investment Justification in Flexible Manufacturing Systems," *Engineering Costs and Production Economics*, 21, 1991, pp. 203-209.

Kaku, B. K., "Fitting Flexible Manufacturing Systems to the Task: An Analysis of Current Practices," *Working chapter series MS/S 92-003*, College of Business and Management, University of Maryland, 1992.

Kamath, M., "Recent Developments in Modeling and Performance Analysis Tools for Manufacturing Systems," *Computer Control of Flexible Manufacturing Systems*, S. B. Joshi and J. S. Smith (Eds.), Chapman and Hall, 1994.

Kamath, M., and J. L. Sanders, "Modeling Operator/Workstation Interference in Asynchronous Automatic Assembly Systems," *Discrete Event Dynamic Systems: Theory and Applications*, Vol. 1, 1991, pp. 93-124.

Kenneth, N. M, F. R. Safayeni, and J. A. Buzacott, "A Review of Hierarchical Production Planning and its Applicability for Modern Manufacturing," *Production Planning and Control*, Vol. 6, No. 5, 1995, pp. 384-394.

Kiesler, S., "New Technology in the Workplace/Robotics: Cause and Effect," *Public Relations Journal*, Vol. 39, No. 12, December 1983, pp.12-16.

Kim, J. and A. A. Desrochers, "Modeling and analysis of semiconductor manufacturing plants using time Petri net models: COT Business Case Study," *Proc. of IEEE Int. Conf. Sys. Man Cybern.*, Orlando, FL, 1997, pp. 3227-3232.

Kimura, O. and H. Terada, "Design and Analysis of Pull System, a Method of Multistage Production Control," *Int'l J. of Production Research*, 19, 1981, pp. 241-253.

King, R. E., T. J. Hodgson, and S. K. Monteith, "Evaluation of Heuristic Control Strategies for AGVs Under Varying Demand and Arrival Patterns," *Progress in Material Handling and Logistics*, Springer Verlag, 1989.

Klahorst, T. H., "Flexible Manufacturing Systems: Combining Elements to Lower Costs, Add flexibility," *Industrial Engineering*, 32(11), 1981, pp.112-117.

Knapp, G. M. and H. P. Wang, "Modeling of Automated Storage/Retrieval Systems Using Petri nets," *Journal of Manufacturing Systems*, 11(2), 1992, pp. 20-29.

Kochikar, V. P. and T. T. Narendran, "On Using Abstract Models for Analysis of Flexible Manufacturing Systems," *Int'l J. of Production Research*, Vol. 32, No. 10, 1994, pp. 2303-2322.

Kochikar, V. P. and T. T. Narendran, "Modeling Automated Manufacturing Systems Using a Modification of Coloured Petri nets," *Robotics and Computer Integrated Manufacturing*, Vol. 9, No. 3, 1992, pp. 181-189.

Koh, I. and F. DiCesare, "Modular transformation methods for generalized Petri nets and their applications in manufacturing automation," *IEEE Trans. on Systems, Man, and Cybernetics*, 21(6), 1991, pp. 963-973.

Krinsky, I., A. Melnez, G. H. Mitenbarg, and B. L. Myers, "Flexible Manufacturing System Evaluation: An Alternative Approach," *Int'l J. of Flexible Manufacturing Systems*, Vol. 3, No. 2, 1991, pp. 237-253.

Krogh, B. H. and C. L. Beck, "Synthesis of place/transition nets for simulation and control of manufacturing systems," *Preprints of 4th IFAC/IFORS Symp. Large Scale Systems*, Zurich, 1986, pp. 661-666.

Krogh, B. H. and R. S. Sreenivas, "Essentially decision free Petri nets for real-time resource allocation," *Proc. of IEEE Int. Conf. Robotics and Automation*, Raleigh, North Carolina. 1987, pp. 1005-1011.

Krogh, B. H., R. Willson, and D. Pathak, "Automated generation and evaluation of control programs for discrete manufacturing processes," *Proc. of 1988 Int. Conf. on Computer Integrated Manufacturing*, Troy, NY, 1988, pp. 92-99.

Kunnathur, A., and P. S. Sundararaghavan, "Issues in FMS Installation: A Field Study and Analysis," *IEEE Trans. on Engineering Management*, Vol. 39, No. 4, 1992, pp. 370-377.

Kwok, S. C., "A Case Report on Integrating FMS and Traditional Machine Tools," *Flexible Manufacturing Systems*, Society of Manufacturing Engineers, Dearborn, MI, 1988.

Lautenbach K., "Linear algebraic techniques for place/transition nets," in *Advances on Petri nets'86 – Part I*, Vol 254 of LNCS, Brauer W., W. Reisig, and G. Rozenberg (eds.), Springer-Verlag, Bad Honnef, Germany, 1987, pp. 142-167.

Lee, C. C. and J. T. Lin, "Deadlock Prediction and Avoidance Based on Petri Nets for Zone-Control Automated Guided Vehicle Systems," *Int'l J. of Production Research*, Vol. 33, No. 12, 1995, pp. 3249-3265.

Lee, L. C., "Parametric Appraisal of the JIT System," *Int'l J. of Production Research*, Vol. 25, No. 10, 1987, pp. 1415-1429.

Lee, C. Y., R. Uzsoy, and L. A. Martin-Vega, "Efficient algorithms for scheduling semiconductor burn-in operations," *Operations Research*, 40, 1992, pp. 764-795.

Lee, D. Y., *Scheduling and Supervisory Control of Flexible Manufacturing Systems Using Petri Nets and Heuristic Search*, Ph.D. Dissertation, Rensselaer Polytechnic Institute, Troy, NY, May 1993.

Lee, D. Y. and F. DiCesare, "Scheduling FMS using Petri nets and heuristic search," *IEEE Trans. on Robotics and Automation*, Vol. 10, No. 2, 1994a, pp. 123-132.

Lee, D. Y. and F. DiCesare, "Integrated Scheduling of Flexible Manufacturing Systems Employing Automated Guided Vehicles," *IEEE Trans. on Industrial Electronics*, 41(6), 1994b, 602-610.

Lee, D. Y. and F. DiCesare, "Petri Net-Based Heuristic Scheduling for Flexible Manufacturing Systems," *Petri Nets in Flexible & Agile Automation*, M. C. Zhou (Ed.), Norwell, MA: Kluwer Academic, 1995, pp. 149-187.

Les, G., Smart Handling Doubles FMS Productivity, *Modern Materials Handling*, Vol. 4, No. 1, January 1990, pp. 64-66.

Lewis, F. L., O. C. Pastravanu, and H. H., Huang, "Controller design and conflict resolution for discrete event manufacturing systems," *Proc. of 32nd IEEE Conf. on Decision and Control*, San Antonio, TX, Dec. 1993, pp. 3288-3293.

Lewis, F. L., H. H., Huang., O. C. Pastravanu, and A. Guerel, "A matrix formulation for design and analysis of discrete event manufacturing systems with shared resources," *Proc. of 1994 IEEE Conf. on Systems, Man and Cybernetics*, San Antonio, TX, October 1994, pp. 1700-1705.

Li, S., T. Takamori, and S. Tadokoro (1995). "Scheduling and re-scheduling of AGVs for flexible and agile manufacturing," *Petri Nets in Flexible and Agile Automation*, M. C. Zhou (Ed.), Kluwer Academic, Boston, MA, pp. 189-205.

Lindemann, C., "Performance Modeling using DSPNexpress," *Models and Techniques for Performance Evaluation of Computer and Communication Systems*, L. Don atiello, R. Nelson (Eds.), Vol. 729, Springer, 1993, pp. 291-306.

Liu, C. M. and F. C. Wu, "Using Petri Nets to Solve FMS Problems *Int'l J. of Computer Integrated Manufacturing*, Vol. 6, No. 3, 1993, pp. 175-185.

Luh, P. B. and D. J. Hoitomt, "Scheduling of manufacturing systems using the Lagrangian relaxation technique," *IEEE Trans. on Automatic Control*, Vol. 38, No. 7, 1993, pp.1066-1079.

Ma, J. M. and M. C. Zhou, "Performance evaluation of discrete event systems via stepwise reduction and approximation of stochastic Petri nets," *Proc.1992 IEEE Int. Conf. on Decision and Control*, Tucson, AZ, 1990, pp. 1210-1215.

Maccarini, G., C. Giardini, L. Zavanella, and A. Bugini, "Different Kind of Tool Room Models for an FMS: A Simulation Approach and Analysis," *International AMSE Conference*, Vol. 4, 1987, pp. 87-89.

Magott, J., "Performance evaluation of concurrent systems using Petri nets," *Inform. Proc. Lett.*, Vol. 18, 1984, pp. 7-13.

Mahadevan, B., and T. T. Narendran, " Design of an automated guided vehicle based material handling system," *Int'l J. of Production Res.*, 28(9), 1990, pp. 1611-1622.

Maimon, O.Z., "Real-time Operational Control of Flexible Manufacturing Systems," *Journal of Manufacturing Systems*, Vol. 6, No. 2, 1987, pp. 125-136.

Maione, B., Q. Semeraro, and B. Turchiano, 1986, "Closed Analytical Formula for Evaluating FMS Performance Measure *Int'l J. of Production Research*, Vol. 24, No. 3, 1986, pp. 583-592.

Malmborg, C. J., "A Model for the Design of Zone Control Automated Guided Vehicle Systems," *Int'l J. of Production Research*, 28(10), 1990, pp. 1741-1758.

Marinov, D. D. and N. Todorov, "Software Development Approach in FMS," *Computers in Industry*, Vol. 10, No. 3, 1988, pp. 171-175.

Martinez, J. and M. Silva, "A Language for the Description of Concurrent Systems Modeled by Coloured Petri Nets: Application to the Control of Flexible Manufacturing Systems," in *Languages for Automation*, S. K. Chang (Ed.), New York: Plenum Press, 1985, pp. 369-388.

Martinez J., H. Alla, and M. Silva, "Petri nets for the specifications of FMSs," In *Modeling and Design of Flexible Manufacturing Systems*, A. Kusiak (Ed.), Elsevier Science Publishers, Amsterdam, 1986, pp. 389-406.

Masory, O., "Monitoring of Tool Wear Using Artificial Neural Networks," *Int'l J. of Material Processing Technology*, Vol. 63, 1990.

Maxwell, W. L. and J. A. Muckstadat, "Design of Automated Guided Vehicle Systems," *IIE Transactions*, Vol. 14, 1982, pp. 114- 124.

May, B., "FMS Control Software Basics," *Proceedings of Flexible Manufacturing Systems*, Chicago, Illinois, 1986.

McPherson, R. F. and K. P. White Jr, "Dynamic Issues in the Planning and Control of Integrated Manufacturing Hierarchies," *Production Planning and Control*, Vol. 6, No. 6, 1995, pp. 544-554.

Menon, S. R., T. J. Quinn, P. M. Ferreira and S. G. Kapoor, "Coordination control of flexible manufacturing systems using colored Petri nets," *Proc. of Int. Conf. on Computer Aided Production Eng.*, 1988, pp. 317-326.

Merabet, A. A., "Synchronization of Operations in a Flexible Manufacturing Cell: The Petri Net Approach," *J. of Manufacturing Systems*, 5(3), 1986, pp. 161-169.

Meyer, B., *Object-oriented Software Construction*, Prentice-Hall, 1988.

Michel, G., *Programmable Logic Controllers: Architectures and Application*, John Wiley and Sons, England, 1990.

Molloy, M. K.. "Performance analysis using stochastic Petri nets," *IEEE Trans. on Computers*, 31(9), 1982, pp. 913-917.

Monarchi, D. E. and G. I. Puhr, "A Research Topology for Object-oriented Analysis and Design," *Communications of the ACM*, Vol. 35, No. 9, 1992, pp. 35-47.

Monden, Y., "How Toyota Shortened Supply Lot Production Time, Waiting Time, and Conveyance Time," *Industrial Engineering*, Vol. 13, No. 9, 1981, pp. 22-30.

Morioka, S. and T. Yamada, "Performance evaluation of marked graphs by linear programming," *Int'l J. of Systems Science*, 22(9), 1991, pp. 1541-1552.

Mullins, P. J., "Feeding Flexible Manufacturing Systems," *Automotive Industry*, Vol. 164, November 1984, pp. 63-64.

Murata T., "Circuit theoretic analysis and synthesis of marked graphs," *IEEE Trans. Circ. and Sys.*, Vol. 27, 1977, pp. 400-405.

Murata, T., "Petri Nets: Properties, Analysis and Applications," *Proceedings of the IEEE*, 77(4), 1989.

Murata, T. and J. Y. Koh, "Reduction and expansion of live and safe marked graphs," *IEEE Trans. Circ. and Sys.*, CAS-27, 1980, pp. 68-70.

Murata, T., N. Komoda, K. Matsumoto, and K. Haruna, "A Petri Net-based Controller for Flexible and Maintainable Sequence Control and its Applications in Factory Automation," *IEEE Trans. on Industrial Electronics*, 33(1), 1986, pp. 1-8.

Murata, T., B. Shenker, and S. Shatz, "Detection of Ada Static Deadlocks Using Petri net Invariants," *IEEE Trans. on Software Engineering*, 15, 1989, pp.314-326.

Narahari, Y., and N. Viswanadham, "A Petri net Approach to the Modeling and Analysis of FMSs," *Annals of Operations Research*, Vol. 30, 1985, pp. 449-472.

Narahari, Y. and N. Viswanadham, "Colored Petri nets: a tool for modeling, validation, and simulation of FMS," *Proc. of IEEE Int. Conf. Robotics and Automation,* Raleigh, NC, 1987, pp. 1985-1990.

Narayanan, S., D. A. Bodner, and C. M. Mitchell, "Object-Oriented Simulation to Support Modeling and Control of Automated Manufacturing Systems," *Proc. of the 1992 Object-Oriented Simulation Conferences*, SCS, 1992, pp.59-63.

Narumol, U. and C.F. Daganzo, "Impact of Parallel Processing on Job Sequences in Flexible Assembly Systems," *Int'l J. of Production Res.*, 27(1), 1986, pp. 73-89.

Naylor, A. W. and R. A. Voltz, "Design of Integrated Manufacturing System Control Software," *IEEE Trans. on Systems, Man, and Cybernetics*, Vol. 17, No. 6, 1987, pp. 881-897.

Nemhauser, G. L., A. H. G. Rinnooy Kan, and M. J. Todd, (Eds.), *Optimization, Vol 1. of Handbooks in Operations Research and Management Science*, Amsterdam, The Netherlands: North-Holland, 1989.

Newton, D., "Simulation Model Helps Determine How Many Automatic Guided Vehicles are Needed," *Industrial Engineering*, Vol. 17, No. 1, 1985, pp. 68-79.

Nof, S. Y., *Handbook of Industrial Robots*, John Wiley and Sons, New York, 1985.

Occena, G. and T. Yokota, "Modeling of an Automated Guided Vehicle System in a Just-in-time Environment," *Int'l J. of Production Res*, 29(3), 1991, pp. 495-511.

Ould-Kaddour, N. and M. Courvoisier, "A Multi-tasking Environment Based on Petri nets With Objects and Modula-2," *Proc. of the 15th Annual Conference of IEEE Industrial Electronics Society*, Philadelphia, PA, 1989, pp. 799-804.

Ozden, M., " A Simulation Study of Multiple Load Carrying Automatic Guided Vehicles in a Flexible Manufacturing System," *Int'l J. of Production Research*, Vol. 26, No. 7, 1988, pp. 1353-1366.

Pearl, J., *Heuristics: Intelligent Search Strategies for Computer Problem Solving*. Reading, MA: Addison-Wesley, 1984.

Pessen, D.W., *Industrial Automation: Circuit Design and Components*, New York: Wiley, 1989.

Pessen, D.W., "Ladder-Diagram Design for Programmable Controllers," *Automatica*, Vol. 25, No. 3, 1989, pp. 407-412.

Peterson, J. L., *Petri Net Theory and the Modeling of Systems*, Prentice, Inc., Englewood Cliffs, NJ,1981

Pimentel, J., *Communication Networks for Manufacturing*, Prentice-Hall Inc., 1990.

Primrose, P. L. and R. Leonard, "Selecting Technology for Investment in Flexible Manufacturing," *Int'l J. of Flexible Manufacturing Systems*, 4(1), 1991, pp. 51-77.

Proth, J. M., "Discrete Manufacturing Systems: From Specification to Evaluation," Invited Lecture, *Proc.s of the 2nd Int. Conf. on Automation Technology*, Taipei, Taiwan, 1992, pp. 7-8.

Proth, J. M. and I. Minis, "Planning and scheduling based on Petri nets," *Petri Nets in Flexible & Agile Automation*, M. C. Zhou (Ed.), Norwell, MA: Kluwer Academic, 1995, pp. 109-148.

Proth, J. M. and X. Xie, *Petri Nets*, John Wiley & Sons, New York, 1996.

Raju, K. R. and O. V. K. Chetty, "Design and Evaluation of Automated Guided Vehicle Systems for Flexible Manufacturing Systems: An Extended Timed-Petri net-based Approach," *Int'l J. of Production Res.*, 31(5), 1993a, pp. 1069-1096.

Raju, K. R. and O. V. K. Chetty, "Priority nets for Scheduling Flexible Manufacturing Systems," *J. of manufacturing systems*, 12, 1993b, pp. 326-340.

Ram, S. S. and G. P. Yash, "Strategic Cost Measurement for Flexible Manufacturing Systems," *Long Range Planning*, Vol. 24, No. 5, 1991, pp. 34-40.

Ramadge, P. J. and W. M. Wonham, "Supervisory control of a class of discrete event processes," *SIAM J. of Control and Optimization*, 25(1), 1987, pp. 206-230.

Ramamoorthy, C. V. and G. S. Ho, "Performance evaluation of asynchronous concurrent systems using Petri nets," *IEEE Trans. on Software Engineering*, vol. SE-6, No. 5, 1980, pp. 440-449.

Ramaswamy, S. E. and S. B. Joshi, "Deadlock-free schedules for automated manufacturing workstations," *IEEE Trans. on Robotics and Automation*, Vol. 12, No. 3, 1996, pp. 391-400.

Ramaswamy, S. and K. P. Valavanis, "Modeling, analysis and simulation of failures in a material handling system using extended Petri nets," *IEEE Trans. on Systems, Man, and Cybernetics*, 24(9), 1994, pp. 1358-1373.

Ramchandani, C., *Analysis of Asynchronous Concurrent Systems by Timed Petri Nets*, Doctoral Dissertation, MIT, Cambridge, MA, 1974.

Ranky, P. G., *The Design and Operation of FMS: Flexible Manufacturing Systems*, IFS Publications Ltd., UK, 1986.

Ranky, P., G., *An Introduction to Concurrent/Simultaneous Engineering, An Engineering Multimedia CD-ROM*, CIMware, 1997.

Reddy, C. E., O. V. K. Chetty, and D. Chaudhuri, "A Petri net Based Approach for Analyzing Tool Management Issues in FMS," *Int'l J. of Production Research*, Vol. 30, No. 6, 1992a, pp. 1427-1446.

Reddy, C. E., O. V. K. Chetty, and D. Chaudhuri, "Design of a Tool Delivery System Using Expert Simulation in FMS," *Proceedings of the 6th Convention of Computer Engineers*, Trichy, India, September, 1990b.

Reddy, C. E., O. V. K. Chetty, and D. Chaudhuri, "Expert Tool in Flexible Manufacturing Systems," *Proceedings of the International Conference on Automation, Robotics and Computer Vision*, Singapore, September, 1990c.

Reddy, C. E., O. V. K. Chetty, and D. Chaudhuri, "Objective SIMTOOL in FMS," *Proceedings of the 5th International Conference on CAD/CAM Robotics and Factories of the Future*, Norfolk, USA, December 2-5, 1990d.

Rembold, U., B. O. Nnaji, and A. Storr, *Computer Integrated Manufacturing and Engineering*, Addison-Wesley, Wokingham, England, 1993.

Resig, W., *Petri Nets*, New York: Springer-Verlag, 1985.

Reveliotis, S. A., M. A. Lawley, and P. M. Ferreira, "Polynomial complexity deadlock avoidance policy for flexible manufacturing systems," *IEEE Trans. Robotics and Automation*, Vol. 13, 1997, pp. 1344-1357.

Righini, G., "Modular Petri nets for Simulation of Flexible Production Systems," *Int'l J. of Production Research*, Vol. 31, No. 10, 1993, pp. 2463-2477.

Rodammer, F. A. and K. P. Jr. White, "A Recent Survey of Production Scheduling," *IEEE Trans. on Systems, Man, and Cybernetics*, 18(6), 1988, pp. 841-851.

Rogers, R.V., "Understanding Implications of Object-oriented Simulation and Modeling," *Proceedings of IEEE International Conference on Systems, Man, and Cybernetics*, Charlottesville, VA, 1991, pp. 285-289.

Rolston, L. J., "Modeling FMSs with MAP/1," *Annals of Operations Research*, Vol. 3, 1985, pp. 189.

Ross, D. T., "Structured analysis SA: A Language for Communicating Ideas," *IEEE Trans. on Software Eng.*, Vol. 3, No. 1, 1977, pp. 16-34.

Rumbaugh, J., M. Blaha, W. Premerlani, F. Eddy, and W. Lorensen, *Object-Oriented Modeling and Design*, Prentice Hall, Englewood Cliffs, 1991.

Sabuncuoglu, I. and L. Hommuertzheim, "An Investigation of Machine and AGV Scheduling Rules in an FMS," *Operation Research Models and Applications*, K.E. Stecke and R. Suri (Eds.), 1989, pp. 261-266.

Sahraoui, A. E. K. and N. Ould-Kaddour, "Control Software Prototyping," *Computers and Industry*, Vol. 20, No. 3, 1992, pp. 327-334.

Salomon, D. P. and J. E. Beigel, "Assessing Economic Attractiveness of FMS Applications in Small-Batch Manufacturing," *Industrial Engineering*, June 1984, pp. 88-96.

Sanchez, A. M., "Adopting Advanced Manufacturing Technologies: Experience from Spain," *J. of Manufacturing Systems*, Vol. 15, No. 2, 1996, pp. 133-140.

Sarkar, B. R. and J. A. Fitzsimmons, "The Performance of Push and Pull Systems: A Simulation and Comparative Study," *Int'l J. of Production Research*, Vol. 27, No. 10, 1989, pp. 1715-1732.

Sarkar, B. R., "Simulating a Just in Time Production System," *Computers and Industrial Engineering*, Vol. 16, No. 1, 1989, pp. 127-130.

Sato, T. and K. Nose, "Automatic generation of sequence control programs via Petri net and logic tables for industrial applications," *Petri Nets in Flexible and Agile Automation*, Zhou, M. C., (Ed.) Kluwer Academic, Boston, MA, 1995, pp. 93-107.

Scalpone, R. W., "Education Process Is Vital to Realization of CIM Benefits, Handling of Pitfalls," *Industrial Engineering*, Vol. 16, No. 10, 1984, pp. 110-116.

Sciomachen, A., S. Grotzinger and F. Archetti, "Petri net-based emulation for a highly concurrent pick-and-place machine," *IEEE Trans. on Robotics and Automation*, 6(2), 1990, pp. 242-247.

Schonberger, J. R., "Applications of Single-card and Dual-card Kanban," *Interfaces*, Vol. 13, No.4, 1983, pp. 56-67.

Schroer B. J. and F.T. Tseng, " Modeling Complex Manufacturing Systems Using Discrete Event Simulation," *Computers and Industrial Engineering*, Vol. 14, 1985, pp. 455-464.

Sifakis, J., "Use of Petri nets for performance evaluation," *Measuring, Modelling, and Evaluating Computer Systems*, Amsterdam: North-Holland, 1977, pp. 75-93.

Silva, M. and S. Velilla, "Programmable logic controller and Petri nets: a comparative study," *Proc. of the IFAC Conf. on Software for Computer Control*, Madrid, Spain, 1982, pp. 38-88.

Silva, M., *Las redes de Petri en la Automatica y la Informatica*, Editorial AC, Madrid, 1985.

Silva, M. and R. Valette, "Petri nets and Flexible Manufacturing," *Advances in Petri Nets 1989*, Springer-Verlag, Heidelberg, 1990, pp. 374-417.

Shen, L., Q. Chen, J. Y. Luh, S. C. Chen, and Z. Zhang, "Truncation of Petri Net Models of Scheduling Problems for Optimun Solutions," *Proc. of Japan/USA Symposium on Flexible Automation,* 1992, pp. 1681-1688.

Shih, H. and T. Sekiguchi, "A timed Petri net and beam search based on-line FMS scheduling system with routing flexibility," *Proc. of the 1991 IEEE Int. Conf. on Robotics and Automation,* Sacramento, CA, April 1991, pp. 2548-2553.

Smith, J. and S.B. Joshi, "Reusable Software Concepts Applied to the Development of FMS Control Software," *Int'l J. of Computer Integrated Manufacturing,* 5(3), 1992, pp. 182-196.

Spearman, L. M., D. L. Woodruff, and W. J. Hopp, "CONWIP: A Pull Alternative to Kanban," *Int'l J. of Production Research,* Vol. 28, No. 5, 1990, pp. 879-894.

Srinivasan, V. S. and M. A. Jafari, "Monitoring and Fault Detection in Shop Floor Using Timed Petri nets," *Proceedings of IEEE Conference on Systems, Man, and Cybernetics,* Charlottesville, VA, 1991, pp. 355-360.

Stanton, M. J., W. F. Arnold, and A. A. Buck, "Modeling and Control of Manufacturing Systems Using Petri Nets," *Preprints of the 13th World Congress of IFAC,* San Francisco, CA, June 30 - July 5, 1996, pp. 329-335.

Stecke, K. E., "Formulation and Solution of Nonlinear Integer Production Planning Problems for Flexible Manufacturing Systems," *Management Science,* Vol. 29, No. 3, March 1983, pp. 273-288.

Stecke, K., "FMS Design and Operating Problems and Solutions," *Proceedings of the 2nd Intelligent Factory Automation Symposium,* ISCIE, 1989, pp. 17-32.

Stecke, K. E. and J. Solberg, "Loading and Control Policies for Flexible Manufacturing System," *Int'l J. of Production Research,* 19(5), 1981, pp. 481-490.

Stefano, D. A. and O. Mirabella, "A Fast Sequence Control Device Based on Enhanced Petri nets," *Microprocessors and Microsystems,* 15, 1991, pp. 179-186.

Stulle, M. and G. Schmidt, "Object-Oriented Petri-Nets for State Reconstruction of Flexible Manufacturing Systems," *Preprints of the 13th World Congress of IFAC,* San Francisco, CA, June 30 - July 5, 1996, pp. 335-340.

Sturzenbecker, M. C., "Building an Object-oriented Environment for Distributed Manufacturing Software," *Proceedings of the 1991 IEEE International Conference on Robotics and Automation,* Sacramento, CA, 1991, pp. 1972-1978.

Sun, T. H., C. W. Cheng, and L. C. Fu, "A Petri net based approach to modeling and scheduling for an FMS and a Case Study," *IEEE Trans. on Industrial Electronics,* 41(6), 1994, pp. 593-601.

Suresh, N., "Towards an Integrated Evaluation of Flexible Automation Investments," *Int'l J. of Production Research,* 28(9), 1990, pp.1657-1672.

Suri, R. and G. W. Diehl, "MANUPLAN: A Precursor to Simulation for Complex Manufacturing Systems," *Proceedings of Winter Simulation Conference,* 1985.

Suri, R. and J. W. Dille, "A Technique for On-line Sensitivity Analysis of Flexible Manufacturing Systems," *Annals of Operations Research,* Vol. 3, 1985, pp. 381.

Suzuki, T., T. Kanehara, A. Inaba, and S. Okuma, "On algebraic and graph structural properties of assembly Petri Net," *Proc. of the IEEE Int. Conf. On Robotics and Automation*, Atlanta, GA, May 2-6, 1993, pp. 507-514.

Suzuki, I. and T. Murata, "A method for stepwise refinements and abstractions of Petri nets," *J. of Comp. and Syst. Sci.*, 27, 1983, 51-76.

Taholakian, A. and W.M.M. Hales, "A Methodology for Designing, Simulating and Coding PLC Based Control Systems Using Petri Nets," *Int'l J. of Production Research*, Vol. 35, No. 6, 1997, pp.1743-1762.

Teng, S. H. and J. T. Black, "Cellular Manufacturing System Modeling: The Petri net approach," *Journal of Manufacturing Systems*, Vol. 9, No. 1, 1990, pp. 41-54.

Thomas, J. P., N. Nissanke, and K. D. Baker, "A hierarchical Petri net framework for the representation and analysis of assembly," *IEEE Trans. on Robotics and Automation*, 12(2), 1996, pp. 268-279.

Tomek, P., "Tooling Strategies Related to FMS Management," *The FMS Magazine*, Vol. 5, No. 4, April 1986, pp.102-107.

Uzma, S., A. D. Robbi, and M. C. Zhou, "Anticipatory real-time scheduling of flexible manufacturing systems," *Proc. of 1995 IEEE Int. Conf. on Systems, Man, and Cybernetics*, Vancouver, Canada, Vol. 5, October 1995, pp. 4131-4136.

Uzsoy, R., C. Y. Lee and L. A. Martin-Vega, "A review of production planning and scheduling models in the semiconductor industry part I: system characteristics, performance evaluation and production planning," *IIE Trans. on Scheduling and Logistics*, Vol. 24, No. 4, 1992, pp. 47-60.

Uzsoy, R., L. A. Martin-Vega, C. Y. Lee and P. A. Leonard, "Production scheduling algorithms for a semiconductor test facility," *IEEE Trans. on Semiconductor Manufacturing*, Vol. 4, 1991, pp. 271-280.

Valavanis, K. P., "On the Hierarchical Modeling, Analysis and Simulation of FMSs with Extended Petri nets," *IEEE Trans. on Systems, Man and Cybernetics*, Vol. 20, No. 1, 1990, pp. 94-110.

Valette, R., "Analysis of Petri nets by stepwise refinements," *J. of Comp. and Syst. Sci.*, 18, 1979, pp. 35-46.

Valette, R., M. Courvoisier, and D. Mayeux, "Control of flexible production systems and Petri nets," *Informatik Fachberichte 66*, Springer Verlag, 1982, pp. 264-267.

Valette, R., M. Courvoisier, J. M. Bigou, and J. Albukerque, "A Petri Net Based Programmable Logic Controller," *Computer Applications in Production and Engineering*, E.A. Warman (editor), North-Holland, IFIP, 1983, pp. 103-115.

Valette, R., M. Courvoisier, H. Demmou, J. M. Bigou, and C. Desclaux, "Putting Petri nets to work for controlling flexible manufacturing systems," *Proc. of Int. Symp. on Circ. & Sys*, Kyoto, Japan, 1985, pp. 929-932.

Valette, R., "Nets in production systems," In W. Brauer, W. Reisig, and G. Rozenberg (eds.), *Advances in Petri Nets 1986*, Vol. 255, Part I, Springer-Verlag, 1987, pp. 191-217.

Veeramani, D., D.M. Upton, and M.M. Barash, "Cutting-Tool Management in Computer-Integrated Manufacturing," *Int'l J. of Flexible Manufacturing Systems*, Vol. 5, N0. 2, 1992, pp. 238-265.

Venkatesh, K., O. V. K. Chetty, and K.R. Raju, "Simulating Flexible Automated Forming and Assembly Systems," *Journal of Material Processing and Technology*, Vol. 24, 1990, pp. 453-462.

Venkatesh, K., *Petri Nets - An Expeditious Tool for Modeling, Simulation and Analysis of Flexible Multi Robot Assembly Systems*, M. Tech. Project Report, Manufacturing Engineering Section, Indian Institute of Technology, Madras, India, 1990.

Venkatesh, K., O. V. K. Chetty, and V. Radhakrishnan, "Software Development for Future Unmanned Industries," *Proc. of Int. Conf. on Design Automation and Computer Integrated Manufacturing*, Coimbatore, India, 1991, pp. 80-91.

Venkatesh, K. and O. V. K. Chetty, "Petri Nets as an Efficient Modeling Tool for Modeling Tool Management in FMSs," *Proc. of the 5th Annual Research Conf. on Recent Advances in Robotics*, Florida Atlantic University, June 1992, pp. 583-594.

Venkatesh, K., O. Masory, and J. Wu, "Simulation and Scheduling of Robots in an Flexible Factory Automated System Operating With JIT Principles Using Timed Petri nets," *Proc. of Int. Conf. on Automation and Technology*, Taipei, Taiwan, R.O.C., July 4-6, 1992, pp. 73-80.

Venkatesh, K. and M. Kaighobadi, "Modeling, Simulation, and Analysis of Flexible Assembly System Using Stochastic Petri Nets," Presented *at Production and Operations Management Society Annual Meeting*, Orlando, FL, October 18-21, 1992.

Venkatesh, K. and E. B. Fernandez, "Object-oriented Simulation and Control of Flexible Manufacturing Systems Using Timed Petri Nets and Ada," *Technical Report No. TR-CSE-92-93*, Florida Atlantic University, Boca Raton, FL, 1993.

Venkatesh, K., C. Subramanian, and O. Masory, "A Sequence Controller Based on Augmented Timed Petri Nets," *Proceedings of the 6th Annual Conference on Recent Advances in Robotics*, Gainsville, Florida, March 19-21, 1993.

Venkatesh, K., M. Kaighobadi, M. C. Zhou, and R. J. Caudill, "Augmented Timed Petri nets for Modeling of Robotic Systems with Breakdowns," *Journal of Manufacturing Systems*, Vol. 13, No. 4, 1994a, pp. 289-301.

Venkatesh, K., M. C. Zhou, and R. J. Caudill, "Comparing Ladder Logic Diagrams and Petri Nets for Sequence Controller Design through a Discrete Manufacturing System," *IEEE Trans. on Industrial Electronics*, 41(6), 1994b, pp. 611-619.

Venkatesh, K., M. C. Zhou, R. J. Caudill, and E. Fernandez, "A Control Software Design Methodology for CIM Systems," *Proc. of Rutgers Conference on CIM in the Process Industries*, East Brunswick, NJ, April 25-26, 1994c, pp. 565-579.

Venkatesh, K., M. C. Zhou, and R. J. Caudill, "Evaluating the Complexity of Petri Nets and Ladder Logic Diagrams to Design Sequence Controllers in Flexible

Automation," *Proceedings of Seiken/IEEE Symposium on Emerging Technologies & Factory Automation*, Tokyo, Japan, Nov. 6-10, 1994d, pp. 428-435.

Venkatesh, K., M. C. Zhou, and R. J. Caudill, "Automatic Generation of Petri Nets from Logic Controller's Sequence Specification," *Proc. of the 4th Rennsselaer's Int. Conf. on Computer-Integrated Manufacturing and Automation Technology,* Troy, NY October 1994e, pp. 242-247.

Venkatesh, K., *A Petri Net Based Approach for Modeling, Simulation, Analysis, and Control of Flexible Manufacturing Systems*, Doctoral Dissertation, Mechanical Engineering Department, New Jersey Institute of Technology, Jan. 1995.

Venkatesh, K., M. C. Zhou, and R. Caudill, "Design of Sequence Controllers Using Petri Net Models," *Proc. of 1995 IEEE Int. Conf. on Systems, Man, and Cybernetics*, Vancouver, Canada, Vol. 5, October 1995, pp. 3469-3474.

Venkatesh, K. and M. Ilyas, "Real-time Petri Nets for Modeling, Controlling, and Simulation of Local Area Networks in Flexible Manufacturing Systems," *Computers and Industrial Engineering*, Vol. 28, No. 1, 1995, pp. 147-162.

Venkatesh, K., M. Kaighobadi, M. C. Zhou, and R. J. Caudill, "A Petri Net Approach to Investigating Push and Pull Paradigms in Flexible Factory Automated Systems," *Int. J. of Prod. Res.*, Vol. 34, No. 3, 1996, pp. 595-620.

Venkatesh, K., M. C. Zhou, and R. J. Caudill, "Design of Artificial Neural Network for Tool Wear Monitoring," *J. of Intelligent Manufacturing*, 8, 1997, pp. 215-226.

Venkatesh, K., M. C. Zhou, and R. J. Caudill, "Object-Oriented Design of FMS Control Software Based on Object Modeling Technique Diagrams and Petri Nets," *Journal of Manufacturing Systems*, Vol. 17, No. 2, 1998, pp. 118-136.

Vernon, M., J. Zahorjan, and E. D. Lazowska, "A Comparison of Performance Petri nets and Queuing Network Models," *Proc. of the Int. Workshop on Modeling Techniques and Performance Evaluation,* Paris, France, 1987.

Viswanadham, N. and Y. Narahari, "Stochastic Petri net models for performance evaluation of automated manufacturing systems," *Information and Decision Technologies*, 14, North-Holland, 1988, pp. 125-142.

Viswanadham, N., Y. Narahari, and L. Johnson, "Deadlock prevention and avoidance in flexible manufacturing systems using Petri net models," *IEEE Trans. on Robotics and Automation*, 6(6), 1990, pp. 713-723.

Voltz, R., T. N. Mudge, and D. Gal, "Using Ada as a Programming Language for Robot-based Manufacturing Cells," *IEEE Trans. on Systems, Man, and Cybernetics*, Vol. 14, No. 6, 1984, pp. 863-878.

Vosniakos, G. C. and A.G. Mamalis, "Automated guided vehicle system design for FMS applications," *Int'l J. of Machine Tools and Manufacture*, 30, 1990, pp. 85-97.

Wang, F. Y. and G. N. Saridis, "A coordination theory for intelligent machines," *Automatica*, 26(5), 1990, pp. 833-844.

Wang, H. P. B. and S. A. Hafeez, "Performance Evaluation of Tandem and Conventional AGV Systems Using Generalized Stochastic Petri nets," *Int. J. of Prod. Res.*, Vol. 32, No. 4, 1994, pp. 917-932.

Wang, J. and S. Jiang, "Stochastic Petri net models of communication and flexible systems," *Petri Nets in Flexible & Agile Automation*, M. C. Zhou (Ed.), Norwell, MA: Kluwer Academic, 1995, pp. 207-238.

Wang, L. C., "The Development of an Object-oriented Petri net Cell Control Model," *Int'l J. of Advanced Manufacturing Technology*, Vol. 11, 1996, pp. 59-69.

Watson III, J. F. and A. A. Desrochers, "Applying GSPN to manufacturing systems containing nonexponential transition functions," *IEEE Trans. on System, Man and Cybernetics*, Vol. 21, No. 5, 1991, pp. 1008-1017.

Wegner, P., "Capital Intensive Software Technology, Part 2: Programming in the large," *IEEE Software*, Vol. 1, No. 3, 1984, pp. 24-32.

Westphal, H. and S. Renganathan, "State dependent observer synthesis – a module for supervisory control of stochastic discrete event dynamic systems," in *Proc. of 6th IEEE Int. Conf. on Emerging Technologies and Factory Automation*, Los Angeles, CA, September 9-12, 1997, pp. 538-542.

Wong, H.-M. and M. C. Zhou, "Automated generation of modified reachability trees for Petri nets," *1992 Regional Control Conference*, Brooklyn, NY, July 24-25, 1992, pp. 119-121.

Wu, B., "Object-Oriented Systems Analysis and Definition of Manufacturing Operations," *Int'l J. of Production Research*, Vol. 33, No. 4, 1995, pp. 955-974.

Wysk, R. A., P. J. Egbelu, C. Zhou, and B. K. Ghosh, "Use of Spreadsheet Analysis for Evaluating AGV Systems," *Material Flow*, Vol. 4, 1987, pp. 53-64.

Wysk, R. A., N. S. Yang, and S. Joshi, "Resolution of deadlocks in flexible manufacturing systems: Avoidance and recovery approaches," *Journal of Manufacturing Systems*, Vol. 13, 1994, pp. 128-138.

Xiang, D. and C. O'Brien, "Cell Control Research - Current Status and Development Trends," *Int'l J. of Production Res.*, 33(8), 1995, pp. 2325-2352.

Xie, X. L., "Superposition properties and performance bounds of stochastic timed-event graphs," *IEEE Trans. Automatic Control,* 39(7), 1994, pp. 1376-1386.

Xing, K., B. Hu, and H. Chen, "Deadlock avoidance policy for flexible manufacturing systems," *Petri Nets in Flexible and Agile Automation*, M. C. Zhou (Ed.), Kluwer Academic Publishers, Boston, MA, 1995, pp. 239-263.

Xiong, H. H., M. C. Zhou, and C. N. Manikopoulos, "Modeling and Performance Analysis of Medical Services Systems Using Petri Nets," *Proc. of 1994 IEEE Conf. on Systems, Man and Cybernetics*, San Diego, TX, 1994, pp. 2339-2342.

Xiong, H. H., M. C. Zhou, and R. J. Caudill, "A hybrid heuristic search algorithm for scheduling flexible manufacturing systems," *Proc. of 1996 IEEE Int. Conf. on Robotics and Automation*, Minneapolis, MN, April 1996, pp. 2793-2797.

Xiong, H. H., *Scheduling and Discrete Event Control of Flexible Manufacturing Systems based on Petri Nets*, Ph.D. Dissertation, Electrical and Computer Engineering Department, New Jersey Institute of Technology, 1996.

Yamada, T. and S. Kataoka, "On some LP problems for performance evaluation of timed marked graphs," *IEEE Trans. on Automatic Control*, 39, 1994, pp. 696-698.

Yilmaz, O. S. and R. P. Davis, "Flexible Manufacturing Systems: Characteristics and Assessment," *Engineering Management*, Vol. 4, 1987, pp. 209-212.

Yim, D. and R. J. Linn, "Push and Pull Rules for Dispatching Automated Guided Vehicles in a Flexible Manufacturing System," *Int'l J. of Pro 'uction Research*, Vol. 31, No. 1, 1993, pp. 43-57.

Zavanella, L., G. C. Maccarini, and A. Bugini, "FMS Tool Supply in a Stochastic Environment: Strategies and Related Reliabilities," *Int'l J. of Machine Tools Manufacturing*, Vol. 30, No. 3, 1990, pp. 389-402.

Zhou, M. C. (Ed.), *Petri Nets in Flexible and Agile Automation*, Norwell, MA: Kluwer Academic, 1995.

Zhou, M. C., "Deadlock Avoidance Methods for a Distributed Robotic System: Petri Net Modeling and Analysis," *J. of Robotic Systems*, 12(3), 1995, pp. 177-187.

Zhou, M. C., H. Chiu, and H. H. Xiong, "Petri net scheduling of FMS using branch and bound method," *Proc. of 1995 IEEE Int. Conf. on Industrial Electronics, Control, and Instrumentation*, Orlando, FL, Nov. 1995, pp. 211-216.

Zhou, M. C. and F. DiCesare, "Adaptive Design of Petri net Controllers for Error Recovery in Automated Manufacturing Systems," *IEEE Trans. on Systems, Man, and Cybernetics*, Vol. 19, No. 5, 1989, pp. 963-973.

Zhou, M. C. and F. DiCesare, "Parallel and sequential mutual exclusions for Petri net modeling for manufacturing systems," *IEEE Trans. on Robotics and Automation*, 7(4), 1991, pp. 515-527.

Zhou, M. C. and F. DiCesare, *Petri Net Synthesis for Discrete Event Control of Manufacturing Systems*, Kluwer Academic, Norwell, MA, 1993.

Zhou, M. C. and F. DiCesare, "Petri Net Modeling of Buffers in Automated Manufacturing Systems," *IEEE Trans. on Systems, Man, and Cybernetics - Part B: Cybernetics*, 26(1), 1996, pp. 157-164.

Zhou, M. C., F. DiCesare, and A. Desrochers, "A Hybrid Methodology for Synthesis of Petri Net Models for Manufacturing Systems," *IEEE Trans. on Robotics and Automation*, Vol. 8, No. 3, 1992a, pp. 350-361.

Zhou, M. C., F. DiCesare, and D. Guo, "Modeling and performance analysis of a resource-sharing manufacturing system using stochastic Petri nets," *Proc. of Fifth IEEE Symp. on Intelligent Control*, Phildephia, PA, 1990, pp. 1005-1010.

Zhou, M. C., F. DiCesare, and D. Rudolph, "Design and Implementation of a Petri net Based Supervisor for a Flexible Manufacturing System," *Automatica*, Vol. 28, No. 6, 1992b, pp. 1999-2008.

Zhou, M. C., D. L. Guo, and F. DiCesare, "Integration of Petri Nets and Moment Generating Function Approaches for System Performance Evaluation," *J. of Systems Integration*, 3(1), 1993, pp. 43-62.

Zhou, M. C. and M. D. Jeng, "Modeling, Analysis, Simulation, Scheduling, and Control of Semiconductor Manufacturing Systems: A Petri Net Approach," to appear in *IEEE Transactions on Semiconductor Manufacturing*, 1998.

Zhou, M. C. and M. C. Leu, "Modeling and Performance Analysis of a Flexible PCB Assembly System Using Petri nets," *Transactions of ASME, Journal of Electronic Packaging*, Vol. 113, 1991, pp. 410-416.

Zhou, M. C. and J. M. Ma, "Reduction and approximation for performance evaluation of stochastic Petri nets with multi-input and multi-output modules," *Preprints of 12th IFAC World Congress*, Sydney, Australia, 4, 1993, pp. 159-163.

Zhou, M. C., K. McDermot, and P. A. Patel, "Petri Net Synthesis and Analysis of Flexible Manufacturing System Cell," *IEEE Trans. on Systems, Man, and Cybernetics*, Vol. 23, No. 2, 1993, pp. 523-531.

Zhou, M. C. and A. D. Robbi, "Applications of Petri net methodology to manufacturing systems," in *Computer Control of Manufacturing Systems*, S. Joshi and G. Smith (eds.), Chapman and Hall, 1994, 307-330.

Zhou, M. C., A. D. Robbi, and R. Zurawski, "Discrete Event Simulation," *Handbook of Industrial Electronic Engineering*, CRC Press, 1996, pp. 694-705.

Zhou, M. C. and G. A. Thorniley, "Performance modeling and optimization for manufacturing systems using Petri nets," *Advances in Manufacturing Systems: Modeling, Design and Analysis*, Elesevier Scientific: Amsterdam, The Netherlands, 1994, pp. 33-38.

Zhou, M. C., H. H. Xiong, and N. Manikopoulos, "Performance Models for Communication Networks in Manufacturing Environment," *Pro. of the 4th Int. Conf. on CIM and Automation Technology*, Troy, NY, Oct. 10-12, 1994.

Zhou, M. C. and E. Twiss, "A Comparison of Relay Ladder Logic Programming and Petri Net Synthesis for Control of Discrete Event Systems," Technical Report #9501, Discrete Event Systems Laboratory, ECE, New Jersey Institute of Technology, Jan. 1995.

Zhou, M. C. and E. Twiss, "Discrete Event Control Methods: A Review," *Preprints of the 13th World Congress of IFAC*, San Francisco, CA, 1996, pp. 401-406.

Zhou, M. C. and E. Twiss, "Design of Industrial Automated Systems via Relay Ladder Logic Programming and Petri Nets," *IEEE Trans. on Systems, Man, and Cybernetics*, 28(1), 137-150, 1998.

Zhou, M. C., B. Zhang, R. J. Caudill, and D. Sebastian, "A cost model for multi-lifecycle engineering design," in *Proc. of 1996 IEEE Int. Conf. on Emerging Technologies and Factory Automation*, Hawaii, November 1996, pp. 385-391.

Zuberek, W. M., "Application of timed Petri nets to modeling the schedules of manufacturing cells," *Proc. of INRIA/IEEE Conf. on Emerging Technologies and Factory Automation,* Paris, France, Vol. 2, 1995, pp. 311-322.

Zuberek, W. M., "Application of timed Petri nets to modeling the schedules of manufacturing cells," *Proc. of 1996 IEEE Int. Conf. on Systems, Man, and Cybernetics,* Beijing, China, 1996, pp. 2990-2995.

Zurawski, R. and T. Dillon, "Modeling and verification of flexible manufacturing systems using Petri nets," *Modern Tools for Manufacturing Systems,* R.Zurawski, T. Dillon (Eds.), Elesevier Scientific, 1993, pp. 237-261.

Zurawski, R., "Systematic construction of functional abstractions of Petri net models of flexible manufacturing systems," *IEEE Trans. on Industrial Electronics,* 41(6), 1994, pp. 584-592.

Zurawski, R. and M. C. Zhou, "Petri Nets and Industrial Applications: A Tutorial," *IEEE Trans. on Industrial Electronics,* 41(6), 1994, pp. 567-583.

Zussman, E., B. Scholz-Reiter, and H. Scharke, "Modeling and Planning of Disassembly Processes," *Proc. of the IFIP WG5.3 Int. Conf. on Life-Cycle Modeling and Processes,* Berlin, 1995, pp. 221 - 232.

Zussman, E. and M. C. Zhou, "A Methodology for Modeling and Adaptive Planning of Disassembly Processes," Technical Report 97-0112, Multi-lifecycle Engineering Research Center, New Jersey Institute of Technology, Newark, NJ, also submitted to *IEEE Trans. on Robotics and Automation* for review, 1997.

INDEX

A* algorithm, 333, 346
AC net, 99, 145
activity, 15, 74
actuating signal, 285
    momentary actuating signal, 285
    pulsed actuating signal, 285
    sustained actuating signal, 285
acyclic Petri net, 105-106
adaptive colored Petri net, 44
admissible, 347, 351
aggregate Petri net, 131-133
aggregation, 312
agile manufacturing, 1-3
agility, 1
AGV, 14, 336
AMG, 112, 145
anticipatory scheduling, 336
APN, 106, 145
ATPN, 223, 225
    application illustration, 230
    example, 229
    model for designing the optimum number of assembly fixtures, 234-235
    modeling of breakdown handling, 232, 235
    simulation and analysis, 236
approximation, 175
arc, 9, 57, 64-65, 74
    deactivation arcs, 226
    secondary arc, 226
assembly Petri net, 106-107, 145
associated Petri net, 142
association, 313

asymmetric choice Petri net, 99-105, 145
asynchronous, 1, 5
augmentation, 83-84
augmented marked graph, 112-117, 145
augmented timed Petri net, see ATPN
automated manufacturing system, 337
automated storage and retrieval system, 45
automatic guided vehicle, 14, 49-50, 56
automaton, 8

B-bounded, 14, 44
Backtracking, 333, 346, 348
basic elements, 262
    for logical AND, 279
    for logical OR, 280
    for sequential modeling, 282
    for timed logical AND, 283
    for logical OR, 284
    for timed sequential modeling, 284
    for emergency stop modeling, 286-287
    for modeling a counter, 287-288
    for modeling a relay, 289
    method to estimate basic elements, 289
basic relation, 118-119
behavioral property, 40-41, 43
best-first, 333, 346
bottom-up design, 125-130, 337
bottom-up design stage, 125-126

bottom-up synthesis, 125-130, 337
boundedness, 41, 44, 70, 74, 82
breakdown, 11
    modeling using ATPNs, 232-234
bridge, 103
buffer, 45, 129, 342, 354
buffer place, 129, 342

CAD/CAM, 4
CASE tool environment, 365-369
centralized control, 56
choice, 91, 93, 99
choice-free Petri net, 95-99, 147
choice-synchronization, 131
choice-synchronization structure, 115-117, 131
CIM, 4, 15
class, 311
    attributes, 311
    name, 311
    operations, 311
    relations among classes, 311
colored Petri net, 44
common path, 97, 128
composition, 128-130
computational intelligence, 334
computer-integrated manufacturing, 2-4, 15
concurrent, 1, 5, 178
concurrent engineering, 3
condition, 9, 59
conflict, 5, 118
conservative, 14, 69
conservativeness, 15, 69, 75, 82
consistency, 14, 73, 75, 82
consistent, 14, 73
control code, 41
Control nets, 51, 112

control specifications, 279
    logic constructs, 279
coordination, 45
coverability graph, 70
CPM/PERT, 335
cycle time, 6, 151
cyclic behavior, 6, 95, 118

deadlock, 6, 11, 41, 48, 72, 93, 110, 115, 125, 139
deadlock avoidance, 6, 48, 109, 376
deadlock-free controller, 48
decomposition, 44, 125
dedicated resource, 126, 128
DEDS, 5
DEEP, 46
design complexity, 261
DES, 1
deterministic, 95
deterministic stochastic Petri net, 166, 175
deterministic timed Petri net, 147-152
directed arc, 9, 74
disassembly Petri net, 105-109, 145
discrete event control, 7, 39
discrete event dynamic system, 5
discrete event simulation, 7, 39, 89, 177
    procedure, 178
    simulation steps, 179
    token game simulation, 187
discrete event system, 5, 39
distributed control, 56
distributed system, 56
DPN, 105, 145
DSPN, 166, 175
dual, 106

Index

elementary circuit, 92
elementary loop, 92, 125, 149
elementary path, 91, 149
enabled, 9-11, 66
enabling rule, 66, 68, 147
error recovery, 46, 73
ESP, 46
ESPN, 166
essentially decision free, 44
event, 9, 59
event-driven, 40, 59
event trace diagram, 314
execution rule, 66, 147
exponential distribution, 156
extended stochastic Petri net, 166, 175
extendibility, 303

fairness, 69
FAS, 201, 230
fault-prone module, 126-127
FC, 99, 145
FIFO, 45
finite state machine, 91-94
fire, 9-11, 66
firing rate, 156-157
firing rule, 147
first-in-first-out, 45
first-level Petri net, 131
flexibility, 1, 51
flexible assembly system, 201, 230
flexible and agile manufacturing, 1
flexible manufacturing cell, 8-13, 133-144, 153-155
flexible manufacturing system, 8-13, 44
flexible route, 345
flexible station, 121-124, 169-174

FMS, 1-3, 8-13, 14, 15-18, 39, 45, 133-144
 applications, 24-25
 automated material handling, 30-31
 control and communication, 32-34
 emerging trends and new demands, 35
 installation, implementation, integration, 20-24
 managerial problems, 26
 performance evaluation, 195
 Petri net model with pull paradigm, 209
 Petri net model with push paradigm, 211
 technical problems, 27
 tool management, 28-30
free choice net, 99-105, 145
functional abstraction, 45

generalized stochastic Petri net, 47, 164-165
generalization, 312
Grafcet, 43, 375
graphical complexity, 262
GreatSPN, 47
GSPN, 47-49, 164

handle, 103
headlight example, 106-107
heuristic function, 347
heuristic dispatching, 334
heuristic search, 333
hierarchical decomposition, 125
hierarchical time-extended Petri nets, 46
high-level Petri net, 183

high-level timed Petri net,
home state, 72-73
hybrid BF-BT, 349, 360-363
hybrid BT-BF, 350, 360-363
hybrid method, 132-133
hybrid strategy, 333, 349, 360-363
hybrid synthesis, 45, 132-133

incidence matrix, 65
inhibitor arc, 66-67
initial marking, 64
input function, 64
intelligent automation, 2
intelligent machine, 45
interaction diagram, 314
interlock, 285
invariant analysis method, 80-82, 90

JIT, 46, 196-197
job-shop, 4
just-in-time, 46, 196

Kanbans, 45-46
knitting technique, 103

ladder logic diagram, see LLD
life-cycle engineering, 372
linear programming, 109, 151-152
live, 41, 44
liveness, 41, 44, 71-72, 75
LLD, 259
    models of logic constructs, 262
    representation of logical AND, 262
    representation of logical OR, 262
    representation of concurrency, 262
    representation of time delay, 262
    representation of synchronization, 262

local control level, 51
logic controller, 45
loop, 92, 95
lot size, 342, 359

maintainability, 51
makespan, 335, 360
manufacturing system, 1, 41
marked graph, 95-99, 125, 145
marked Petri net, 13
marking, 64
Markov chain, 6
Markov process, 6
material flow, 142
mathematical programming, 334
methodology
    for FMS control software development, 307-309
    for using the analytical formulas, 289
MG, 102, 145
model, 1, 182
modeling, 44-46
modeling methodology, 118-124
modifiability, 303
modified Petri net, 44
moment generating function, 175
monitoring, 42
moving lot size (MLS), 202
multi-lifecycle engineering, 372-375
mutual exclusion, 5, 118
mutually exclusive, 118

node, 8, 40, 91
non-determinism, 5

object, 311

Index

object modeling technique diagram, see OMT
object-oriented design, see OOD
object-oriented models, 184
OOD, 303
    fundamentals, 309
OMT, 303
    class definition, 312
    diagram for aggregation, 313
    diagram for association, 313
    diagram for generalization, 312
operation place, 132
optimal, 333, 336
optimality, 360
ordinary Petri net, 91
output function, 64
overflow, 70, 75, 90, 125

•p, 91
p•, 91
P-invariant, 80-82
parallel mutual exclusion, 118
performance, 46-49, 157-159
Periodically-maintained resource /operation module, 126-127, 129-130
Petri net, 1, 40, 64-65
    advantage, 41-42
    augmented-timed Petri net, see ATPN
    CASE tools, 192
    design, 120
    industrial application, 43
    models of logic constructs, 262
    representation of logical AND, 262
    representation of logical OR, 262
    representation of concurrency, 262
    representation of time delay, 262
    representation of synchronization, 262
    real-time Petri net, see RTPN
    various methods of sequence control, 245
Petri net execution algorithm, 41
Petri net language, 371
Petri net model, 9
    conventions of modeling, 199
    for pull paradigm, 202, 204-207
    for push paradigm, 207-208
    procedure for analysis of pull paradigm, 212
Petri net simulator, 187, 192
place, 8, 40, 64, 74
    deactivation place, 226
planning, 7, 55-58, 333
PLC, 7, 41, 43
PN, 1, 67
post-set, 91
PPN, 109-111, 145
pre-set, 91
priority, 69
priority module, 126-127, 129-130
probability density function, 156
process planning, 55-58
production rate, 6, 159
production process net, 109-111, 145
productivity, 336
programmable logic controller, 7, 41, 43
Prot net, 195
production log size (PLS), 202
push paradigm, 195
    Petri net modeling, 197-198
push system, 197
pull paradigm, 195
    Petri net modeling, 197-198

procedure for Petri net modeling, 212
pull system, 197

queuing models, 182
queuing networks, 335

rapid prototyping, 3
reachability, 70, 75
reachability analysis method, 75-80, 90
reachability set, 70
reachability tree, 70
reachable, 66, 70
reduction method, 82-88, 90
reduction rule, 83-84
refinement, 131, 135, 143
repetitive, 72, 75, 82, 95
resource, 39
resource allocation, 41, 44
resource place, 105, 110
resource-operation module, 125-126, 129-130
resource-sharing, 112, 128, 145
resource-utilization, 158
reusability, 303
reversibility, 41, 72-73
reversible, 41, 72
rework module, 127-130
robotic assembly system, 45, 59-64
robotic control software, 54
route, 345
RTPN, 241-243
   application example using an assembly system, 248
   case study using an electro-pneumatic system, 251
   comparison with earlier PN techniques, 246-248
   controlling using RTPN controller, 245
   input signal vector, 242
   output signal vector, 242
   procedure for formulating RTPN based controller, 244
   sequence controller design, 253
   software description to execute RTPNs, 255-257

•S, 91
S•, 91
safeness, 41, 70, 75
scheduling, 7, 11, 333, 335, 369-370
self-loop, 40, 68, 112
semiconductor fabrication process, 46, 128-130, 356, 369-370
semiconductor test facility, 356
sequence, 118
sequential condition, 281
sequential control, 189
sequential function charts, 301
sequential mutual exclusion, 118, 142
sequential relation, 118
shared buffer, 112
shared resource, 118-119
simulation, 89-90
siphon, 100
   basis, 100
   minimal, 100
SM, 102
SM-component, 102
software package
   description to execute RTPNs, 255-257

description to execute TPNs, 189-192
specification, 39
SPN, 157
SPNP, 47-48, 167-169
stability, 6
state, 59
state diagram, 314
state explosion, 61, 64
state machine, 59-64
state machine Petri net, 91-94, 125
state transition diagram, 314
state transition models, 182
steady state probability, 158
stepwise refinement, 117, 146
stochastic Petri net, 6, 157
strongly connected, 92
structural property, 40-41, 64-75
structured colored adaptive Petri net, 44
sub-Petri net, 125-130
subnet, 125-130
sub-optimal, 333, 336
supervisory control, 5, 371-372
supervisory controller, 50-54
symbolic performance solution, 175
synchronic distance, 69
synchronization, 39, 64, 115-117
synthesis, 125-144
system deadlock, 6, 11, 41, 48, 72, 93, 110, 115, 125, 139
systematic synthesis method, 125-133

•t, 91
t•, 91
T-invariant, 80-82
throughput , 159
throughput rate, 6, 159
timed event-graph, 47
timed Petri net, 6, 147-152, 187
time arc Petri net, 147-152
timed place Petri net, 147-152
timed transition Petri net, 147-152
token, 8, 40, 74
token game simulation, 187
   algorithm used for software package, 190
token player, 187
top-down design, 131-132
top-down synthesis, 131-132
transfer line, 44
transfer station, 16, 17
transition, 8, 40, 64, 74
   deactivation transition, 226
transition rate matrix, 158
trap, 100
   basis, 100
   minimal, 100
tree marked graph, 105

work-in-process, 336
workstation, 8
WS, 8

Z, 91
Z', 91